Bogdan Gabrys, Kauko Leiviskä, Jens Strackeljan (Eds.)

Do Smart Adaptive Systems Exist?

Studies in Fuzziness and Soft Computing, Volume 173

Editor-in-chief
Prof. Janusz Kacprzyk
Systems Research Institute
Polish Academy of Sciences
ul. Newelska 6
01-447 Warsaw
Poland
E-mail: kacprzyk@ibspan.waw.pl

Further volume of this series can be found on our homepage: springeronline.com

Vol. 158. K.K. Dompere
Cost-Benefit Analysis and the Theory of Fuzzy Decisions – Identification and Measurement Theory, 2004
ISBN 3-540-22154-9

Vol. 159. E. Damiani, L.C. Jain, M. Madravia (Eds.)
Soft Computing in Software Engineering, 2004
ISBN 3-540-22030-5

Vol. 160. K.K. Dompere
Cost-Benefit Analysis and the Theory of Fuzzy Decisions – Fuzzy Value Theory, 2004
ISBN 3-540-22161-1

Vol. 161. N. Nedjah, L. de Macedo Mourelle (Eds.)
Evolvable Machines, 2005
ISBN 3-540-22905-1

Vol. 162. R. Khosla, N. Ichalkaranje, L.C. Jain
Design of Intelligent Multi-Agent Systems, 2005
ISBN 3-540-22913-2

Vol. 163. A. Ghosh, L.C. Jain (Eds.)
Evolutionary Computation in Data Mining, 2005
ISBN 3-540-22370-3

Vol. 164. M. Nikravesh, L.A. Zadeh, J. Kacprzyk (Eds.)
Soft Computing for Information Prodessing and Analysis, 2005
ISBN 3-540-22930-2

Vol. 165. A.F. Rocha, E. Massad, A. Pereira Jr.
The Brain: From Fuzzy Arithmetic to Quantum Computing, 2005
ISBN 3-540-21858-0

Vol. 166. W.E. Hart, N. Krasnogor, J.E. Smith (Eds.)
Recent Advances in Memetic Algorithms, 2005
ISBN 3-540-22904-3

Vol. 167. Y. Jin (Ed.)
Knowledge Incorporation in Evolutionary Computation, 2005
ISBN 3-540-22902-7

Vol. 168. Yap P. Tan, Kim H. Yap, Lipo Wang (Eds.)
Intelligent Multimedia Processing with Soft Computing, 2005
ISBN 3-540-22902-7

Vol. 169. C.R. Bector, Suresh Chandra
Fuzzy Mathematical Programming and Fuzzy Matrix Games, 2005
ISBN 3-540-23729-1

Vol. 170. Martin Pelikan
Hierarchical Bayesian Optimization Algorithm, 2005
ISBN 3-540-23774-7

Vol. 171. James J. Buckley
Simulating Fuzzy Systems, 2005
ISBN 3-540-24116-7

Vol. 172. Patricia Melin, Oscar Castillo
Hybrid Intelligent Systems for Pattern Recognition Using Soft Computing, 2005
ISBN 3-540-24121-3

Vol. 173. Bogdan Gabrys, Kauko Leiviskä, Jens Strackeljan (Eds.)
Do Smart Adaptive Systems Exist?, 2005
ISBN 3-540-24077-2

Bogdan Gabrys
Kauko Leiviskä
Jens Strackeljan (Eds.)

Do Smart Adaptive Systems Exist?

Best Practice
for Selection and Combination
of Intelligent Methods

Bogdan Gabrys
Bournemouth University
School of Design
Engineering & Computing
Poole House
Talbot Campus, Fern Barrow
Poole, BH12 5BB
U.K.
E-mail: bgabrys@bournemouth.ac.uk

Jens Strackeljan
Universität Magdeburg
Institut für Mechanik
Lehrstuhl für Technische Dynamik
Universitätsplatz 2, 39106 Magdeburg
Germany
E-mail:jens.strackeljan@mb.uni-magdeburg.de

Kauko Leiviskä
University Oulu
Department of Process Engineering
Control Engineering Laboratory
P.O.Box 4300, 90014 Oulu
Finland
E-mail: kauko.leiviska@oulu.fi

ISSN print edition:1434-9922
ISSN electronic edition: 1860-0808
ISBN-10   3-540-24077-2 Springer Berlin Heidelberg New York
ISBN-13   978-3-540-24077-8 Springer Berlin Heidelberg New York

Library of Congress Control Number: 2004116662

This work is subject to copyright. All rights are reserved, whether the whole or part of the material is concerned, specifically the rights of translation, reprinting, reuse of illustrations, recitation, broadcasting, reproduction on microfilm or in any other way, and storage in data banks. Duplication of this publication or parts thereof is permitted only under the provisions of the German Copyright Law of September 9, 1965, in its current version, and permission for use must always be obtained from Springer. Violations are liable for prosecution under the German Copyright Law.

Springer is a part of Springer Science+Business Media
springeronline.com

© Springer-Verlag Berlin Heidelberg 2005
Printed in The Netherlands

The use of general descriptive names, registered names, trademarks, etc. in this publication does not imply, even in the absence of a specific statement, that such names are exempt from the relevant protective laws and regulations and therefore free for general use.

Typesetting: by the authors and TechBooks using a Springer LATEX macro package
Cover design: E. Kirchner, Springer Heidelberg
Printed on acid-free paper   89/3141/jl- 5 4 3 2 1 0

Preface

At first sight, the title of the present book may seem somewhat unusual, since it ends with a question: Do Smart Adaptive Systems Exist? We have deliberately chosen this form for the title, because the content of this book is intended to elucidate two different aspects: First of all, we wish to define what is meant by the term "Smart Adaptive Systems". Furthermore, the question asked in the title also implies that the applications described must be critically examined to determine whether they satisfy the requirements imposed on a smart adaptive system. Many readers will certainly have an intuitive notion of the concepts associated with the terms "smart" and "adaptive". Some readers will probably also think of applications from their own field of work which they consider to be both smart and adaptive.

Is there any need for a book of this kind? Is an attempt to provide a definition of terms and to describe methods and applications in this field necessary at all? Two years ago, we answered this question with an unambiguous "yes" and also started the book project for this reason. The starting point was the result of joint activities among the authors in the EUNITE network, which is dedicated to the topic of smart adaptive systems. EUNITE, the European Network on Intelligent Technologies for Smart Adaptive Systems, was the European Network of Excellence that started 2001 and ended mid of 2004. It concerned with intelligent technologies, including neural networks, fuzzy systems, methods from machine learning, and evolutionary computing, that have recently lead to many successful industrial applications. Terms and definitions have been the subject of intensive discussions within the scope of this network. These discussions were necessary because the existence of a generally accepted definition as a working basis is a prerequisite for joint activity among the members of a network consisting of scientists and representatives from industry. Finding such a definition proved to be quite difficult, especially because of the multiplicity of highly personal opinions which could not be expressed in concise and conclusive form without contradiction. We hope that this book will provide an idea of the consensus which has been reached within EUNITE. Since a large number of European experts in the fields of computer science, engineering and economic sciences, mathematics, biology, and medicine were active in this network, these terms also have a broad interdisciplinary basis. Adaptivity is a highly essential property which has steadily gained importance for solving current problems in the fields of process control, process and condition monitoring, and finance. Standard solutions without adaptation to match changing ambient conditions or customer groups, etc. are no longer competitive nowadays. Among other factors, this progress is due to the availability of highly efficient sensors at a reasonable price during recent years. These sensors function as vital interfaces for perception of the external world. Furthermore, current developments in the field of multimedia and use of the internet require adaptations to match special user profiles. In this context, the developments in designing of adaptive search engines may be cited as an example.

In addition, this book has received the subtitle: "A Best Practice Guideline". We know that the demands imposed by this subtitle are quite exacting. Best Practice – these two words imply no more and no less than a presentation of the best-possible techniques available today for realising concrete applications. Is there really nothing better than those methods which are described in this book as examples of smart adaptive systems? We cannot guarantee this with 100-percent certainty; nevertheless, we are sure that the present book contains a very good collection of methods and applications. In the corresponding chapters, the authors have implemented the "best practice" concept with the integration of sections entitled "Hints and Tips". These sections can provide valuable suggestions to readers from the industrial field as well as those from the research sector for solving typical problems of their own. Moreover, a Best Practice Guideline is always referred to concrete fields of application. Consequently, we have decided to subdivide the book into two essential parts. The first part is devoted to methods for applications and is concentrated on the theoretical foundation which is necessary for understanding adaptive methods within the field of intelligent methods. This part considers both individual intelligent technologies (i.e. neural networks, fuzzy systems, evolutionary algorithms and machine learning approaches) and hybrid methods (i.e. neuro-fuzzy systems, genetic-neuro-fuzzy approaches etc.) that are in the focus of EUNITE. It discusses the areas where particular methods have been applied and how the main features of particular methods respond to the need for adaptation.

In the second part, on the other hand, applications based on the methods presented in the first part are described. It focuses on applications that require some degree of adaptation; it describes the way from applications to methods, while the previous part goes from methods to applications. Here authors' experiences are reviewed with respect to the question of how to design an application requiring adaptation and how to select tools for it. The book can be read in two different ways: We recommend the classical procedure of reading from cover to cover, since we consider this approach advantageous. If the reader prefers, however, he can begin with the second part and then read the necessary background information on special methods in the first part as required.

We hope that our readers will find it a pleasure to read this book, and that this subject will be every bit as fascinating to them as it has been for us.

January 2005
Jens Strackeljan
Kauko Leiviskä
Bogdan Gabrys

Contents

1 Do Smart Adaptive Systems Exist? – Introduction
B. Gabrys 1

2 Problem Definition – From Applications to Methods
K. Leiviskä 19

3 Data Preparation and Preprocessing
D. Pyle 27

Part I From Methods To Applications

4 Artificial Neural Networks
C. Fyfe 57

5 Machine Learning and Reinforcement Learning
M. van Someren and S. ten Hagen 81

6 Fuzzy Expert Systems
J.M. Garibaldi 105

7 Learning Algorithms for Neuro-Fuzzy Systems
D.D. Nauck 133

8 Hybrid Intelligent Systems: Evolving Intelligence in Hierarchical Layers
A. Abraham 159

9 Evolving Connectionist Systems with Evolutionary Self-Optimisatio
N. Kasabov, Z. Chan, Q. Song, and D. Greer 181

Part II From Applications To Methods

10 Monitoring
J. Strackeljan..205

11 Examples of Smart Adaptive Systems in Model-Based Diagnosis
K. Leiviskä...233

12 Design of Adaptive Fuzzy Controllers
K. Leiviskä and L. Yliniemi...251

13 Optimal Design Synthesis of Component-Based Systems Using Intelligent Techniques
P.P. Angelov, Y. Zhang, and J.A. Wright............................267

14 Intelligent Methods in Finance Applications: From Questions to Solutions
M. Nelke...285

15 Neuro-Fuzzy Systems for Explaining Data Sets
D.D. Nauck...305

16 Fuzzy Linguistic Data Summaries as a Human Consistent, User Adaptable Solution to Data Mining
J. Kacprzyk and S. Zadrożny..321

17 Adaptive Multimedia Retrieval: From Data to User Interaction
A. Nürnberger and M. Detyniecki....................................341

1

Do Smart Adaptive Systems Exist? – Introduction

B. Gabrys

Computational Intelligence Research Group, School of Design, Engineering and Computing, Bournemouth University
bgabrys@bournemouth.ac.uk

This chapter serves as an introduction to the book and especially to its first part entitled "From methods to applications". It begins with a description of the motivations and driving forces behind the compilation of this book and work within European Community concerned with smart adaptive systems as the main theme of EUNITE Network of Excellence. This will be followed by a short account of individual intelligent technologies within the scope of EUNITE and their potential combinations which are perceived as good candidates for constructing systems with a degree of adaptiveness and intelligence. The chapters in the first part of the book cover some of these intelligent technologies like artificial neural networks, fuzzy expert systems, machine and reinforcement learning, evolutionary computing and various hybridizations in more detail. As it has proved throughout the life span of EUNITE it was not an easy task to agree on the definitions of what adaptive and smart systems are and therefore some compromise have had to be reached. As a result the definitions of three levels of adaptivity and some interpretations of the word smart adopted within EUNITE are given first. We then look at more general requirements of intelligent (smart) adaptive systems of the future and discuss some of the issues using the example of Evolutionary Connectionist Systems (ECOS) framework. Within this general framework and the scope of the book in the remaining sections we then concentrate on short description of existing methods for adaptation and learning, issues to do with model selection and combination and conflicting goals of having systems that can adapt to new/changing environments and at the same time have provable stability and robustness characteristics. The pointers to the relevant chapters discussing the highlighted issues in much greater detail are provided throughout this introductory chapter.

1.1 Introduction

Rapid development in computer and sensor technology not only used for highly specialised applications but widespread and pervasive across a wide range of business and industry has facilitated easy capture and storage of immense amounts of data. Examples of such data collection include medical history data in health care, financial data in banking, point of sale data in retail, plant monitoring data based on instant availability of various sensor readings in various industries, or airborne hyperspectral imaging data in natural resources identification to mention only a few. However, with an increasing computer power available at affordable prices and the availability of vast amount of data there is an increasing need for robust methods and systems, which can take advantage of all available information.

In essence there is a need for intelligent and smart adaptive methods but do they really exist? Are there any existing intelligent techniques which are more suitable for certain type of problems than others? How do we select those methods and can we be sure that the method of choice is the best for solving our problem? Do we need a combination of methods and if so then how to best combine them for different purposes? Are there any generic frameworks and requirements which would be highly desirable for solving data intensive and unstationary problems? All these questions and many others have been the focus of research vigorously pursued in many disciplines and some of them will be addressed in this book.

One of the more promising approaches to constructing smart adaptive systems is based on intelligent technologies including artificial neural networks [5, 25, 45], fuzzy systems [4, 31, 41, 46, 54, 55], methods from machine learning [8, 14, 15, 35, 44], parts of learning theory [51] and evolutionary computing [16, 24, 26, 30] which have been especially successful in applications where input-output data can be collected but the underlying physical model is unknown. The incorporation of intelligent technologies has been used in the conception and design of complex systems in which analytical and expert systems techniques are used in combination. Viewed from a much broader perspective, the above mentioned intelligent technologies are constituents of a very active research area called soft computing (SC) (the terms computational intelligence and hybrid intelligent systems are also frequently used) [1, 3, 6, 27, 28, 29, 32, 33, 34, 36, 39, 50, 56]. According to Zadeh [56], who coined the term soft computing, the most important factor that underlies the marked increase in machine intelligence nowadays is the use of soft computing to mimic the ability of the human mind to effectively employ modes of reasoning that are approximate rather than exact. Unlike traditional hard computing based on precision, certainty, and rigor, soft computing is tolerant of imprecision, uncertainty and partial truth. The primary aim of soft computing is to exploit such tolerance to achieve tractability, robustness, a high level of machine intelligence, and a low cost in practical applications. Although the fundamental inspirations for each of the constituent intelligent technologies are quite different, they share the common ability to improve the intelligence of systems working in an uncertain, imprecise and noisy environment, and since they are complementary rather than competitive, it is frequently advantageous to employ them in combination rather than exclusively.

This realisation has also been reflected in various European initiatives, which have concentrated on bringing together communities working with and potentially benefiting from the intelligent technologies. The identification of the potential benefits from integration of intelligent methods within four thematic networks of excellence for machine learning (MLNet), neural networks (NeuroNet), fuzzy logic (Erudit) and evolutionary computing (EvoNet) initially led to conception of the cluster on Computational Intelligence and Learning (CoIL) [50] and subsequent launch of EUNITE.

But before continuing with the theme of integration and hybridization within the context of smart adaptive systems let us take a brief look at those intelligent technologies and their frequently nature inspired origins.

1.2 Intelligent Methods

The intelligent technologies considered in this book have frequently exhibited very desirable characteristics by direct modelling or taking advantage in other ways of both the natural structures (i.e. human brain organisation) and processes (i.e. natural selection, human's approximate reasoning and ways of thinking, learning etc.) occurring in nature.

So let us have a very brief look at what are these intelligent methods we are going to cover in some detail in later chapters and what are their main weak and strong points which could prove useful or otherwise in building SAS.

1.2.1 Artificial Neural Networks

Artificial neural networks (ANN) [5, 25, 45] have been inspired by biological neural systems and the earliest ideas date back to the 1943 when McCalloch and Pits introduced a model of an artificial neuron. Since then many researchers have been exploring artificial neural networks as a novel and very powerful non-algorithmic approach to information processing. The underlying presumption of investigations involving ANNs has been the notion that by simulating the connectionists architectures of our brains supplemented by various learning algorithms can lead to the emergence of mechanisms simulating intelligent behaviour. As highlighted in Chap. 4 discussing ANNs in much greater detail, they are not truly SAS despite being based on models of truly adaptive system – the brain. However, without oversimplifying, in the context of SAS probably the most important features of ANNs are the various learning algorithms and Chap. 4 discusses a number of such algorithms developed for different types of ANNs and falling into two main groupings: supervised and unsupervised learning algorithms. On the other hand, one of the most often quoted drawbacks of ANNs is their lack of transparency or decision explanation ability.

1.2.2 Machine Learning

Machine learning [8, 14, 15, 35, 44] has its roots in the artificial intelligence and one of its central goals from the conception was the reproduction of human learning performance and modelling intelligence. While modelling human learning seems to be a secondary issue in the current stream of ML research, the recent focus has been on the analysis of very large data sets known as Data Mining and Information Retrieval using varying sources and types of data (please see Chaps. 5 and 17 for further details). What distinguishes ML approaches and can be considered as one of their main strengths are the rigorous, principled and statistically based methods for model building, selection and performance estimation which are related to the most fundamental issue of data overfitting versus generalisation when building models directly from data. More details can be found in Chap. 5 and some additional discussion will be provided in further sections of this chapter. Chapter 5 will also provide some discussion of reinforcement learning particularly useful in non-stationary problems and dynamically changing environments. Reinforcement learning has been frequently quoted as the third major group of learning algorithms in addition to supervised and unsupervised approaches.

1.2.3 Fuzzy and Expert Systems

One of the basic traits of intelligent behaviour is the ability to reason and draw conclusions. Conventional AI research focused on capturing and manipulating knowledge in a symbolic form. One of the most successful practical outcomes of conventional AI research were knowledge based or expert systems which proved especially useful in some narrowly defined areas. The modelling of human reasoning and uncertainty were also the main motivations of developing fuzzy set theory by Zadeh [54] and subsequent modifications and extensions by a large number of other researchers [4, 31, 41, 46]. In contrast to the classical set theory based on two value logic (true or false) fuzzy set theory is based on degrees of truth (membership values) and provides a systematic framework and calculus for dealing with uncertain and ambiguous data and information. Fuzzy sets provided a powerful extension to the classical expert systems. The fuzzy inference systems (FIS) proved to be extremely powerful and successful tool for effective modelling of human expertise in some specific applications. While one of the main advantages of fuzzy expert systems is their interpretability and ability to cope with uncertain and ambiguous data, one of their serious drawbacks, as far as SAS are concerned, is a complete lack of adaptability to changing environments. Further details concerning fuzzy expert systems can be found in Chap. 6.

1.2.4 Evolutionary Computing

One of the nature's most powerful, though admittedly not the quickest or particularly clever mechanisms for adaptation are based on biological evolution with the reproduction, mutation and survival of the fittest stages playing a prominent role. Similarly to natural evolution the evolutionary computing [16, 24, 26, 30] is based on the evolving population of solutions which ideally should represent better solutions from one generation to another according to some fitness criterion. The evolutionary techniques like genetic algorithms have been particularly useful in hard combinatorial optimization problems and applications where search spaces are large and it is difficult to reduce them. While some of the gradient based learning, adaptation and optimization techniques could be quite efficient when the gradient of the cost functions can be calculated the evolutionary algorithms provide a very useful complementary, global search technique which additionally have been shown to offer some advantages in multicriteria optimization problems. Since the performance of many fuzzy, neural network or machine learning techniques often critically rely on a number of user defined parameters, finding the optimal or at least very good set of such parameters for a given problem could be a very tedious and time consuming process. Moreover, if the models are to be deployed in a dynamically changing environment there is a need for continuous adaptation of both models and their parameters. Evolutionary algorithms and some other stochastic search techniques have proven extremely useful for such applications [40, 53] though the computational complexity involved in those searches is a definite drawback of these techniques. Some more details on how they can be used with other intelligent techniques towards achieving more flexible, smart and adaptive solutions are included in Chaps. 8 and 9.

1.2.5 Hybrid Intelligent Systems

As we can see from the above short overview of some of a wide range of intelligent technologies in existence today there is a large potential in capturing some of the desired characteristics deemed necessary for intelligent behaviour. However, it has also been frequently observed that none of the individual technologies as developed over the last half a century have been powerful enough to be considered as a clear leader on their own.

It is not surprising that there is been a strong trend in pursuing combined or hybrid approaches [1, 3, 6, 17, 27, 28, 29, 32, 33, 34, 36, 39, 40, 42, 53] that would overcome the shortcomings of individual methods while taking advantage of their complementary characteristics.

As the term hybrid (intelligent) systems is very broad it would be very difficult to cover all the possible combinations and aspects forming today's very dynamic research agenda of extended artificial intelligence [11]. We will highlight some of the most prominent hybridizations and some further examples will also be discussed in Chaps. 7, 8 and 9 and in the context of specific applications in Chaps. 10 to 17.

Neuro-fuzzy systems [2, 18, 19, 20, 21, 22, 23, 27, 34, 37, 38] are probably the most extensively researched combination of intelligent techniques with a number of demonstrated successes. The main motivation for such combination is the neural networks learning ability complimenting fuzzy systems' interpretability and ability to deal with uncertain and imprecise data. Further details on learning algorithms for fuzzy systems are covered in Chap. 7.

Some of the other hybridizations commonly used include a use of evolutionary algorithms for optimization of structures and parameters of both neural networks (evolutionary neural networks [53]) and fuzzy systems (evolutionary fuzzy systems [40]). A combination of three or more techniques in the quest for smart adaptive systems is not uncommon as also illustrated in Chaps. 8 and 9.

However, as correctly pointed out in [1] hybrid soft computing frameworks are relatively young, even comparing to the individual constituent technologies, and a lot of research is required to understand their strengths and weaknesses. Nevertheless hybridization and combination of intelligent technologies within a flexible open framework like ECOS [28] discussed in the following sections and Chap. 9, seem to be the most promising direction in achieving the truly smart and adaptive systems today.

1.3 Smart Adaptive Systems – Definitions

But what exactly do we mean by smart adaptive systems? To put some more context behind this quite broad phrase which can mean many different things in different situations, let us give some definitions which though clearly not ideal and potentially debatable will hopefully clarify and focus a little bit more the material covered in the following chapters.

As part of the work within EUNITE the following formal definitions of the words "adaptive" and "smart" have been adopted in the context of smart adaptive systems.

Due to the fact that there are systems with different levels of adaptivity in existence, for our purposes the word "adaptive" has been defined on the following three different levels which also imply an increasingly challenging applications:

1. *Adaptation to a changing environment.* In this case the system must adapt itself to a drifting (over time, space, etc.) environment, applying its intelligence to recognise the changes and react accordingly. This is probably the easiest concept of adaptation for which examples abound: customers preferences in electronic commerce system, control of non stationary systems (e.g. drifting temperature), telecommunication systems with varying channel or user characteristics.
2. *Adaptation to a similar setting without explicitly being ported to it.* Here the accent is more on the change of the environment itself than on a drift

of some features of the environment. Examples are systems that must be ported from one plant to another without a need to explicitly perform a porting of its main parameters, or a financial application that must be ported from a specific market to a similar one (e.g. a different geographical location).
3. *Adaptation to new/unknown application.* This level is the most futuristic one, but its open problems have been addressed already by a number of researchers, especially in the Machine Learning and Hybrid Intelligent Systems fields where, starting from very limited information about the problem it is possible to build a system through incremental learning.

The working definition adopted within EUNITE is that "smart" implies that intelligent techniques must be involved in the adaptation of a system. That is, to be a smart adaptive system, a system must be adaptive to one of the levels defined above and must utilise one or more of the intelligent techniques mentioned in the previous section. This, though very practical, is somewhat a very narrow definition of the word "smart" and therefore systems conforming to this definition are denoted as narrowly smart.

On the other hand, on a more general level, the word "smart" can be taken to have a meaning closer to a traditional dictionary definition of "intelligent": i.e. *the ability to learn or understand or to deal with new or trying situations; or the ability to apply knowledge to manipulate one's environment or to think abstractly as measured by objective criteria (as tests)* [The Merriam-Webster's Dictionary]. This is beyond even level 3 adaptation as defined above and we denote it as generally smart.

While examples of levels 1 and 2 applications can be found in abundance and many of such applications are discussed in the second part of the book (Chaps. 10 to 17) the level 3 problems and issues are much more challenging.

So do any attempts have been made to construct such systems? Have any general requirements/desirable characteristics for such systems been specified? What are the challenges and problems?

Let us now focus on an example of one such general framework and illustrate some of the main issues and attempt to answer some of the above questions.

1.4 General Framework for SAS

While searching for answers to the question whether smart adaptive systems exist we have come across the following definition. Paraphrasing Kasabov who was describing knowledge engineering on-line and emerging intelligence, we could also say that a smart adaptive system is a system "that develops its structure and functionality in a continuous, self-organised, adaptive, interactive way. It can use many sources of incoming information and can perform intelligent tasks, such as adaptive pattern recognition, concept formation, language learning, reasoning with uncertainty, decision making, and many more".

And in fact Kasabov [27] listed seven major requirements of future intelligent systems and proposed an open framework called Evolving Connectionist System (ECOS) [27, 28], some advanced aspects of which are discussed in Chap. 9. We will now use both the requirements and the ECOS framework as the basis for our discussion as, though with a number of open problems still to be solved, they are the closest yet to the level 3 smart adaptive system as defined in the previous section.

Seven major requirements for future intelligent/smart adaptive systems:

1. SAS should have an open, extendable and adjustable structure. This means that the system has to have the ability to create new inputs and outputs, connections, modules etc. while in operation. It should also be able to accommodate in an incremental way all the data that is know and will be known in the future about the problem.
2. SAS should adapt in an on-line, incremental, life-long mode where new data is used as soon as it is available.
3. SAS should be able to learn fast from large amount of data ideally in an "one-pass" training mode.
4. SAS should be memory-based with an ability to add, retrieve and delete individual pieces of data and information.
5. SAS should be able to improve its performance via active interaction with other systems and with the environment in a multi-modular, hierarchical fashion.
6. SAS should represent adequately space and time at their different scales, inner spatial representation, short- and long-term memory, age, forgetting, etc.
7. SAS should be able to self-improve, analyse its own performance and explain what it has learnt about the problem it is solving.

While many of the above issues have been acknowledged and addressed by researchers working with intelligent technologies since their conceptions, the focus tends to be on the individual or small subsets of the above listed points. The reason for this is that any of the issues listed are often difficult enough in themselves and either the structure, learning or the knowledge representation of the existing techniques are not adequate or flexible enough to cover all of them. And as correctly pointed out in [27] it is not likely that a truly smart adaptive system can be constructed if all of the above seven requirements are not met and therefore radically new methods and systems are required.

One such attempt at proposing a model or a general framework that addresses, at least in principle, all seven requirements is the ECOS framework. ECOS are multi-level, multi-modular, open structures which evolve in time through interaction with environment.

There are five main parts of ECOS which include:

1. *Presentation part.* In this part an input filtering, preprocessing, feature selection, input formation etc. are performed. What is interesting and something that would naturally be a consequent of acquiring new data and

information is that the dimensionality of inputs can be variable. This usually can pose serious problems with many existing statistical, neural or machine learning techniques.
2. *Representation and memory part.* In the original ECOS this part is formed by a multi-modular, evolving structure of neural network modules used for storing input patterns (information). In fact this part could consist of various functional modules from various domains as long as they could be created and optimised in an on-line manner (i.e. decision trees, neuro-fuzzy modules etc.).
3. *Higher level decision part.* This part consisting of several modules is quite critical as it assumes that a suitable feedback from environment is received and decisions about the functioning and adaptation of the whole ECOS are made here. Many unresolved research issues and important questions are still to be tackled in this part and will pose significant challenge in the near future.
4. *Action part.* In this part the decisions from the decision module are passed to the environment.
5. *Self-analysis and rule extraction modules.* This part is used for the extraction of the information from the representation and decision modules and converts it into different forms of rules, associations etc. which may be useful in the process of interpretability and transparency of the functionality and knowledge stored in ECOS model.

As mentioned in Chap. 9 there are a number of realisations of ECOS framework especially evolving, hierarchically structured neuro-fuzzy systems with the ability to grow and shrink as required. An example of such a system also include the general fuzzy min-max neural networks (GFMM) [18, 19, 20, 21, 22, 23] in which a number of requirements for SAS have been addressed and are being further developed and investigated.

An example of the process model for developing adaptive systems using machine and reinforcement learning with some components which are similar to the above ECOS framework are discussed in Chap. 5 while another example of typical general steps or functional elements required for SAS application are highlighted in Chap. 2.

So does this mean that the problem of smart adaptive systems has been solved? The answer is NO. Though significant advances have been made there are still important issues and challenges requiring intensive research.

Let us now try a take a closer look at some of those critical issues and challenges within the context of the seven requirements and the listed ECOS framework components.

1.5 Achievements, Critical Issues and Challenges

1.5.1 Data Preparation and Preprocessing

One of the most underrated but absolutely critical phases of designing SAS are data preparation and preprocessing procedures [43]. In order to highlight the importance of this stage, as part of the introductory material to the main two parts of the book, Chap. 3 has been entirely dedicated to these issues.

Why are they so important? Since the changing environments requiring adaptation usually manifest themselves in changes in the number and type of observed variables (see requirement 1 and presentation part of ECOS), it is particularly important/critical to know the implications it could have on preprocessing (e.g. dealing with noise, missing values, scaling, normalisation etc.) and data transformation techniques (e.g. dimensionality reduction, feature selection, non-linear transformations etc.) which in turn have an enormous impact on the selection and usage of appropriate modelling techniques. It is even more critical as this stage in a SAS application should be performed in an automatic fashion while currently in practise a vast majority of the data cleaning, preprocessing and transformation is carried out off-line and usually involves a considerable human expertise.

1.5.2 Adaptation and Learning

Not surprisingly the adaptation and learning algorithms should play a prominent role in SAS and in fact in one form or another are represented in all seven requirements.

There are currently two main groups of search, adaptation and learning algorithms: local (often gradient based) search/learning algorithms and global (stochastic) approaches.

The main learning algorithms in the first group have been developed over the years in statistics, neural networks and machine learning fields [4, 5, 13, 44, 45, 52]. Many of the methods are based on the optimization/minimization of a suitable cost, objective or error function. Often those algorithms also require the cost function to be differentiable and many gradient based optimization techniques have been used to derive learning rules on such basis. Unfortunately such differentiable cost functions do not always exist.

Evolutionary algorithms and other stochastic search techniques [16, 24, 26, 30] provide a complementary powerful means of a global search often used where gradient based techniques fail. As mentioned earlier these global searches usually are less computationally effective then the gradient based learning techniques.

Due to their complementary nature it is more and more common that both types of algorithms are used in combination for SAS as illustrated in Chaps. 8 and 9.

Another distinction between different learning algorithms which all are needed in SAS general framework is based on the availability and type of the training feedback signal and fall into three broad categories:

- *Supervised learning* [5, 45] – also known as learning with a teacher where both the input patterns and the desired target signal are available.
- *Unsupervised learning* [5, 45] – which does not assume the presence of the feedback signal and attempts to group the input patterns on the basis of a suitably chosen similarity measure.
- *Reinforcement learning* [49] – in which the external feedback signal is restricted to a discrete, delayed reward and punishment signals and the correct output is not known.

While a lot of research has been carried out in each of the above mentioned groups and further details and examples of some representative algorithms can be found in Chaps. 4 and 5, the nature of real problems requiring adaptation makes it also necessary that these different types of algorithms be combined [17, 18, 22, 42]. The combination of these learning algorithms is a vigorously pursued research area on its own and they are particularly relevant to the SAS requirements 1, 2, 3, 5 and 6.

With respect to the general issue of adaptive learning there are two pertinent problems which have received a lot of attention in the literature:

- *Stability/plasticity dilemma* which is associated with on-line, incremental model parameter learning and adaptation. In simple terms the problem relates to the ability of SAS to follow the changes and react quickly to such changes versus ability to filter out noisy data and exhibit stable, good overall performance.
- *Bias/variance dilemma* which is associated with model structure learning and finding and maintaining the right model complexity. In other words if the model is too simple it will not be able to capture the complex nature of the problem (often resulting in stable but over-generalising models with high bias) while too complex model could result in data overfitting and poor generalisation characteristics (models generated are often too flexible which manifests itself in a high variance error component).

With regard to adaptive parameter learning and stability/plasticity dilemma one can distinguish two main learning modes present in ECOS/SAS frameworks:

- *On-line/incremental learning* - which refers to the ability to learn new data without destroying old data and learning while system is in operation (dominated by various supervised, unsupervised and reinforcement techniques from ANN and ML [5, 13, 28, 45, 44]).
- *Life long learning* - which refers to the ability of the system to learn during its entire existence. Growing and pruning (techniques from ML, ANN and FS [18, 22, 28, 44, 52]) are the primary mechanisms in this type of learning.

The issues addressed in adaptive parameter learning procedures are particularly relevant to the SAS requirements 1, 2, 3, 4 and 6.

With regard to the structure learning/adaptation and bias/variance dilemma there are currently three main learning modes present in ECOS/SAS frameworks:

- *Constructivism* - which starts with a simple initial model structure that is grown over time [22, 28]. In these approaches the growth is controlled by similarity or output error measures.
- *Selectivism* - which starts with complex system and applies pruning procedures to avoid data overfitting [20, 28]. A criterion for pruning procedures to operate is required.
- *Hybrid approaches* - the most popular approach is to use evolutionary algorithms for structure optimization [2, 28, 53]. A suitable fitness function which is related to the optimized system performance is required.

The issues addressed in the structure learning procedures are particularly relevant to the SAS requirements 1, 5, 6 and 7.

Probably the most challenging issue within such flexible open frameworks as ECOS is the development of robust procedures for synchronisation and optimal deployment of the different types and modes of learning while ensuring the overall stability of the system.

All the above learning and adaptation approaches and modes of operation are also very tightly connected to the issues of model representation (flat or hierarchical models, rule based or connectionist etc.) and model selection/combination (single or multiple version systems etc.) briefly discussed in the following sections.

1.5.3 Model Representation

As highlighted above the different forms and modes of adaptation and learning are pivotal in creating SAS and span all seven requirements. However, all those learning procedures are closely associated with the chosen SAS model representation. This is reflected in requirements 1, 4, 5, 6 and 7 focusing on the need for open, extendable and adjustable structures, memory based, multi-modular, hierarchical and interpretable representations of the data and knowledge stored and manipulated in SAS frameworks.

Though the representation and memory part of the ECOS framework do not stipulate the use of any particular models and in principle any number and mixture of ANN, ML or HS based modules could be used, in an attempt to cover all those requirements the neuro-fuzzy models emerged as the most prominent and often used realisations of the ECOS framework.

One reason is highlighted in Chap. 7 as the neuro-fuzzy approaches with their rule based structure lead to interpretable solutions while the ability to incorporate expert knowledge and learn directly from data at the same time allows them to bridge the gap between purely knowledge-based (like the ones discussed in Chap. 6) and purely data driven methods (like for instance ANNs covered in Chap. 4).

There are many other desirable features of the fuzzy set/rule base representation with respect to the requirements for SAS discussed in Chaps. 7, 8 and 9.

Having said that, there are many methods from statistics, ANN, ML, FS and HS that have equivalent functional characteristics (i.e. classification methods are one typical example) and choosing one of them for different applications could pose a considerable problem.

So are there any good guidelines or criteria for selection of suitable methods? A number of examples can be found in Chaps. 10 to 17 which attempt to answer exactly this question from different application domain points of view but let us now consider some important issues and procedures that can be applied for model selection and combination.

1.5.4 Model Selection and Combination

One of the main problems with flexible structure models which need to be used in the SAS framework is the suitable selection of appropriate models, their combination and fusion. There are a number of potential criteria that can be used for model selection but probably the most commonly used is the performance of the models as estimated during the overall selection process following by the second criterion of interpretabilty or transparency of the models. Please see Chap. 15 for some more details of the issue of the trade-off between accuracy and interpretabilty.

One of the strongest, principled and statistically based approaches for model error estimation are various methods based on statistical resampling and cross-validation techniques [52] which are commonly used in the ML community and have not been as rigourously applied with some other intelligent methods.

The performance on the data for which models have been developed (the training data) can be often misleading and minimization of the error for the training sets can lead to the problem of data overfitting and poor generalisation performance on unseen future data. This is quite critical but since the cross-validation approaches can be used off-line with a known labelled data sets only there can be problems with direct application to the dynamically changing environments and some compromises would have to be made.

Some of the most common problems and challenges with ECOS framework and its flexible structure is that a large number of parameters need to be adjusted continuously (see Chap. 9) for the model to work well. These adjusting of the parameters is not a trivial but in fact a very critical process.

One approach to reduce the influence of individual parameters and improve the overall stability and performance of the system is to use multi-component, aggregated or combined methods like the ensemble of neural networks discussed in Chap. 4 or a number of other multi-classifier, multi-predictor etc.

methods [7, 9, 12, 18, 48]. The adaptation in such multi-component structure can be much more difficult then with individual models though.

In general, however, the combination of models can be seen as a good way of generating more robust less parameter sensitive solutions.

Another significant challenge is the fusion of different types of models (e.g. models for recognition of face, gestures, sound etc.) and the adaptation of the fusion methods themselves.

Self optimization of the whole hierarchies and multiple-units at the same time is currently the topic of challenging and interesting research and some examples are illustrated in Chaps. 8 and 9.

1.6 Conclusions

So what is the answer to our original question: Do smart adaptive systems exist? Within the context of this book and state-of-the-art of current research we could give a quantified YES answer to this question but there is still a long way to constructing truly smart adaptive systems and some significant challenges will have to be overcome first.

Much of the improvement of current intelligent systems stems from a long and tedious process of incremental improvement in existing approaches (i.e. neural networks, fuzzy systems, evolutionary computing techniques etc.). Extracting the best possible performance from known techniques requires more work of this kind, but exploration of new and combined approaches supplies additional opportunities. However, while a number of combined techniques have been very successful in certain application domains, they are usually constructed in an ad hoc manner. Another major weakness of most of the hybrid systems, as well as individual intelligent techniques, is that their successful performance heavily (sometimes critically) rely on a number of user-specified parameters. In order for such systems to be adopted as everyday tools by unskilled users the focus of the future research has to be on (semi-) automatic settings of such parameters. This is also required if such systems are to be fully adaptable to changing environments and operating conditions as required in SAS applications. Therefore, apart from the engineering challenge of building complex, hybrid, smart adaptive systems capable of accomplishing a wide range and mixture of tasks, one of the major scientific challenges consists of providing integrated computational theories that can accommodate the wide range of intellectual capabilities attributed to humans and assumed necessary for nonhuman intelligences.

As the flexibility of the smart adaptive systems increases there will also be a greater need for more sophisticated complexity control mechanisms. Integration of intelligent technologies is today vigorously pursued along several dimensions: integrating systems that support different capabilities, combining theories and methodologies that concern different facets of intelligence, and reconciling, accommodating and exploiting ideas from various disciplines. All of these dimensions still pose significant scientific and engineering challenges.

Despite all the challenges it is however unquestionable that smart adaptive intelligent systems and intelligent technology have started to have a huge impact on our everyday life and many applications can already be found in various commercially available products as illustrated in the recent report very suggestively titled: "Get smart: How intelligent technology will enhance our world" [10].

References

1. A. Abraham, "Intelligent Systems: Architectures and Perspectives", *Recent Advances in Intelligent Paradigms and Applications*, Abraham A., Jain L. and Kacprzyk J. (Eds.), Studies in Fuzziness and Soft Computing, Springer Verlag Germany, Chap. 1, pp. 1-35, Aug. 2002
2. A. Abraham, "EvoNF: A Framework for Optimization of Fuzzy Inference Systems Using Neural Network Learning and Evolutionary Computation", *2002 IEEE International Symposium on Intelligent Control (ISIC'02)*, Canada, IEEE Press, pp. 327-332, 2002
3. B. Azvine and W. Wobcke, "Human-centered Intelligent Systems and Soft Computing", *BT Technology Journal*, vol. 16, no. 3, July 1998
4. J.C. Bezdek, *Pattern Recognition with Fuzzy Objective Function Algorithms*, Plenum, New York, 1981
5. C.M. Bishop, *Neural Networks for Pattern Recognition*, Clarendon Press: Oxford, 1995
6. P. Bonissone, Y-T. Chen, K. Goebel, P.S. Khedkar, "Hybrid Soft Computing Systems: Industrial and Commercial Applications", *Proc. Of the IEEE*, vol. 87, no. 9, Sep. 1999
7. L. Breiman, "Bagging predictors", *Machine Learning*, vol. 24, no. 2, pp. 123-140, 1996
8. T.G. Dietterich, "Machine Learning Research: Four Current Directions", *AI Magazine*, vol. 18, no. 4, pp. 97-136, 1997
9. T.G. Dietterich, "An Experimental Comparison of Three Methods for Constructing Ensembles of Decision Trees: Bagging, Boosting, and Randomization", *Machine Learning*, vol. 40, pp. 139-157, 2000
10. C. Doom, "Get Smart: How Intelligent Technology Will Enhance Our World", *CSC's Leading Edge Forum Report* (www.csc.com), 2001
11. J. Doyle, T. Dean et al., "Strategic Directions in Artificial Intelligence", *ACM Computing Surveys*, vol. 28, no. 4, Dec. 1996
12. H. Drucker, C. Cortes, L.D. Jackel, Y. LeCun, and V. Vapnik, "Boosting and other ensemble methods", *Neural Computation*, vol. 6, pp. 1289-1301, 1994
13. R.O. Duda, P.E. Hart, D.G. Stork, *Pattern Classification*, John Wiley and Sons, Inc., 2000
14. U.M. Fayyad, G. Piatetsky-Shapiro, P. Smyth, and R. Uthurusamy, Eds., *Advances in Knowledge Discovery and Data Mining*, Menlo Park, CA: AAAI/MIT Press, 1996

15. U. Fayyad, G.P. Shapiro, and P. Smyth, "TheKDDprocess for extracting useful knowledge from volumes of data", *Commun. ACM*, vol. 39, pp. 27-34, 1996
16. D.B. Fogel, Evolving neural networks, *Biol. Cybern.*, vol. 63, pp. 487-493, 1990
17. B. Gabrys and L. Petrakieva, "Combining labelled and unlabelled data in the design of pattern classification systems", *International Journal of Approximate Reasoning*, vol. 35, no. 3, pp. 251-273, 2004
18. B. Gabrys, "Learning Hybrid Neuro-Fuzzy Classifier Models From Data: To Combine or not to Combine?", *Fuzzy Sets and Systems*, vol. 147, pp. 39-56, 2004
19. B. Gabrys, "Neuro-Fuzzy Approach to Processing Inputs with Missing Values in Pattern Recognition Problems", *International Journal of Approximate Reasoning*, vol. 30, no. 3, pp. 149-179, 2002
20. B. Gabrys, "Agglomerative Learning Algorithms for General Fuzzy Min-Max Neural Network", the special issue of the *Journal of VLSI Signal Processing Systems* entitled "Advances in Neural Networks for Signal Processing", vol. 32, no. $^{1}/_{2}$, pp. 67-82, 2002
21. B. Gabrys, "Combining Neuro-Fuzzy Classifiers for Improved Generalisation and Reliability", *Proceedings of the Int. Joint Conference on Neural Networks*, (IJCNN'2002) a part of the WCCI'2002 Congress, ISBN: 0-7803-7278-6, pp. 2410-2415, Honolulu, USA, May 2002
22. B. Gabrys and A. Bargiela, "General Fuzzy Min-Max Neural Network for Clustering and Classification", *IEEE Transactions on Neural Networks*, vol. 11, no. 3, pp. 769-783, 2000
23. B. Gabrys and A. Bargiela, "Neural Networks Based Decision Support in Presence of Uncertainties", *ASCE J. of Water Resources Planning and Management*, vol. 125, no. 5, pp. 272-280, 1999
24. D.E. Goldberg, *Genetic Algorithms in Search, Optimization and machine Learning*, Addison-Wesley, Reading, MA, 1989
25. S. Haykin, *Neural Networks. A Comprehensive Foundation*, Macmillan College Publishing Company, New York, 1994
26. J.H. Holland, *Adaptation in natural and artificial systems*, The University of Michigan Press, Ann Arbor, MI, 1975
27. N. Kasabov and R. Kozma (Eds.), *Neuro-Fuzzy Techniques for Intelligent Information Systems*, Physics Verlag, 1999
28. N. Kasabov, *Evolving connectionist systems – methods and applications in bioinformatics, brain study and intelligent machines*, Springer Verlag, London-New York, 2002
29. R. Khosla and T. Dillon, *Engineering Intelligent Hybrid Multi-Agent Systems*, Kluwer Academic Publishers, 1997
30. J.R. Koza, *Genetic Programming*, MIT Press, 1992
31. L.I. Kuncheva, *Fuzzy Classifier Design*, Physica-Verlag Heidelberg, 2000
32. C.T. Leondes (Ed.), *Intelligent Systems: Technology and Applications* (vol. 1 – Implementation Techniques; vol. 2 – Fuzzy Systems, Neural Networks, and Expert Systems; vol. 3 – Signal, Image and Speech Processing; vol. 4 – Database and Learning Systems; vol. 5 – Manufacturing, Industrial, and Management Systems; vol. 6 – Control and Electric Power Systems), CRC Press, 2002
33. L.R. Medsker, *Hybrid Intelligent Systems*, Kluwer Academic Publishers, 1995
34. S. Mitra and Y. Hayashi, "Neuro-Fuzzy Rule Generation: Survey in Soft Computing Framework", *IEEE Trans. on Neural Networks*, vol. 11, no. 3, pp. 748-768, May 2000

35. T.M. Mitchell, "Machine learning and data mining", *Commun. ACM*, vol. 42, no. 11, pp. 477-481, 1999
36. S. Mitra, S.K. Pal and P. Mitra, "Data Mining in Soft Computing Framework: A Survey", *IEEE Trans. on Neural Networks*, vol. 13, no. 1, pp. 3-14, 2002
37. D. Nauck and R.Kruse, "A neuro-fuzzy method to learn fuzzy classification rules from data", *Fuzzy Sets and Systems*, vol. 89, no. 3, pp. 277-288, 1997
38. D. Nauck and R. Kruse. "Neuro-fuzzy systems for function approximation", *Fuzzy Sets and Systems*, vol. 101, pp. 261-271, 1999
39. S. Pal, V. Talwar, P. Mitra, "Web Mining in Soft Computing Framework: Relevance, State of the Art and Future Directions", To appear in IEEE Trans. on Neural Networks, 2002
40. W. Pedrycz (Ed.), *Fuzzy Evolutionary Computation*, Kluwer Academic Publishers, USA, 1997
41. W. Pedrycz and F.Gomide, *An Introduction to Fuzzy Sets: Analysis and Design*, The MIT Press, 1998
42. L. Petrakieva and B. Gabrys, "Selective Sampling for Combined Learning from Labelled and Unlabelled Data", In the book on *"Applications and Science in Soft Computing"* published in the Springer Series on Advances in Soft Computing, Lotfi, A. and Garibaldi, J.M. (Eds.), ISBN: 3-540-40856-8, pp. 139-148, 2004
43. D. Pyle, *Data Preparation for Data Mining*, Morgan Kaufman Publishers, 1999
44. J.R. Quinlan, *C4.5: Programs for Machine Learning*, Morgan Kaufman, San Mateo, CA, 1993
45. B.D. Ripley, *Pattern Recognition and Neural Networks*, Cambridge University Press, 1996
46. E.H. Ruspini, P.P. Bonissone, and W. Pedrycz (Eds.), *Handbook of Fuzzy Computation*, Institute of Physics, Bristol, UK, 1998
47. D. Ruta, and B. Gabrys, "An Overview of Classifier Fusion Methods", *Computing and Information Systems* (Ed. Prof. M. Crowe), University of Paisley, vol. 7, no. 1, pp. 1-10, 2000
48. D. Ruta, and B. Gabrys, "Classifier Selection for Majority Voting", Special issue of the journal of *INFORMATION FUSION* on Diversity in Multiple Classifier Systems, vol. 6/1 pp. 63-81, 2004
49. R.S. Sutton and A.G. Barto, *Reinforcement Learning: An Introduction*, MIT Press, 1998
50. G. Tselentis and M. van Someren, "A Technological Roadmap for Computational Intelligence and Learning", Available at www.dcs.napier.ac.uk/coilweb/, June 2000
51. V. Vapnik, *The nature of statistical learning theory*, Springer Verlag, New York, 1995
52. S.M. Weiss and C.A. Kulikowski, *Computer Systems That Learn: Classification and Prediction Methods from Statistics, Neural Nets, Machine Learning, and Expert Systems*, Morgan Kaufman Publishers, Inc., 1991
53. X. Yao, "Evolving Artificial Neural Networks", *Proc. of the IEEE*, vol. 87, no. 9, pp. 1423-47, Sep. 1999
54. L. Zadeh, "Fuzzy sets", *Information and Control*, vol. 8, pp. 338-353, 1965
55. L. Zadeh, "Fuzzy logic and the calculi of fuzzy rules and fuzzy graphs: A precis". *Int. J. Multiple-Valued Logic*, vol. 1, pp. 1-38, 1996
56. L. Zadeh, "Soft Computing and Fuzzy Logic", *IEEE Software*, pp. 48-56, Nov. 1994

2

Problem Definition – From Applications to Methods

K. Leiviskä

University of Oulu, Department of Process and Environmental Engineering,
P.O. Box 4300, 90014 Oulun yliopisto, Finland
Kauko.Leiviska@oulu.fi

Are there any general features that make some application especially suitable for Smart Adaptive Systems (SAS)? These systems are now technologically available, even though commercial applications are still few. And what things are essential when starting to decide, define and design this kind of application? What makes it different from the usual IT-project? Does adaptation itself mean increasing complexity or need for special care? This chapter tries to answer these questions at the general level under the title "Problem Definition – From Applications to Methods" which also serves as an introduction to the second part of the book. Each case has its own requirements, but some common themes are coming out. The author's own experience is in industrial systems, which is, inevitably, visible in the text.

Section 2.1 looks at adaptation from the Control Engineering point of view; it is an area where practical applications have existed for four decades and some analogies exist with other application areas. Also intelligent methods are used there and therefore we can speak about genuine SAS applications in control. Section 2.2 looks at the need for adaptation in different disciplines and tries to give characteristics for fruitful application areas together with economical potential. Section 2.3 lists some topics that should be kept in mind when defining the SAS problem. Section 2.4 is a short summary.

2.1 Introduction

Many applications of intelligent methods, including fuzzy logic, neural networks, methods from machine learning, and evolutionary computing, have been recently launched especially in cases were an explicit analytical model is difficult to obtain. Examples come from the domains of industrial production, economy, healthcare, transportation, and customer service applications in communications, especially Internet. While being powerful and contributing to the increased efficiency and economy, most solutions using intelligent methods lack one important property: they are not adaptive (or not adaptive enough), when the application environment changes. In other words, most such systems have to be redesigned when the system settings or parameters change significantly.

This may look strange, because at the first sight adaptivity seems to be the central issue for intelligent technology. The typical "learning by example" concept combined with techniques inspired by biological concepts is supposed to provide enhanced adaptivity. However, after looking more carefully, it becomes clear that most techniques for "learning" are used for one-time estimation of models from data, which remain fixed during routine application. "True" online learning, in most cases, has not reached a level mature enough for industrial applications.

Adaptation is not a new concept in Systems Science and early definitions origin from the 1950's. Control Engineering has also taken the advantage of adaptive systems, when introducing adaptive controllers; controllers with adjustable parameters and a mechanism for adjusting them. Further on, Control Engineering applications include the main features of adaptivity that are, in a way, comparable to the three levels of Smart Adaptive Systems (SAS) defined earlier in this book:

1. Gain scheduling, that adapts the controller to the known changes in the environment. The effects of changes must, of course, be modelled and the system must include an adaptive mechanism that can compensate for these effects.
2. Model Reference Adaptive Control that adapts the system parameters so that the system response follows the recommended model behaviour.
3. Self-Tuning Control that adapts the controller so that the system response remains optimal in the sense of an objective function under all circumstances.

However, it should be remembered that the abovementioned Control Engineering methods are mostly concerned with parameter adaptation, which is only one alternative when speaking about Intelligent Systems. Structural adaptation is also needed in many real-world cases.

2.2 Why to Use SAS?

Why to use or not to use adaptive systems? What is the reason for introducing adaptation in existing constant parameter (or structure) systems? There are two main reasons to evaluate: the need and the effect.

Need for adaptation comes from many reasons and it varies from one application to another, but some general features can be given. In industrial production, customer requirements vary and lead to several products and grades that require their own ways of production and control. This may require adaptation in the production system and especially in its quality monitoring and control systems together with process diagnosis. Production in short series that is becoming more and more common in every industry sets the most stringent requirements.

In many economic systems, e.g. in banking and insurance, the varying customer demands and profiles have the same kind of effect on the need to adapt to changing situations. Also the security reasons, e.g. fighting against fraud, come into the picture, here. In healthcare, each human is an own case requiring own medication and care in different stages of life. Safety seems to work to two opposite directions in healthcare: increasing safety would require adaptation, but in many cases instead of automatic adaptation a tailored system for each patient is needed.

All abovementioned needs seem to come together in Internet. There are many customers and customer groups with different needs. Every individual would like to have the system tailored for his/her own needs. And the safety aspect must be taken into account in every application.

The effect can be approached from many directions, but only the economical effects are accounted for here. Usually they offer the strongest criteria for the selection of adaptive systems, usually also coming before the need. Of course, the effects to users, customers, image, etc. cannot be underestimated.

The economical benefits of SAS seem to come from the same sources as other IT investments:

- Increased throughput, better quality and higher efficiency that effect directly on the economy of the company and can usually be calculated in money.
- Better quality of operations and services that effect on the market position and bring along economical benefits.
- Improved working conditions and safety that increase the staff motivation and also improve the company image.
- Improved overall performance.

There are some common features in companies and actors that can gain in using SAS[1]:

- Tightening quality requirements characteristic to high quality products and services, together with requirements to customise the products depending on the customer [1, 2, 3, 4, 5, 6, 7, 8, 9, 10, 11, 12, 13, 14].
- Increasing throughput meaning high capacity requirements to the process or mass productisation, and services offered to big audiences (e.g. in Internet) or to a high number of customers (air traffic) [15, 16, 17, 18, 19, 20].
- Complicated functional structures e.g. industry with a high number of processes, mills, customers, and products or services offered to high number of customers with varying customer profiles, needs and capabilities in using IT applications [21, 22, 23, 24, 25, 26, 27, 28, 29, 30].
- Capital-intensive investments including high economical risk in production decisions [31, 32, 33].

---

[1] The references given here are from three Eunite Annual Symposiums

- Rapidly changing markets that lead to needs to adapt to changing environment, customer needs, and economical situation [34, 35].
- Solving difficult diagnosis problems in industry, medicine and economy, especially safety critical applications including a high technological risk in production, risks to humans in medicine, high economical risk in offering services, and possibilities for misuse and fraud [39, 40, 41, 42, 43, 44, 45, 46, 47, 48, 49, 50, 51].
- Innovative companies/company image that leads to using the up-to-date technology.
- Market push from technology companies or need for intelligent products.

In some areas, there is a general push to intelligent technologies from system providers and vendors. However, the attitude of companies varies. Vendor companies are more or less selling results, not technologies as such. SAS have more a role of tools; they have been embedded in the system and concerned only if there is some doubt of the results. Companies depending on SAS as products are, of course, more actively marketing also this technology.

2.3 How to Build SAS?

Defining the SAS problem includes the specification of at least five functional elements or tasks that together form an SAS. To put them shortly, the elements are: DETECT, ADAPT, IMPLEMENT, INDICATE and EVALUATE. Next these elements are considered in some detail.

2.3.1 DETECT

DETECT means monitoring of the signals (in a wide sense) that tell when the adaptation activity is triggered. First, the signals must be named. In technical applications, this means the definition of external measurements, grade change times, etc. In other applications, e.g. the development of certain features in customer profiles is followed. The procedures to assure for data quality and security must be defined in the same connection together with possible thresholds or safety limits that must be exceeded before triggering the adaptation.

Adaptation may also be triggered in the case of deterioration in the system performance. The performance factor should be defined together with the limits that cannot be exceeded without adaptation in the system structure or parameters.

2.3.2 ADAPT

ADAPT means the mechanism used in changing the system. The calculation routines or methods must be defined in the early stage of the design. Adaptation is in many cases based on models, and the conventional stages of model design must be followed. In data-intensive cases, a lot of data pre-processing is usually required and these routines must be defined. If expert knowledge is needed, its acquisition and updating must be clarified in the start. If modelling requires experimental work, as usual in technical applications, systematic experiment design must be used.

In choosing the methods, rule-based and data-based paradigms compete and no general guidelines are possible. Application determines the method and some tips and hints are given in the further chapters. Of course, the availability and costs play a big role here.

2.3.3 IMPLEMENT

IMPLEMENT means the ways, how the actual change in the system is carried out and these mechanisms must be defined carefully case by case. In technical systems, IMPLEMENT is usually integrated with ADAPT, but in service applications in web it would mean that information shown or activities allowed are adapted based on the changes in the customer structure or in the services required. It is quite curious to see that safety plays a remarkable role in both applications. In technical systems the question is to secure the system integrity and prevent changes that would be harmful or impossible to realize. In other systems, the changes that would endanger e.g. customer privacy should not be allowed.

2.3.4 INDICATE

The act of adaptation should be transparent to the user. He/she should know when the changes are made and how they effect on the system performance. This is especially important in cases where the user can interfere the system operation and via own activity cancel the changes made by adaptive systems. This element calls for careful user interface design and user involvement in the early design phases of the whole systems. It also underlines the need for user training.

The ways to indicate are many. In simple process controllers, green lamp light is showing that the adaptive controller is on. In other system, the changes may be recorded in the system log. In some cases, the user acceptance may also be asked before implementing the change.

2.3.5 EVALUATE

The system performance should be continuously evaluated to recognize the effects of adaptation. The performance criteria are agreed in the start and developed, if necessary, when more experience is available. The correct criteria are, of course, especially important, if they are used in triggering and guiding the adaptation.

2.4 Summary

This chapter introduced some special themes necessary in defining the SAS problem. Many points of view are possible in deciding, defining and designing Smart Adaptive Systems. Need for adaptivity and its costs and benefits come in the first line. Next, some practical aspects that make SAS project different from usual IT-projects must be recognized.

One last comment is necessary. The availability of systems and tools is crucial for gaining further ground for smart adaptive systems. Pioneers are, of course, needed, but established, in-practice-proven, commercial systems are needed in the market. This fact is clearly visible when looking at the life cycle of any new paradigm from an idea to the product.

References[2]

1. Garrigues, S., Talou, T., Nesa, D. (2001): Neural Networks for QC in Automotive Industry: Application to the Discrimination of Polypropylene Materials by Electronic Nose based on Fingerprint Mass Spectrometry. Eunite 2001.
2. Rodriguez, M.T., Ortega, F., Rendueles, J.L., Menendez, C. (2002): Determination of the Quality of Hot Rolled Steel Strips With Multivariate Adaptive Methods. Eunite 2002.
3. Oestergaard, J-J. (2003): FuzEvent & Adaptation of Operators Decisions. Eunite 2003.
4. Rodriguez, M.T., Ortega, F., Rendueles, J.L., Menéndez, C. (2003): Combination of Multivariate Adaptive Techniques and Neural Networks for Prediction and Control of Internal Cleanliness in Steel Strips. Eunite 2003.
5. Leiviskä, K., Posio, J., Tirri, T., Lehto, J. (2002): Integrated Control System for Peroxide Bleaching of Pulps. Eunite 2002.
6. Zacharias, J., Hartmann, C., Delgado, A. (2001): Recognition of Damages on Crates of Beverages by an Artificial Neural Network. Eunite 2001.
7. Linkens, D., Kwok, H.F., Mahfouf, M., Mills, G. (2001): An Adaptive Approach to Respiratory Shunt Estimation for a Decision Support System in ICU. Eunite 2001.
8. Chassin, L. et al. (2001): Simulating Control of Glucose Concentration in Subjects with Type 1 Diabetes: Example of Control System with Inter-and Intra-Individual Variations. Eunite 2001.
9. Mitzschke, D., Oehme, B., Strackeljan, J. (2002): User-Dependent Adaptability for Implementation of an Intelligent Dental Device. Eunite 2002.
10. Ghinea, G., Magoulas, G.D., Frank, A.O. (2002): Intelligent Multimedia Transmission for Back Pain Treatment. Eunite 2002.

---
[2] Note: References are to three Eunite Annual Symposium CD's: Eunite 2001 Annual Symposium, Dec. 13–14, 2001, Puerto de la Cruz, Tenerife, Spain. Eunite 2002 Annual Symposium, Sept. 19–21, 2002, Albufeira, Algarve, Portugal. Eunite 2003 Annual Symposium, July 10–11, 2003, Oulu, Finland.

11. Emiris, D.M., Koulouritis, D.E. (2001): A Multi-Level Fuzzy System for the Measurement of Quality of Telecom Services at the Customer Level. Eunite 2001.
12. Nürnberger, A., Detyniecki, M. (2002): User Adaptive Methods for Interactive Analysis of Document Databases. Eunite 2002.
13. Raouzaiou, A., Tsapatsoulis, N., Tzouvaras, V., Stamou, G., Kollias, S. (2002): A Hybrid Intelligence System for Facial Expression Recognition. Eunite 2002.
14. Balomenos, T., Raouzaiou, A., Karpouzis, K., Kollias, S., Cowie, R. (2003): An Introduction to Emotionally Rich Man-Machine Intelligent Systems. Eunite 2003.
15. Arnold, S., Becker, T., Delgado, A., Emde, F., Follmann, H. (2001): The Virtual Vinegar Brewer: Optimizing the High Strength Acetic Acid Fermentation in a Cognitive Controlled, Unsteady State Process. Eunite 2001.
16. Garcia Fernendez, L.A., Toledo, F. (2001): Adaptive Urban Traffic Control using MAS. Eunite 2001.
17. Hegyi, A., de Schutter, B., Hellendoorn, H. (2001): Model Predictive Control for Optimal Coordination of Ramp Metering and Variable Speed Control. Eunite 2001.
18. Niittymäki, J., Nevala, R. (2001): Traffic Signals as Smart Control System – Fuzzy Logic Based Traffic Controller. Eunite 2001.
19. Jardzioch, A.J., Honczarenko, J. (2002). An Adaptive Production Controlling Approach for Flexible Manufacturing System. Eunite 2002.
20. Kocka, T., Berka, P., Kroupa, T. (2001): Smart Adaptive Support for Selling Computers on the Internet. Eunite 2001.
21. Macias Hernandez, J.J. (2001): Design of a Production Expert System to Improve Refinery Process Operations. Eunite 2001.
22. Guthke, R., Schmidt-Heck, W., Pfaff, M. (2001): Gene Expression Based Adaptive Fuzzy Control of Bioprocesses: State of the Art and Future Prospects. Eunite 2001.
23. Larraz, F., Villaroal, R., Stylios, C.D. (2001): Fuzzy Cognitive Maps as Naphta Reforming Monitoring Method. Eunite 2001.
24. Woods, M.J. et al. (2002): Deploying Adaptive Fuzzy Systems for Spacecraft Control. Eunite 2002.
25. Kopecky, D., Adlassnig, K-P., Rappelsberger, A. (2001): Patient-Specific Adaptation of Medical Knowledge in an Extended Diagnostic and Therapeutic Consultation System. Eunite 2001.
26. Palaez Sanchez, J.I., Lamata, M.T., Fuentes, F.J. (2001): A DSS for the Urban Waste Collection Problem. Eunite 2001.
27. Fernandez, F., Ossowski, S., Alonso, E. (2003): Agent-based decision support services. The case of bus fleet management. Eunite 2003.
28. Hernandez, J.Z., Carbone, F., Garcia-Serrano, A. (2003): Multiagent architecture for intelligent traffic management in Bilbao. Eunite 2003.
29. Sterrit, R., Liu, W. (2001): Constructing Bayesian Belief Networks for Fault Management in Telecommunications Systems. Eunite 2001.
30. Ambrosino, G., Logi, F., Sassoli, P. (2001): eBusiness Applications to Flexible Transport and Mobility Services. Eunite 2001.
31. Koulouriotis, D.E., Diakoulakis, I.E., Emiris, D.M. (2001): Modeling Investor Reasoning Using Fuzzy Cognitive Maps. Eunite 2001.
32. Danas, C.C. (2002): The VPIS System: A New Approach to Healthcare logistics. Eunite 2002.

33. Heikkilä, M. (2002): Knowledge Processing in Real Options. Eunite 2002.
34. Nelke, M. (2001): Supplier Relationship Management: Advanced Vendor Evaluation for Enterprise Resource Planning Systems. Eunite 2001.
35. Carlsson, C., Walden, P. (2002): Mobile Commerce: Enhanced Quests for Value-Added Products and Services. Eunite 2002.
36. Biernatzki, R., Bitzer, B., Convey, H., Hartley, A.J. (2002): Prototype for an Agent Based Auction System for Power Load Management. Eunite 2002.
37. Dounias, G.D., Tsakonas, A., Nikolaidis, E. (2003): Application of Fundamental Analysis and Computational Intelligence in Dry Cargo Freight Market. Eunite 2003.
38. Riedel, S., Gabrys, B. (2003): Adaptive Mechanisms in an Airline Ticket Demand Forecasting Systems. Eunite 2003.
39. Adamopoulos, A. et al. (2001): Adaptive Multimodel Analysis of the Magnetoencephalogram of Epileptic Patients. Eunite 2001.
40. Adamopoulos, A., Likothanassis, S. (2001): A Model of the Evolution of the Immune System Using Genetic Algorithms. Eunite 2001.
41. Tsakonas, A., Dounias, G.D., Graf von Keyserlingk, D., Axer, H. (2001): Hybrid Computational Intelligence for Handling Diagnosis of Aphasia. Eunite 2001.
42. Poel, M., Ekkel, T. (2002): CNS Damage Classification in Newborn Infants by Neural Network Based Cry Analysis. Eunite 2002.
43. Hernando Perez, E., Gomez, E.J., Alvarez, P., del Pozo, F. (2002): Intelligent Alarms Integrated in a Multi-Agent Architecture for Diabetes Management. Eunite 2002.
44. Castellano, G., Fanelli, A.M., Mencar, C (2003): Discovering Human Understandable Fuzzy Diagnostic Rules From Medical Data. Eunite 2003.
45. Janku, L., Lhotska, L., Kreme, V. (2002): Additional Method for Evaluation of the Training Capturing: Intelligent System for the Estimation of the Human Supervisory Operator Stress Level under the Normal Safety Critical Circumstances. Eunite 2002.
46. Marino, P. et al. (2002): Development of a Hybrid System for a Total In-line Inspection. Eunite 2002.
47. Juuso, E., Gebus, S. (2002): Defect Detection on Printed Circuit Boards using Lingustic Equations Applied to Small Batch Processes. Eunite 2002.
48. Kivikunnas, S., Koskinen, J., Lähdemäki, R., Oksanen, P. (2002): Hierarchical Navigation and Visualisation System for Paper Machine. Eunite 2002.
49. Vehi, J., Mujica, L.E. (2003): Damage Identification using Case Based Reasoning and Self Organizing Maps. Eunite 2003.
50. Döring, C., Eichhorn, A., Klose, A., Kruse, R. (2003): Classification of Surface Form Deviations for Quality Analysis. Eunite 2003.
51. Christova, N., Vachkov, G. (2003): Real Time System Performance Evaluation By Use of Mutually Excluding Relations Analysis. Eunite 2003.
52. Strackeljan, J., Dobras, J. (2003): Adaptive Process and Quality Monitoring using a new LabView Toolbox. Eunite 2003,
53. Ahola, T., Kumpula, H., Juuso, E. (2003): Case Based Prediction of Paper Web Break Sensitivity. Eunite 2003.
54. Zmeska, Z. (2002): Analytical Credit Risk Estimation Under Soft Conditions. Eunite 2002.

3
Data Preparation and Preprocessing

D. Pyle

Consultant, Data Miners Inc.,
77 North Washington Street,
9th Floor Boston, MA 02114, United States
dpyle@modelandmine.com

The problem for all machine learning methods is to find the most probable model for a data set representing some specific domain in the real world. Although there are several possible theoretical approaches to the problem of appropriately preparing data, the only current answer is to treat the problem as an empirical problem to solve, and to manipulate the data so as to make the discovery of the most probable model more likely in the face of the noise, distortion, bias, and error inherent in any real world data set.

3.1 Introduction

A model is usually initially fitted to (or discovered in) data that is especially segregated from the data to which the model will be applied at the time of its execution. This is usually called a training data set. This is not invariably the case, of course. It is common practice to fit a model to a data set in order to explain that specific data set in simplified and summarized terms – those of the model. Such a model is a simplification and summarization of the data that is more amenable to human cognition and intuition than the raw data. However, adaptive systems are generally not intended primarily to model a specific data set for the purposes of summary. Adaptive systems, when deployed, are (among other things) usually intended to monitor continuous or intermittent data streams and present summarized inferences derived from the training data set that are appropriate and justified by the circumstances present in the execution, or run-time data. Thus, any modification or adjustment that is made to the training data must equally be applicable to the execution data and produce identical results in both data sets. If the manipulations are not identical, the result is operator-induced error – bias, distributional nonstationarity, and a host of unnecessary problems – that are introduced directly as a result of the faulty run-time data manipulation. Transformations of raw data values are generally needed because the raw algorithms prevalent today cannot "digest" all types of data. For instance, in general most popular neural network algorithms require the data to be presented to the neurons as numbers since what is happening inside the neurons is often a form of logistic regression. Similarly, many types of decision trees require data to be presented as categories and this entails a requirement to convert numbers into categories if numeric variables are to be used. And regardless of whether these, or other types of transformations are needed, it is crucial that any transformation follow Hippocrates' admonition: "First, do no harm". That, in fact, is the ultimate purpose of preparing data – so that after preparation it will not harm the chosen algorithm's ability to discover the most probable model. In fact, in the best case it should enhance the algorithm's ability.

The necessary transformations can be grouped into 5 general topics:

- Modifying range.
- Modifying distributions.
- Finding an appropriate numeric representation of categories.
- Finding an appropriate categorical representation of numeric values, either continuous or discrete.
- Finding an appropriate representation for nulls or missing values.

Following this section, the chapter provides a separate section for each of these topics followed by a short section of practical considerations and a summary section. One or more – or conceivably none – of the transformations mentioned may be needed, and the need for transformation depends principally on the requirements of the algorithm that is to be applied in fitting or discovering the needed model. Further, any such data manipulations need to consider the computational complexity involved in applying them at run time since many applications are time sensitive, and too much complexity (i.e., too slow to produce) may render even optimal transformations impractical or irrelevant by the time they are available. They may also need to be robust in the face of greater variance than that experienced in the training data set.

3.2 Transmitting Information

A data set is a device for transmitting information. In a supervised model (one where the outcome is known in all cases in the training data), the collection of variables that forms the input battery transmits information so that for a given input uncertainty about the value of the output is reduced. In an unsupervised model (where no outcome is specified), information is transmitted about the various characteristic, or typical, states of the data set as a whole, and the modeler then evaluates those typical states for relevance to world states of interest. Thus, there are always output states of interest, but the discussion is more concrete for supervised models when the outcomes of interest are explicitly represented in the data set, and they are the main focus of this discussion for that reason.

Information theory presents a method of measuring information transmission that is insensitive to the mode of transmission. Yet the modeling tools available today are far from insensitive to transmission mode. Information can be regarded as being transmitted through 3 modes, usefully labeled as:

- Linear modes
  These include actually linear relationships and transformably linear relationships (those which are curvilinear but can be expressed in linear form).
- Functional modes
  These include non-linear and other relationships that can be expressed as mathematical functions. (A function produces a unique output value for a unique set of input values.) These also include information transmitted through interaction effects between the variables of the input battery.
- Disjoint modes
  These include relationships that cannot be represented by functions. Examples of disjoint modes include non-convex clusters and relationships that are non-contiguous, and perhaps orthogonal or nearly so, in adjacent parts of the input space.

This characterization is useful only because machine-learning (and statistical modeling) algorithms are generally sensitive to these differences in the data structures corresponding to these modes. Almost all algorithms can easily characterize linear mode relationships. Many can characterize functional mode relationships, especially if they are known to exist and the algorithm is conditioned appropriately. Few current algorithms can characterize disjoint mode relationships even when they are known to be present. Thus, if information transmitted through a more difficult to characterize mode can be transformed into an easier to characterize mode through data preparation, while at the same time neither removing useful information from the data set nor adding spurious information, then data preparation makes the modelers' job far easier – moreover, sometimes it is only through adequate data preparation that a satisfactory model can be discovered at all. Information theory has a nomenclature of its own, none of which, useful as it is, is needed to understand the effects of data preparation or preprocessing. To make this discussion concrete, general remarks and comments are illustrated by examples that are taken from real world data sets that are available for exploration[1]. The essential intent of data preparation is to modify the data in ways that least disturb the information content of a data set while most facilitating transmission of information through the relevant transmission mode(s) dependant on the needs of the tool.

---

[1] The data sets discussed in this chapter are available for download from the author's personal web site, www.modelandmine.com

3.3 Modifying Range

Many algorithms require the values of numeric variables to fall into some specific range, often 0 to 1 or +1 to −1. Range modification is univariate, which is to say that each variable is adjusted for range, if necessary, considering only the values of that variable and not in relationship to other variables. As far as making such modifications in the training data set is concerned, little comment is needed since the procedure is simple and straightforward – so long as it is only in the training data set that range modification is needed. However, if the final model is to be applied to run-time data and if it is possible that the run-time data may contain values that fall outside the limits discovered in the training data, then some accommodation has to be made for any run-time out-of-range values. Some algorithms require that any numeric values are turned into categories, and this is usually done through the process of binning. Binning is the assignment of ranges of values of numeric variables to particular categories. However, there are a wide variety of binning methods, some of which are discussed in more detail later in this chapter. Regardless of which binning method is adopted, it is a common practice and also a good practice of data modeling, to create 2 more bins than the binning method requires – one for assigning run-time values that fall below the range of values found in the training data, the other for those that fall above. Since any model constructed in the training data will, by definition, never be exposed to out-of-range values, the model cannot validly make inferences for such values. What action the modeler chooses to take depends on the needs of the model and the situation and is outside the realm of data preparation. When the modeling algorithm of choice requires variables in a continuous numeric form, out-of-range values still present a problem. One way of dealing with such values is to "squash" them back into the training range. There are several methods for squashing values, and a popular one is softmax scaling. The expression for softmax scaling is shown here in two steps for ease of presentation:

$$x = \frac{(\nu - \nu_{mean})}{\lambda \left(\frac{\sigma_\nu}{2\pi}\right)} \tag{3.1}$$

$$s = \frac{1}{1 + e^{-x}} \tag{3.2}$$

where ν_{mean} is the mean of ν and σ the standard deviation.

Figure 3.1 shows the effect of softmax scaling. Similarly to the logistic function, regardless of the value of the input, the output value lies between 0 and 1. The value of λ determines the extent of the linear range. For purposes of illustration the figure shows a much shorter linear section of the softmax curve than would be used in practice. Usually 95%–99% of the curve would be linear so that only out-of-training-range values would be squashed.

Squashing out-of-range values, using this or some similar technique, places them in the range over which the model was trained – but whether this is a

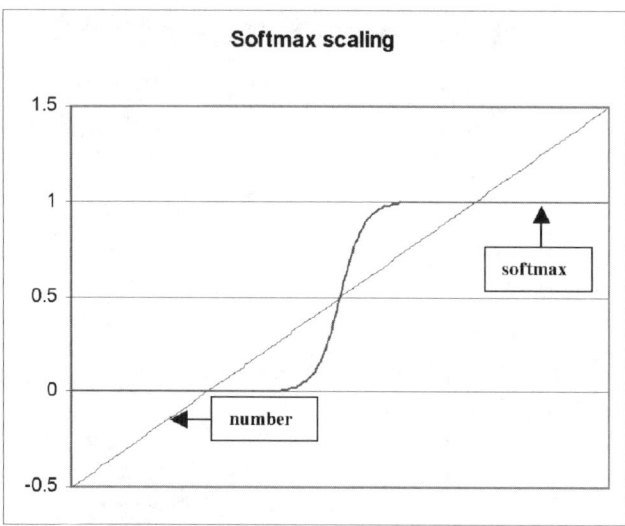

Fig. 3.1. The effect of softmax scaling

valid approach to dealing with the issue depends on the needs of the model. Out-of-range values are still values that were not present in the training data, and are values which the model has not characterized during its creation, whether the values are squashed back inside the training range or not. Range manipulation, including the use of out-of-range bins and squashing, has no effect on the information content of the data set. Measures of information are, in general, insensitive to the information transmission mode whether it be linear or not, and introducing potential curvilinearities or nonlinearities makes no difference to the total information transmitted. It does, of course, change the mode of information transmission, and that may – almost certainly will – have an impact on the modeling tool chosen.

3.4 Modifying Distributions

Modeling tools identify the relationships between variables in a data set – and in the case of supervised modeling, specifically the relationship between the variables in the input battery and the variable(s) in the output battery. In a sense any such modeling tool has to characterize the multivariate distribution of the input battery and relate features in that distribution to features in the (if necessary multivariate) distribution of the output battery. Such tools, therefore, provide an exploration, characterization and summarization of the multivariate distributions. Since this is the case, any data preparation must be careful to retain undistorted any multivariate distributional relationships. However, and perhaps surprisingly, modifying the univariate distributions of the variables that comprise the input and output batteries does not necessarily

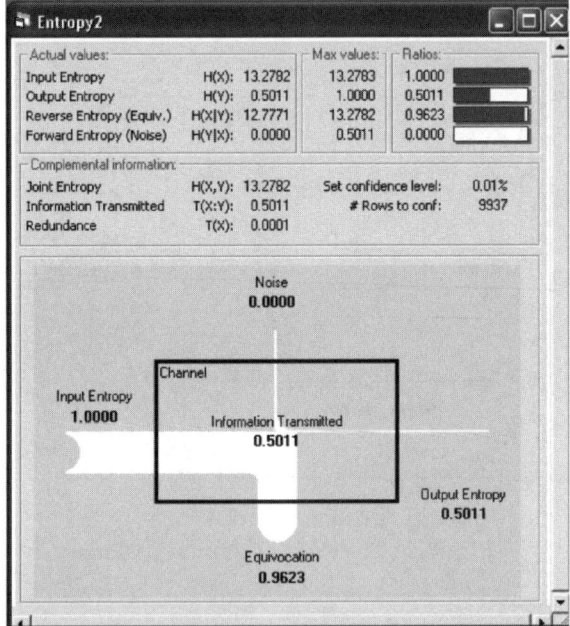

Fig. 3.2. Information transmission in unmodified CA data set

adversely affect the relationships within the multivariate distribution. However, linearizing the univariate distributions while retaining the multivariate relationships can have an enormous impact on improving final model quality, ease of modeling, and speed in generating results. The reason that univariate distribution "flattening" can have a useful impact is that most data modeling tools can characterize linear relationships more easily than functional relationships, and functional relationships more easily than disjoint relationships. Univariate distribution flattening has the effect of moving information transmission from more complex modes to simpler ones. As a concrete example, in the unmodified CA data set (see note 1), information is transmitted as shown in Fig. 3.2.

All information transmission characteristics and quantification in this chapter are produced using the data survey component of Powerhouse[2]. Figure 3.2 shows that this data set contains far more information than is needed to specify the output – most of it unused as "equivocation". However, the most salient point is that the amount of information transmitted, 0.5011 bits, is sufficient to perfectly define the output values since that requires 0.5011 bits. This is discovered because in Fig. 3.2, H(Y), the amount of

---

[2] Powerhouse is a discovery suite that includes a data survey tool that, in part, measures the information content of, and information transmitted by a data set. More information is available at www.powerhouse-inc.com

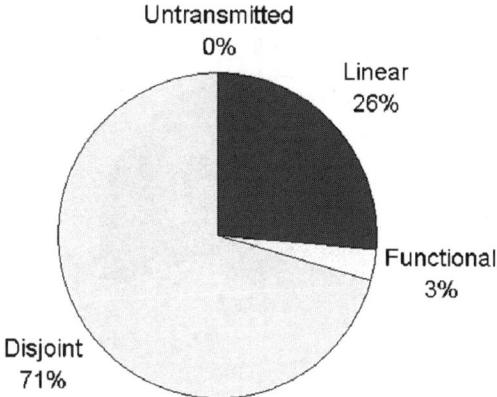

Fig. 3.3. Amount of information transmitted through the 3 modes for the unmodified CA data set

information required to completely specify the output states, equals 0.5011 and the amount of information transmitted, T(X:Y) also equals 0.5011. These measurements are in bits, although the unit is unimportant in this discussion. However, while the output is perfectly defined, the information transmission mode is equally, if not more important to most modeling algorithms than the total amount of information transmitted. Figure 3.3 illustrates the amount of information transmitted by each mode for the unmodified CA data set.

Notice in Fig. 3.3 that the amount of information transmitted is: linear mode 26%, functional mode 3%, disjoint mode 71%. Also note that since the amount of information is sufficient to perfectly define the output, none (or 0%) is shown as "untransmitted", which would represent the quantity of additional information required to completely define the output states. Most of the information is transmitted through the disjoint mode, and information transmitted in this mode is most difficult for most algorithms to characterize. Although the data set transmits sufficient information to completely define the output states, it will be really, really tough in practice to create a perfect model in this data set! And this discussion so far ignores the presence of what is colloquially called "noise" in data modeling, although that is discussed later. The total amount of information transmitted through modes most accessible to data modeling tools is 29%. It would be easier to create a model if more of the information present in the data set were transmitted through the linear and functional modes – and that is what distribution flattening achieves as shown in Fig. 3.4.

Figure 3.4 shows the change in information transmission mode distribution as a result of distribution flattening. This would still be a hard model to make since 62% of the information is transmitted through the disjoint mode, but linear and functional modes now account for 38% of the information transmitted.

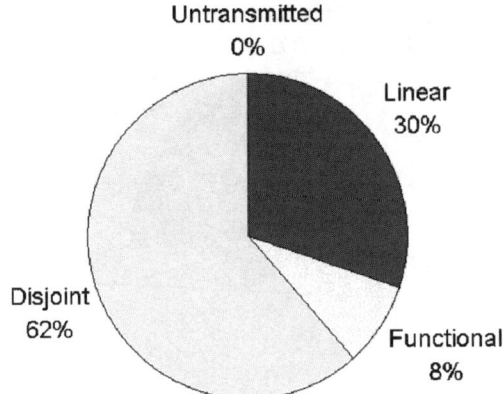

Fig. 3.4. Information transmission modes after distribution flattening in the CA data set

3.4.1 Continuous Redistribution

Continuous redistribution optimally flattens the distribution of a variable and produces, as nearly as the values in the variable permit, a completely flat, or rectangular distribution. The effect of continuous redistribution is shown in Fig. 3.5. As can be seen in Fig. 3.5, perfect rectangularity is not always possible and depends on the availability of sufficient discrete values in suitable quantities. Continuous redistribution does achieve the optimal distribution flattening possible and results in increasing the linearity of relationship in a data set as much as is possible without distorting the multivariate distribution.

However, optimal as it is, and useful at moving information from more complex to less complex-to-model modes as it is, continuous redistribution requires a moderately complex algorithm and it will turn out later in this chapter that equivalent results can be obtained with far less complexity. Thus, the algorithm is not further discussed here, and continuous redistribution

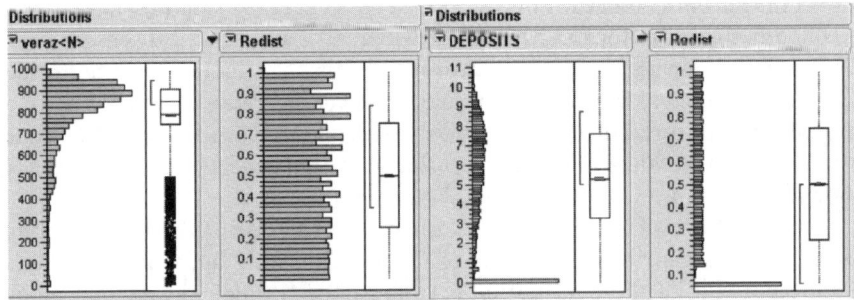

Fig. 3.5. Distribution histograms for the two variables', "veraz" and "deposits", before (*left*) and after (*right*) continuous redistribution

is discussed only to see the effect of this optimal transform on information transmission modes, which can be seen by comparing the difference between Fig. 3.3 and Fig. 3.4.

3.5 Numerating Categories

Some algorithms require that the input variables be presented as numeric values. When this is the case, and there are categorical values present in the input battery, it is necessary to convert the categories into a numerical representation. Just as with most data preparation techniques, so too for numerating categories there are arbitrary and principled ways of performing the numeration. The examples in this section make use of the "cars" data set. (See note 1.) The cars data set contains several categorical variables as well as some numeric variables. These examples exclusively make use of the categorical variables, ignoring the numeric variables – except for "miles per gallon" that is used as the output battery variable. The data set represents data describing various features of 404 cars available from 1971 through 1983 in the USA.

Unlike numeric variables, categorical variables are considered to pass all of their information through the disjoint mode. Since each category is considered as an unordered and distinct class, it is not appropriate to characterize categories as having linear or functional components. However, once numerated, they are usefully described in this way, exactly as if they were numeric variables. Indeed, numeration of categorical variables is, in a sense, the process of turning categorical variables into numeric variables – or at least, variables that can usefully be treated as numeric.

Figure 3.6 shows the information transmitted exclusively by the categorical variables in the cars data set. Note that the information transmitted is insufficient to completely define the output. (Note that the entropy of the output, $H(Y)$ is 5.5148 bits and the information transmitted $T(X:Y)$ is 5.0365 bits – less than the entropy of the output.) Thus, whatever else is expected, the information transmission modes will include a portion of untransmitted information.

3.5.1 Coding

One method of dealing with categorical variables is to code them. Coding explodes each category into a separate variable, and for a variable with few categories, this might work satisfactorily. In cars, the variable "origin" has 3 values, "USA", "Europe" and "Japan". Three binary variables with a "1" or "0" as appropriate might work. However, the variable "model" with 294 distinct categories would increase the variable count by 294 variables, a distinctly bad practice. It not only adds no information to the data set (and makes no difference to the information measures at all), but many algorithms

36   D. Pyle

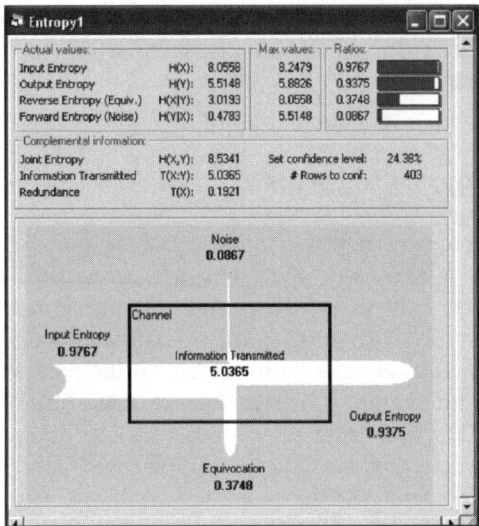

Fig. 3.6. Information transmitted by the categorical variables in the cars data set

suffer tremendously in the presence of increased variable count (for many well founded reasons not discussed here), and this violates the first principle of data preparation – first, do no harm. For that reason alone, category coding is not further discussed here.

3.5.2 Numeration

An alternative to coding is to assign numerical values to each category and to use the numbers in place of the category label. And this is where the arbitrary and principled assignment of values occurs.

Arbitrary Numeration

This is a fairly common method of numeration and assigns values to categories without particular reference to the rest of the data set. Perhaps the numbering is in order of category occurrence, or perhaps alphabetically, or perhaps simply random. Such assignment makes no difference to the information transmitted, of course. Replacing one label with another has no impact at all on information content. However, if there is in fact an implicit order to the categories (and in any data set there almost always is – invariably in the author's experience) then any such arbitrary assignment creates a potentially exceedingly intricate numerical relationship between the arbitrarily numerated variable and the output variable – and indeed with the other variables in the data set. Figure 3.7(a) shows the information transmission modes for the arbitrarily numerated categorical variables in the cars data set.

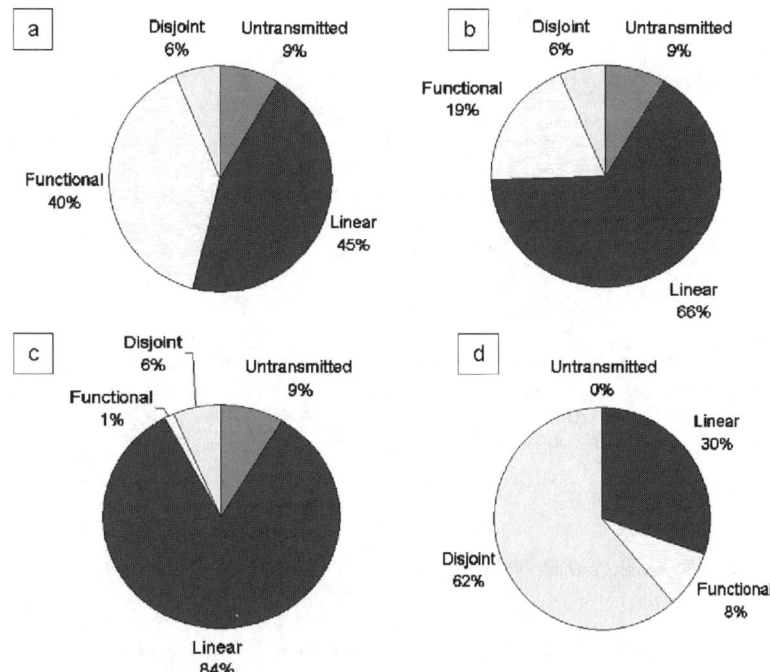

Fig. 3.7. (a) Information transmission modes for arbitrary category numeration in cars data. (b) Information transmission modes in the cars data set after unsupervised numeration of the categorical variables. (c) Information transmission modes in the cars data set after supervised numeration of the categorical variables. (d) Amount of information transmitted through the 3 modes after optimally binning the CA data set

Principled Numeration

There are two methods of principled numeration, unsupervised and supervised. Unsupervised numeration reflects the appropriate order for categories from all of the other variables in the data set except for any output battery variables. Thus, the numeration is made without reference to the output values. This may lead to little better numeration than offered by arbitrary numeration in some cases, but it has the advantage that it offers a rationale for category ordering (which can be explained in terms of the data set if necessary), and it can be used when unsupervised models are required and there is no explicit output battery. Figure 3.8 shows the information transmission modes for the same variables in the cars data set as before, but this time after unsupervised numeration of the categories. What is clear in Fig. 3.7(b) is that there is a sharp increase in linear mode and a decrease in functional mode information transmission. However, to be sure, this change is more by luck than judgment since the arbitrary assignment was made at random – and by chance could have been as good as or better than this. (Except that there are usually

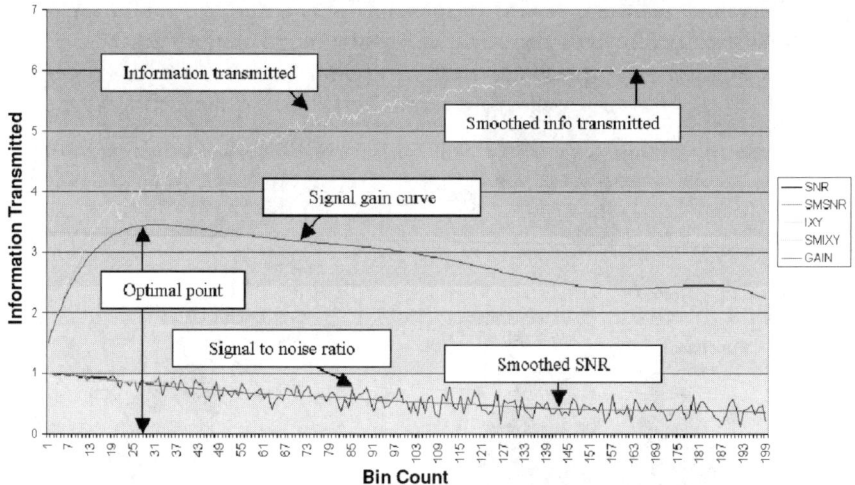

Fig. 3.8. Signal to Noise optimization through bin count selection

more ways to get it wrong than there are to get it right, so chance favors this sort of outcome.) Supervised numeration of categories, the second method of principled numeration, results in the information transmission modes shown in Fig. 3.7(c).

The improvement – that is, the increase in linear mode information transmission – is no accident in Fig. 3.7(c). In this case, it would almost be possible to use linear regression to fit a model. There is an apparently irreducible 6% of information transmitted through the disjoint mode, and 9% of the information required for a perfect model is not available at all, but the functional mode information has been reduced to almost nothing (1%).

3.6 Categorizing Numbers

One common method of assigning numbers to categories is binning. Binning also can be used to effectively modify variables' distributions, and while it includes distribution flattening effects (discussed earlier in the chapter), it also has other useful characteristics for a data modeler. Binning is a more complex topic than commonly perceived and unless carefully carried out, binning can reduce (sometimes very significantly) the useful information transmitted. The problem is that binning replaces several to many values in the original distribution with a single value. Regardless of how it is done, binning reduces the information content available in any variable. However, although the total amount of information available for transmission is inevitably reduced, if there is still the same amount of information transmitted to the output after binning as before, then the overall information reduction can be beneficial – especially if, among other effects, it transmits more information through easier to characterize modes. And that, of course, is the effect illustrated in Fig. 3.7(d).

One of the notable features that can be seen by comparing Fig. 3.7(d) with Fig. 3.4 is that they appear identical in the proportion of information transmitted through each mode. Indeed, by using optimal binning techniques, which are significantly less complex than those required for continuous redistribution, equivalent results can more easily be produced, plus additional benefits of principled binning (discussed later in this section) are also available. Binning requires consideration of two separate issues: (i) selecting an appropriate number of bins, and (ii) determining where to best place the bin boundaries. In addition, binning may be unsupervised (made considering only a single variable at a time) or supervised (considering the binning effect on another variable). In general arbitrary choice of bin count and bin boundary location produces poor, sometimes damaging results, and this section looks at some alternatives. Also, supervised binning techniques are generally better than unsupervised techniques.

3.6.1 Bin Count

There are many ways of selecting an appropriate bin count. Many, if not most ways of selecting the number of bins seem to rely on intuition and are not designed to produce a specifically measurable effect. While some methods may produce good estimates for a useful bin count, these are essentially arbitrary choices since there is no measurable effect that the binning is intended to produce, or more appropriately, there is no optimizable effect. For instance, binning to get a better model gives no indication of how to measure when the best model is achieved since it may always be possible that another bin count may produce a better model, and there is no way to tell without actually trying them all! In contrast, a few methods of selecting bin count do rely on measurable and optimizable metrics. This section briefly discusses some representative options:

Purely Arbitrary

Purely arbitrary bin count selection relies essentially on the opinion of the modeler. It seems common practice to choose 10 to 20 bins: although some recommend more, few arbitrary choices seem to reach 100. This count seems to be chosen for ease of interpreting the "meaning" of the bins in terms of the model, and low bin count seems to be chosen to make the explanation easier to understand.

Principled Arbitrary

This approach uses some usually statistical parameter describing the variable to determine a bin count. Variance based bin count determination is one example where the number of bins is chosen as: Round($3.5 \cdot \sigma \cdot n^{(1/3)}$). This produces a bin count that depends on the variance of a variable (σ) and the number of records (n) available. However, although the bin count is determined by characteristics of the variable, there is no measure of how "good" this binning is, nor even what it means to have a good, better, or best binning.

Principled

This approach uses a criterion to objectively determine the quality of the bin count. SNR bin count is an example of this approach. SNR is an acronym for "Signal to Noise Ratio". Information theory provides an approach to measure the "Signal" and the "Noise" in a variable. Binning, as noted earlier, reduces the amount of information available for transmission. Noise tends to be present in small-scale fluctuations in data. Increasing bin count transmits more information and also more noise: thus, with increasing bin count, the total amount of information transmitted increases and the signal to noise ratio decreases. With increasing bin count from 1 bin comes an optimum balance between the two, and this is illustrated in Fig. 3.8 for one of the variables in the CA data set.

Figure 3.8 shows the various components involved in determining the SNR bin count. Smoothing reduces variance in information transmitted and in the signal to noise ratio, and it is the smoothed versions that are used to determine the signal gain. For the illustrated variable, the optimal point (that is the point at which the maximum signal is passed for the minimum amount of noise included) is indicated at 29 bins. However, other variables in this data set have different optimal bin counts. It is worth comparing the SNR bin count range with the number of bins indicated as suitable using the variance-based binning. For the variable illustrated, "deposits", which has a standard deviation of 2.95 and with 9,936 records, this is:

$$3.5 \cdot \sigma \cdot n^{\frac{1}{3}} = 3.5 \cdot 2.95 \cdot 9936^{\frac{1}{3}} = 222 \; bins \,. \tag{3.3}$$

Although Fig. 3.8 does not show a bin count of 222 bins, it is intuitively apparent that such a number of bins is almost certainly a poor choice if optimizing signal gain (maximum signal for minimum noise) is an objective.

3.6.2 Bin Boundaries

For a given number of bins, it is possible to place the boundaries within the range of a numeric variable in an infinite number of ways. However, some arrangements work better than others[3]. Typically, bin boundaries are arranged using unsupervised positioning. And as with discovering an appropriate bin count, there are arbitrary, principled arbitrary and principled methods of bin boundary positioning.

3.6.3 Unsupervised Bin Boundary Placement

Arbitrary

An arbitrary method would involve distributing the boundaries at random intervals across the variables' range. There is something to be said for this practice, particularly if multiple random assignments are used and the best one chosen. However, other methods have proven to be far more effective.

Principled Arbitrary

These bin boundary assignment methods are fairly popular, and they can work well. Typical methods include:

- Equal Range, where the bins each have the same size, which is 1/n of the range where n is the number of bins. Bins may not contain the same number of records, and in extreme cases (for example the presence of extreme outliers) many or most of the bins may be unpopulated.
- Equal Frequency, where each bin has the same number of records as far as possible, so that each bin contains 1/r of the records, where r is the number of records. Bin boundaries may not span similarly sized sub-ranges of a variable's total range.

Principled

An example of a principled approach is k-means binning. The term "k-means" is usually associated with the multivariate technique of k-means clustering. However, there is no hindrance to applying the technique to a single variable. The technique divides a single variable's distribution into bins (that would be called clusters in a multivariate context) that represent the k most dense associations of values in the distribution. Or, it splits the distribution of values at the least populated k-1 points in the variable's distribution.

3.6.4 Supervised Bin Boundary Placement

Supervised bin boundary placement must be principled since it uses the principle of supervision to determine optimum positioning.

---

[3] "work better" is taken to mean transmits more signal information while rejecting noise information at the same time as transmitting maximum information through the least complex modes.

Principled

A principled approach places the bin boundaries so as to optimize the value of the binning; for example, to maximize the total amount of information transmitted. When used to maximize transmitted information this approach is called Least Information Loss binning (LIL). (It is possible to place the bin boundaries so as to maximize the amount of Linear, Functional, or Linear and Functional information transmission modes, but space limitations prevent further consideration of those options here.) LIL is so called because binning unavoidably involves the loss of some amount of information in the system of the data set, but it loses as little as possible of the information in a variable that characterizes the output variable(s) states. It is important to note that LIL binning is only univariately optimal. This does not justify an assumption of multivariate optimality. In other words, although for each variable considered individually LIL binning assures that for a given bin count the variable transmits maximum information, this does not assure that any selection of LIL binned variables considered together will transmit the maximum amount of information about an output of which the selected variables are capable. This is because any given variable may transmit some information that is identical to information transmitted by another variable (redundant information) while not transmitting unique information that it possesses that would supplement the information transmitted by the first. A fully optimal binning requires LIL-M binning (multivariate LIL binning), where the bin boundaries for any variable are set to transmit the maximum possible non-redundant information given the information transmitted by other variables in the data set.

The LIL binning algorithm is straightforward:

```
CurrentNumberOfBin = 1
do until desired bincount
    for each bin
        inspect all potential splitting points discover splitting
        point that loses least information
    discover least information loss split point in all bins
    insert new bin boundary at global least loss split point
    CurrentNumberOfBin = + 1
```

Note that potential splitting points are all of the values that fall between the actually occurring values for a variable. However, this algorithm may not produce an actual optimal LIL binning since earlier placed boundaries may no longer be in optimal positions with the placement of subsequent boundaries. A second stage of relaxation of the bin boundaries can be applied to ensure optimality, although in practice the improvement in information transmission is usually minute, and the relaxation process is slow, so not usually worth implementing for the minute improvement involved.

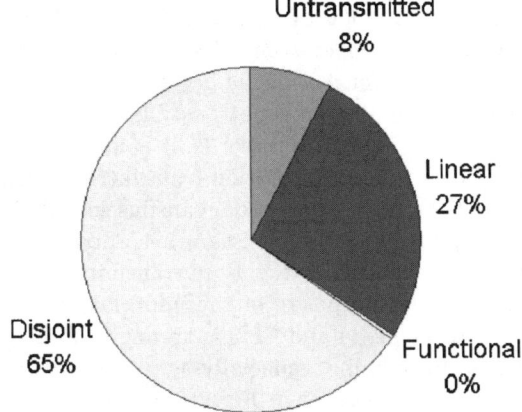

Fig. 3.9. Arbitrary equal range binning in the CA data set

Relaxation:

iterate until no change in bin boundary positioning
    select a contiguous pair of bins at random
    position the bin boundary optimally

Discussion of calculating the information gain at each potential splitting point would consume more space than is available here. It is discussed in detail in Chap. 11 of [3] and again in [4]. Also, in order to allow experimentation, a free demonstration binning tool is available that allows experimentation with equal range, equal frequency, and LIL binning[4]. The effect of LIL binning is shown in Fig. 3.7(d) above and results in the transmission of more information through the linear and functional modes than all types of unsupervised binning, and many other types of supervised binning, although Fig. 3.7(d) shows the effect for the data set as a whole and not just for an individual variable.

Binning Summary

Binning is a useful and valuable tool in the armory of data preparation. However, the ease of binning is not an unmixed blessing because it is very easy to do more harm than good with binning. See, for instance, Fig. 3.9, which shows the result of one recommended method of binning applied to the CA data set, and compare the results with Figs. 3.4 and 3.7(d).

---

[4] A free version of the demonstration binning and numerating tool is available for download from the author's personal web site, www.modelandmine.com. This is a reduced capability version that has been taken from Powerhouse to allow experimentation with some of the techniques discussed in this chapter.

The binning has done a lot of damage. Compared with original data (Fig. 3.3), the data set no longer transmits sufficient information to define the output completely – in fact, 8% of the necessary information is not transmitted. Thus, using this binning, it is not possible, even in principle, to completely define the output – or in other words, it is no longer possible to get a perfect model. The linear mode information transmitted has increased slightly from 26% to 27%, but the functional mode transmission has fallen from 3% to a fraction of 1%. Disjoint mode transmission has apparently decreased from 71% to 65%. However, recall that 8% is untransmitted using this binning practice. To all intents and purposes, of the information that is transmitted 29% is through the linear mode and 71% is through the disjoint mode. This is altogether not a useful binning, especially as compared with what a better binning practice achieves as shown in Fig. 3.7(d). Yet, the binning strategy adopted in this example is taken from the recommended practices published by a major statistical and mining tool vendor for use with their tools.

When carefully applied, and when appropriate techniques for discovering an appropriate bin count and for placing bin boundaries are used, binning reduces noise, increases signal, moves information transmission from more to less complex modes of transmission, and promotes ease of modeling. When inappropriately or arbitrarily applied, binning may increase noise, reduce signal, or move information transmission into more complex modes. As with so many tools in life, binning is a two-edged sword. Careful and principled application is needed to get the most benefit from it.

3.7 Missing Values

Missing values, or nulls, present problems for some tools, and are considered not to for others. However, the issue is not so straightforward. Tools that have a problem usually require a value to be present for all variables in order to generate an output. Clearly, some value has to be used to replace the null – or the entire record has to be ignored. Ignoring records at the very least reduces the information available for defining output states, and may result in introducing very significant bias, rendering the effectively available residual data set unrepresentative of the domain being modeled. From an information transmission perspective, the replacement value should be the one that least disturbs the existing information content of a data set. That is, it should add no information and should not affect the proportion of information transmitted through any mode. This is true because, from one perspective, null values are explicitly not there and should, thus, explicitly have no contribution to make to information content. That being the case, any replacement value that is used purely for the convenience of the modeling tool should likewise make no contribution to the information content of the data set. All such a replacement should do is to make information that is present in other variables accessible to the tool. However, that is not the whole story. Usually (invariably in the author's experience) nulls do not occur at random in any data set. Indeed, the presence of a null very often depends on the values of the other variables in a data set, so even the presence of a null is sensitive to the multivariate distribution, and this is just as true for categorical variables as much for as numeric variables. Thus, it may be important to explicitly capture the information that could be transmitted by the pattern with which nulls occur in a data set – and it is this pattern that is potentially lost by any tool that ignores missing values. By doing so, it may ignore a significant source of information.

For instance, the "credit" data set (see note 1) has many nulls. In fact, some variables are mainly null. The output variable in this data set is "buyer" – an indication of offer acceptance. Ignoring completely any non-null values and considering exclusively the absence/presence of a value results in the information transmission shown in Fig. 3.10. (In other words, a data set is created that has the same variable names and count as in the original, but where the original has a value the new data set has a "1", and where the original has a null the new data set has a "0", except for the output variable that is retained in its original form.) The figure shows that, perhaps surprisingly, the mere absence or presence of a value in this data set transmits 33% of the information needed to define the output – information that might be ignored

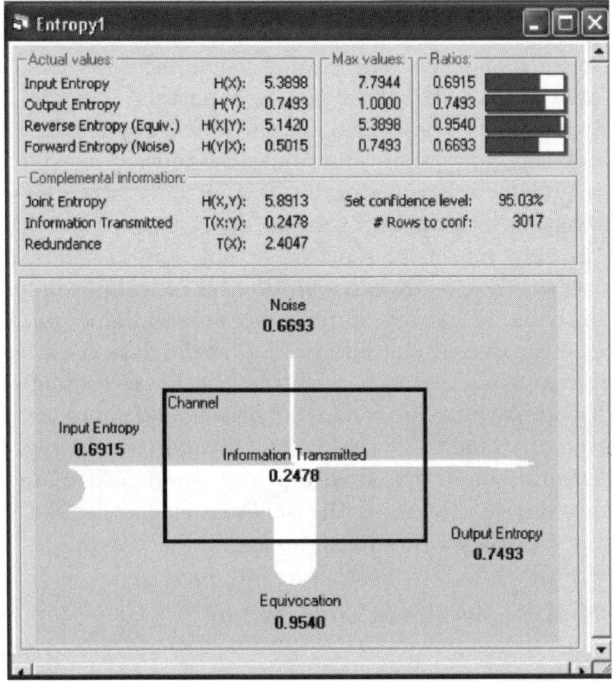

Fig. 3.10. Information transmitted considering only the presence/absence of values in the credit data set

by an algorithm that ignores nulls. (The 33% is determined by noting that (Y), the information required to define the output states, requires 0.7493 bits to specify those states and that T(X:Y), the actual information transmitted, is 0.2478 bits. And 0.2478/.07493 = 0.33 or 33%.)

Thus, dealing with missing values (nulls) requires two distinct considerations. First, some replacement value may need to be used that neither adds to the information present in a data set nor modifies the modes of transmission of existing information, but is used only to make information available in otherwise unusable records. Second, any information that is available in the actual pattern of which values are present and absent has to be at least explained, and possibly captured and explicitly included in the input battery.

Experience shows that purely arbitrary null replacement methods have not proved to have any value, and none have been developed, so none are commented on here.

Principled Arbitrary

However, many if not most current methods for null replacement suggested as good practice are principled arbitrary methods. One such practice, for instance, is to replace a numeric variable null with the mean value of the variable, or the most common class. The idea behind this method is that using the mean is the most representative univariate value. Unfortunately for any method that recommends any single value replacement of nulls in a variable, it is not the univariate distribution that is important. What is important is not distorting the multivariate distribution – and any single value replacement disregards the multivariate distribution – indeed, it explicitly changes the multivariate distribution. Naturally, this introduces new and completely spurious information into the multivariate distribution. Now the credit data set, whether missing values are replaced or not, has so much variability already present that it is already carrying the theoretical maximum amount of information in the input battery, so replacing missing values makes no apparent difference to the quantity of information in the data set. Metaphorically, the bucket is already full, and measuring the level of the content of the bucket before and after garbage has been added makes no difference to the level. Full is full. However, the quality of the content is affected. As Fig. 3.11 shows, a null replacement value imputed to maintain the multivariate distribution (right) is more effective than using the univariate mean (left) at transmitting information through the less complex modes.

Principled Null Replacement Imputation

This requires using multivariate imputation – in other words, taking the values of the variables that are not null into account when imputing a replacement value. Linear multivariate imputation works well – and curvilinear multivariate imputation works better. However, these are far more complex techniques

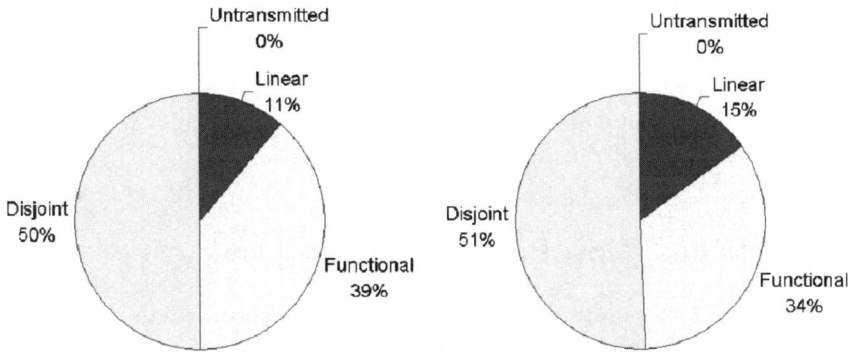

Fig. 3.11. Affect on information transmission mode in credit data set of replacing missing numeric values with mean (*left*) and multivariate sensitive value (*right*)

than using a univariate mean, and in a run-time data set the more complex the technique used, generally the longer the null replacement takes to generate. The optimal value (from an information transmission perspective) is a value that least disturbs the information content of a data set. In numeric variables, linear multivariate imputation is a reasonable compromise that can be applied on a record-by-record basis, and is reasonably fast. For categorical variables, the equivalent of the univariate mean is the most frequent category. The categorical equivalent of multivariate imputation is the selection of the most representative category given the values that are not null, whether numeric or categorical, in the other variables. The combination of including the null value pattern information along with a principled null value imputation makes additional information available. Figure 3.12 shows the information transmission modes for the combined data set.

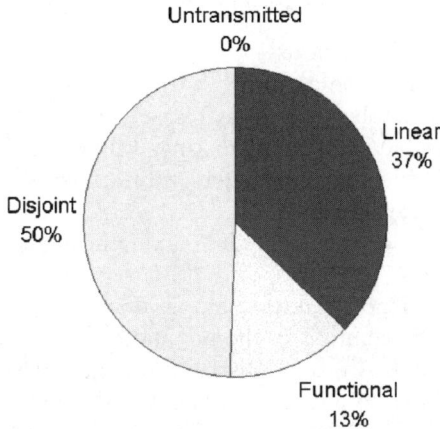

Fig. 3.12. Combined missing value pattern information with principled information neutral null replacements in the credit data set

What is clear in Fig. 3.12 is that, at least in the credit data set, the pattern with which nulls occurs is certainly something that should be investigated since it has such a large effect on the modes of information transmission. Of course, there are many questions raised: why should this be, is it significant, is it valid and can it be used, for instance.

3.8 Hints and Tips: Practical Preparation

This chapter has mainly discussed the effects of various data preparation techniques and practices. It is very important to know that data preparation practices should be applied in a principled way, and to know the effects of different techniques. However, simply knowing of such techniques is clearly not enough. "Know how" is just as crucial as "know what". Unfortunately any single chapter is insufficient to describe in full the details of data preparation. This section attempts to provide a short checklist for a practical approach to preparing data and a few pointers to sources for further exploration. Figure 3.13 provides an outline sketch of a suggested method for approaching data preparation.

Figure 3.13 is necessarily only a brief sketch since fully describing the principles and practice of data preparation requires a far fuller treatment. (See, for instance, [3].) The notes in this section amplify on Fig. 3.13, but should be taken only as a rough guide to introduce these main topics.

3.8.1 Checking Variables

First, consider the univariate characteristics of each variable. An example is shown in Fig. 3.14 for a small data set (cars).
For each variable consider the following characteristics:

1. What type of variable is in use?
2. How is the variable to be used?
3. If a categorical variable: How many classes exist?
4. If a numeric variable: How many distinct values exist?
5. If a numeric variable: How many zero values exist?
6. How many nulls are present?
7. If a numeric variable: What are the maximum, minimum, mean, median and quartile values?
8. What is the confidence that the sample univariate distribution is representative of the distribution in the population?
9. Inspect a histogram: Are there distribution and outlier problems?
10. If tools are available to measure it: What is the appropriate bin count?
11. If tools are available to measure it: What is the univariate noise level?

This set of questions is crucial, if only to provide a sort of "sanity check" that the data is at least within the expected bounds and format.

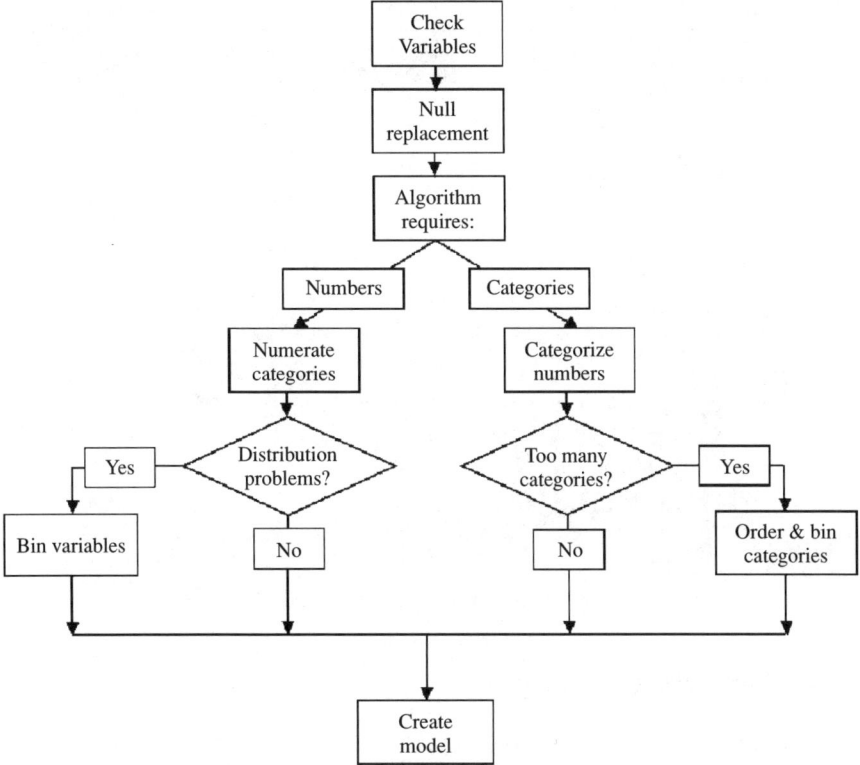

Fig. 3.13. Sketch approach to preparing data

3.8.2 Null Replacement

If nulls exist in any variable it is necessary to make a decision about dealing with them. Using any single value (say the mean, median or modal values) as a replacement is poor practice, potentially damaging to the patterns in the data set. For a brief but thorough discussion of missing data from one perspective see [1]. The most conservative choice is to ignore any rows that contain nulls, although no method of dealing with nulls is foolproof. However, in checking the basic stats for variables it is important to look for surrogate nulls and deal with those too. A surrogate null is a value that is entered in place of a null. That's why it's worth counting the occurrence of zero in a numeric variable since that often is used in place of a null value. All nulls have to be considered, actual and surrogate. Moreover, the decision about handling nulls may be driven by the required output from the model. It may be that the final model should indeed produce a "no prediction" type of output when a null is detected in the input battery. On the other hand, it may be that at run time, and when a null is present in the input battery, the model should produce its best estimate using the values of variables that are present.

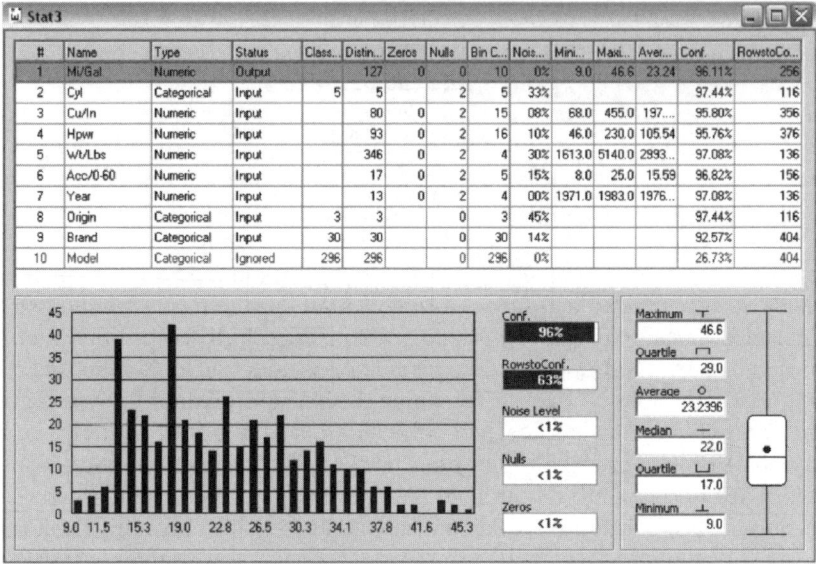

Fig. 3.14. Basic statistics for a small data set

3.8.3 Numerating Categories

One accepted method of coding categories is to create a binary variable for each category in the original variable. This only works when there are few categorical variables and few categories in each. Another practice is to arbitrarily assign numeric values to categories. This only works if the assignment is based on some inherent knowledge and reflects an appropriate relationship to the other variables since a purely arbitrary value assignment asserts a spurious relationship. A better practice is to assign appropriate numeric values to each category if they are known. If they are not known use tools (commercial or otherwise) or techniques (see [2, 3], for instance) for reflecting the categories existing relationships with the other variables in the data set as values for the categories. It's worth noting that when an apparently numeric variable has few discrete values (the variable "Cyl", for instance in Fig. 3.14 which is the number of cylinders with 5 discrete values) then the variable may actually be more appropriately treated as a set of categories. This depends, of course, on the needs of the model.

3.8.4 Problems with Distribution

Although any variable has a distribution of its values, whether categories or numbers, redistribution is usually only applied to numeric variables, and then mainly to deal with distribution problems. (Validly numerated categorical variables will not have correctable distribution problems since the category numeration process produces as nearly an optimal distribution of numeric values as the data set allows.) Many, if not most, algorithms benefit from numeric distributions that are evenly balanced. Well balanced distributions in this sense, for instance, include normal or uniform distributions of values. Distributions with severe outliers, or "lumps" (multi modal distributions) may prevent an algorithm form achieving its best performance. There are several well known ways of dealing with some of these problems, but in general appropriate binning works well. Use supervised binning such as least information loss (best) if tools are available, or unsupervised binning techniques such as equal entropy or equal frequency binning. (Equal range binning is inappropriate for dealing with distribution problems, see [4] for more details.) Bin counts can be determined empirically if automated tools are not available, although this can be somewhat time consuming. Variables need to be individually assigned an appropriate number of bins. It's worth noting that the bin counts shown in Fig. 3.14 are automatically optimized for greatest signal to noise ratio and use least information loss binning. However, this is a univariate optimization. Some modeling algorithms may benefit from a small increase in the bin count which may allow them to capture more information – but only if the chosen algorithm is inherently noise rejecting. If the algorithm is subject to learning noise (sometimes called overtraining) using the optimal signal to noise ratio benefits the model.

3.8.5 Categorizing Numbers

When the chosen algorithm requires to be presented with a number of categories it becomes necessary to represent any continuous values as a logical series of ranges of values. One convenient way to turn numbers into categories is to bin the numbers. Here again supervised binning is preferable. If possible chose least information loss binning with the appropriate number of bins to optimize the signal to noise ratio. This results in numeric ranges that are very often interpretable as being in some way significant (this is a colloquial usage of the word "significant") or meaningful about the to-be-predicted values. Once again, if supervised binning is not available use equal entropy binning if possible or equal frequency binning. Also too, use tools or empirically discover the appropriate number of bins for each numeric variable to maximize signal and minimize noise (garbage).

3.8.6 Problems with Categories

It may happen that there are simply too many categories for the chosen algorithm to handle. Few category oriented tools can handle ZIP codes, for instance, or text messages coded into flagged states but which can run into hundreds or thousands of codes. A useful option is to reduce the number of categories by associating similar categories together into bins. Some binning methods can directly associate categories together, or it may be possible to use a model to make the associations. However, a useful alternative is to numerate the categories and then use least information loss, equal entropy or equal frequency binning on the resulting (now) numerically represented variable.

3.8.7 Final Preparations

Unfortunately space permits only a brief sketch of the preparation options and this section summarizes a possible approach based on the earlier content of the chapter. In fact there remains a lot for a modeler to do after preparation and before modeling, Fig. 3.13 notwithstanding. For sure any modeler will need to survey the final data set, determine the confidence that the data set represents the multivariate distribution, determine how much of the necessary information is transmitted, select a subset of variables for the actual model, and many other tasks. Discussing those tasks is not itself part of preparing the data, and so beyond the scope of this chapter. However, each of those steps may well cause a return to one of the preparation stages for further adjustments. To some degree preparation is an iterative process, and without question it is an integrated part of the whole modeling process even though it is discussed separately here.

3.9 Summary

This chapter has used information theory as a lens to look at the effects of different data preparation techniques and practices. It is a useful tool inasmuch as it is possible to use information transmission measurements that directly relate to the ease (even the possibility in many cases) of creating models of data using available algorithms. If the algorithms available were insensitive to the mode of transmission of information, data preparation – at least of the types discussed here – would not be needed. However, at least currently, that is not the case.

The chapter has provided a brief look at five of the main issues that face any practitioner trying to create a model of a data set: modifying range, modifying distributions, representing categories as numbers, representing numbers as categories and replacing nulls. In all cases, there are poor practices and better practices. In general, the better practices result in increasing the proportion of the information in a data set that is transmitted through the less complex modes. However nomenclature may deceive since what were yesterdays "best practices" may be better than those they replaced, but although still perhaps labeled "best practices" they are not necessarily still the best available option. Time, knowledge and technology move on. Applying the best practices is not always straightforward, and indeed as just mentioned, what constitutes a current best practice has changed in the light of recent developments in data modeling tools, and not all practitioners are aware of the effects on model performance that different practices have. Given the pace of development in all areas of life, data preparation is itself part of an adaptive system – one that adapts to take advantage of developing knowledge and tools. Whatever developments are yet to be discovered, applying principled approaches to data preparation provides the modeler who uses them with a key ingredient in empirically fitting the most probable model to the data available.

References

1. Allison P.D (2002): Missing Data. Sage Publications.
2. Henderson, Th. C. (1990) Discrete Relaxation Techniques, Oxford University Press.
3. Pyle D. (1999) Data Preparation for Data Mining. Morgan Kaufmann Publishers.
4. Pyle D. (2003) Business Modeling and Data Mining. Morgan Kaufmann Publishers.

Part I

From Methods To Applications

4

Artificial Neural Networks

C. Fyfe

Applied Computational Intelligence Research Unit,
The University of Paisley, Scotland
fyfe-ci0@paisley.ac.uk

In this chapter, we give an overview of the most common artificial neural networks which are being used today. Perhaps the crucial thing which differentiates modern searches for artificial intelligence from those of the past is that modern approaches tend to emphasise adaptivity. This is done in artificial neural networks with learning and we define two of the major approaches to learning as supervised (or learning with a teacher) and unsupervised. We illustrate both techniques with standard methods. Now the artificial neural networks which are in use today are not truly "smart adaptive systems"; even though they are based on models of a physical entity – the brain – which is truly a smart adaptive system, something seems to be missing. We conjecture that a combination of some of our modern searches for intelligence might be closer to a smart adaptive system and so we investigate a combination of statistical and neural techniques at the conclusion of this chapter.

4.1 Introduction

Artificial Neural Networks is one of a group of new methods which are intended to emulate the information processors which we find in biology. The underlying presumption of creating artificial neural networks (ANNs) is that the expertise which we humans exhibit is due to the nature of the hardware on which our brains run. Therefore if we are to emulate biological proficiencies in areas such as pattern recognition we must base our machines on hardware (or simulations of such hardware) which seems to be a silicon equivalent to that found within our heads.

First we should be clear about what the attractive properties of human neural information processing are. They may be described as:

- Biological information processing is robust and fault-tolerant: early on in life[1], we have our greatest number of neurons yet though we daily lose

---

[1] Actually several weeks before birth.

many thousands of neurons we continue to function for many years without an associated deterioration in our capabilities
- Biological information processors are flexible: we do not require to be reprogrammed when we go into a new environment; we adapt to the new environment, i.e. we learn.
- We can handle fuzzy, probabilistic, noisy and inconsistant data in a way that is possible with computer programs but only with a great deal of sophisticated programming and then only when the context of such data has been analysed in detail. Contrast this with our innate ability to handle uncertainty.
- The machine which is performing these functions is highly parallel, small, compact and dissipates little power.

Therefore we begin with a look at the biological machine we shall be emulating.

4.1.1 Biological and Silicon Neurons

A simplified neuron is shown in Fig. 4.1 (left). Information is received by the neuron at synapses on its dendrites. Each synapse represents the junction of an incoming axon from another neuron with a dendrite of the neuron represented in the figure; an electro-chemical transmission occurs at the synapse which allows information to be transmitted from one neuron to the next. The information is then transmitted along the dendrites till it reaches the cell body where a summation of the electrical impulses reaching the body takes place and some function of this sum is performed. If this function is greater than a particular threshold the neuron will fire: this means that it will send a signal (in the form of a wave of ionisation) along its axon in order to communicate with other neurons. In this way, information is passed from one part of the network of neurons to another. It is crucial to recognise that synapses are

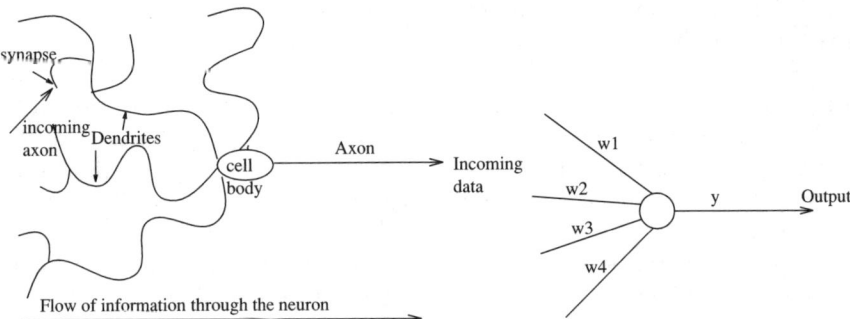

Fig. 4.1. *Left*: A diagrammatic representation of a real neuron. *Right*: The artificial neuron. The weights model the synaptic efficiencies. Some form of processing not specified in the diagram will take place within the cell body

thought to have different efficiencies and that these efficiencies change during the neuron's lifetime. We will return to this feature when we discuss learning.

We generally model the biological neuron as shown in Fig. 4.1 (right). The inputs are represented by the input vector **x** and the synapses' efficiencies are modelled by a weight vector **w**. Therefore the single output value of this neuron is given by

$$y = f\left(\sum_i w_i x_i\right) = f(\mathbf{w}.\mathbf{x}) = f(\mathbf{w}^T \mathbf{x}) \tag{4.1}$$

Notice that if the weight between two neurons is positive, the input neuron's effect may be described as excitatory; if the weight between two neurons is negative, the input neuron's effect may be described as inhibitory.

Therefore we can see that the single neuron is an extremely simple processing unit. The power of neural networks is believed to come from the accumulated power of adding many of these simple processing units together – i.e. we throw lots of simple and robust power at a problem. Again we may be thought to be emulating nature, as the typical human has several hundred billion neurons. We will often imagine the neurons to be acting in concert in layers such as in Fig. 4.2.

In this figure, we have a set of inputs (the input vector, **x**) entering the network from the left-hand side and being propagated through the network via the weights till the activation reaches the output layer. The middle layer is known as the hidden layer as it is invisible from outside the net: we may not affect its activation directly.

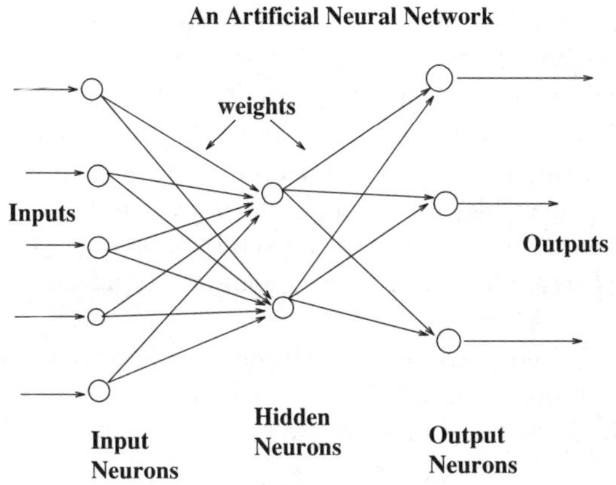

Fig. 4.2. A typical artificial neural network consisting of 3 layers of neurons and 2 connecting layers of weights

4.2 Typical Problem Areas

The number of application areas in which artificial neural networks are used is growing daily. Here we simply produce a few representative types of problems on which neural networks have been used

Pattern completion: ANNs can be trained on sets of visual patterns represented by pixel values. If subsequently, a part of an individual pattern (or a noisy pattern) is presented to the network, we can allow the network's activation to propagate through the network till it converges to the original (memorised) visual pattern. The network is acting like a content-addressable memory. Typically such networks have a recurrent (feedback as opposed to a feedforward) aspect to their activation passing. You will sometimes see this described as a network's *topology*.

Classification: An early example of this type of network was trained to differentiate between male and female faces. It is actually very difficult to create an algorithm to do so yet an ANN has been shown to have near-human capacity to do so.

Optimisation: It is notoriously difficult to find algorithms for solving optimisation problems. A famous optimisation problem is the Travelling Salesman Problem in which a salesman must travel to each of a number of cities, visiting each one once and only once in an optimal (i.e. least distance or least cost) route. There are several types of neural networks which have been shown to converge to "good-enough" solutions to this problem i.e. solutions which may not be globally optimal but can be shown to be close to the global optimum for any given set of parameters.

Feature detection: An early example of this is the phoneme producing feature map of Kohonen: the network is provided with a set of inputs and must learn to pronounce the words; in doing so, it must identify a set of features which are important in phoneme production.

Data compression: There are many ANNs which have been shown to be capable of representing input data in a compressed format losing as little of the information as possible; for example, in image compression we may show that a great deal of the information given by a pixel to pixel representation of the data is redundant and a more compact representation of the image can be found by ANNs.

Approximation: Given examples of an input to output mapping, a neural network can be trained to approximate the mapping so that a future input will give approximately the correct answer i.e. the answer which the mapping should give.

Association: We may associate a particular input with a particular output so that given the same (or similar) output again, the network will give the same (or a similar) output again.

Prediction: This task may be stated as: given a set of previous examples from a time series, such as a set of closing prices for the FTSE, to predict the next (future) sample.

Control: For example to control the movement of a robot arm (or truck, or any non-linear process) to learn what inputs (actions) will have the correct outputs (results).

4.3 Learning in Artificial Neural Networks

There are two modes in artificial neural networks:

1. activation transfer mode when activation is transmitted throughout the network
2. learning mode when the network organises usually on the basis of the most recent activation transfer.

We will now consider the learning mode.

We stated that neural networks need not be programmed when they encounter novel environments. Yet their behaviour changes in order to adapt to the new environment. Such behavioural changes are due to changes in the weights in the network. We call the changes in weights in a neural network learning. The changes in weights in an artificial neural network are intended to model the changing synaptic efficiencies in real neural networks: it is believed that our learning is due to changes in the efficiency with which synapses pass information between neurons.

There are 2 main types of learning in a neural network:

Supervised learning: with this type of learning, we provide the network with input data and the correct answer i.e. what output we wish to receive given that input data. The input data is propagated forward through the network till activation reaches the output neurons. We can then compare the answer which the network has calculated with that which we wished to get. We can now adjust the weights to ensure that the network is more likely to give the correct answer in future if it is again presented with the same (or similar) input data.
Unsupervised learning: with this type of learning, we only provide the network with the input data. The network is required to self-organise (i.e. to teach itself) depending on some structure in the input data. Typically this structure may be some form of redundancy in the input data or clusters in the data.

As with other machine learning paradigms, the interesting facet of learning is not just that the input patterns may be learned/classified/identified precisely but that this learning has the capacity to generalise. That is, while learning will take place on a set of *training patterns* an important property of the learning is that the network can generalise its results on a set of *test patterns* which it has not seen during learning. One of the important consequences here is that there is a danger of overlearning a set of training patterns so that new patterns which are not part of the training set are not properly classified.

4.4 Supervised Learning

We first set the scene with two early networks and then discuss the most common current supervised networks, the multilayered perceptron and radial basis function networks.

The simple perceptron was first investigated by Rosenblatt. For a particular input pattern, $\mathbf{x}^P$, we have an output o^P and target t^P. The algorithm is:

1. begin with the network in a randomised state: the weights between all neurons are set to small random values (between -1 and 1).
2. select an input vector, $\mathbf{x}^P$, from the set of training examples
3. propagate the activation forward through the weights in the network to calculate the output $o^P = \mathbf{w}.\mathbf{x}^P$.
4. if $o^P = t^P$, (i.e. the network's output is correct) return to step 2.
5. else change the weights according to $\Delta w_i = \eta x_i^P (t^P - o^P)$ where η is a small positive number known as the **learning rate**. Return to step 2.

Thus we are adjusting the weights in a direction intended to make the output, o^P, more like the target value, t^P, the next time an input like $\mathbf{x}^P$ is given to the network.

Another important early network was the Adaline (ADAptive LINear Element) from Widrow and co-workers. The Adaline calculates its output as $o = \sum_j w_j x_j + \theta$, with the same notation as before. You will immediately note the difference between this network and the perceptron is the lack of thresholding. The interest in the network was partly due to the fact that it is easily implementable as a set of resistors and switches.

Then the sum squared error from using the Adaline on all training patterns is given by

$$E = \sum_P E^P = \frac{1}{2} \sum_P (t^P - o^P)^2 \tag{4.2}$$

where the fraction is included due to inspired hindsight. Now, if our Adaline is to be as accurate as possible, we wish to minimise the squared error. To minimise the error, we can find the gradient of the error with respect to the weights and move the weights in the opposite direction. Formally $\Delta_P w_j = -\gamma \frac{\partial E^P}{\partial w_j}$. This rule is called the Least Mean Square error (LMS) or Delta rule or Widrow-Hoff rule. Now, for an Adaline with a single output, o,

$$\frac{\partial E^P}{\partial w_j} = \frac{\partial E^P}{\partial o^P} \cdot \frac{\partial o^P}{\partial w_j} \tag{4.3}$$

and because of the linearity of the Adaline units, $\frac{\partial o^P}{\partial w_j} = x_j^P$. Also, $\frac{\partial E^P}{\partial o^P} = -(t^P - o^P)$, and so $\Delta_P w_j = \gamma(t^P - o^P).x_j^P$. Notice the similarity between this rule and the perceptron learning rule; however, this rule has far greater applicability in that it can be used for both continuous and binary neurons.

This has proved to be a most powerful rule and is at the core of many current supervised learning methods.

The Perceptron (and Adeline) proved to be powerful learning machines but there were certain mappings which were (and are) simply impossible using these networks. Such mappings are characterised by being linearly inseparable. Now it is possible to show that many linearly inseparable mappings may be modelled by multi-layered perceptrons; this indeed was known in the 1960s but what was not known was a rule which would allow such networks to learn the mapping. Such a rule appears to have been discovered independently several times but has been spectacularly popularised by the PDP(Parallel Distributed Processing) Group [30] under the name backpropagation.

An example of a multi-layered perceptron (MLP) is shown in Fig. 4.2. Activity in the network is propagated forwards via weights from the input layer to the hidden layer where some function of the net activation is calculated. Then the activity is propagated via more weights to the output neurons. Now two sets of weights must be updated – those between the hidden and output layers and those between the input and hidden layers. The error due to the first set of weights is clearly calculable by the previously described LMS rule; however, now we require to propagate backwards that part of the error due to the errors which exist in the second set of weights and assign the error proportionately to the weights which cause it.

We may have any number of hidden layers which we wish since the method is quite general; however, the limiting factor is usually training time which can be excessive for many-layered networks. In addition, it has been shown that networks with a single hidden layer are sufficient to approximate any continuous function (or indeed any function with only a finite number of discontinuities) provided we use (non-linear) differentiable activation functions in the hidden layer.

4.4.1 The Backpropagation Algorithm

Because it is so important, we will repeat the whole algorithm in a "how-to-do-it" form:

1. Initialise the weights to small random numbers
2. Choose an input pattern, $\mathbf{x}^P$, and apply it to the input layer
3. Propagate the activation forward through the weights till the activation reaches the output neurons
4. Calculate the δs for the output layer $\delta_j^P = (t_j^P - o_j^P)f'(Act_j^P)$ using the desired target values for the selected input pattern.
5. Calculate the δs for the hidden layer using $\delta_i^P = \sum_{j=1}^{N} \delta_j^P w_{ji}.f'(Act_i^P)$
6. Update all weights according to $\Delta_P w_{ij} = \gamma.\delta_i^P.o_j^P$
7. Repeat steps 2 to 6 for all patterns.

4.4.2 Hints and Tips

Batch vs On-line Learning

The backpropagation algorithm is only theoretically guaranteed to converge if used in batch mode i.e. if all patterns in turn are presented to the network, the total error calculated and the weights updated in a separate stage at the end of each training epoch. However, it is more common to use the on-line (or pattern) version where the weights are updated after the presentation of each individual pattern. It has been found empirically that this leads to faster convergence though there is the theoretical possibility of entering a cycle of repeated changes. Thus in on-line mode we usually ensure that the patterns are presented to the network in a random and changing order.

The on-line algorithm has the advantage that it requires less storage space than the batch method. On the other hand the use of the batch mode is more accurate: the on-line algorithm will zig-zag its way to the final solution. It can be shown that the expected change (where the expectation is taken over all patterns) in weights using the on-line algorithm is equal to the batch change in weights.

Activation Functions

The most popular activation functions are the logistic function, $\frac{1}{1+\exp(-t)}$, and the tanh() function. Both of these functions satisfy the basic criterion that they are differentiable. In addition, they are both monotonic and have the important property that their rate of change is greatest at intermediate values and least at extreme values. This makes it possible to saturate a neuron's output at one or other of their extreme values. The final point worth noting is the ease with which their derivatives can be calculated:

- if $f(x) = \tanh(bx)$, then $f'(a) = b(1 - f(a) * f(a))$;
- similarly, if $f(x) = \frac{1}{1+\exp(-t)}$ then $f'(a) = bf(a)(1 - f(a))$.

It is folk wisdom that convergence is faster when tanh() is used rather than the logistic function. Note that in each case the target function must be within the output range of the respective functions. If you have a wide spread of values which you wish to approximate, you must use a linear output layer.

Initialisation of the Weights

The initial values of the weights will in many cases affect the final converged network's values. Consider an energy surface with a number of energy wells; then, if we are using a batch training method, the initial values of the weights constitute the only stochastic element within the training regime. Thus the network will converge to a particular value depending on the basin in which the original vector lies. There is a danger that, if the initial network values are sufficiently large, the network will initially lie in a basin with a small basin of attraction and a high local minimum; this will appear to the observer as a network with all weights at saturation points (typically 0 and 1 or +1 and −1). It is usual therefore to begin with small weights uniformly distributed inside a small range. Reference [22], page 162, recommends the range $(-\frac{2.4}{F_i}, +\frac{2.4}{F_i})$ where F_i is the number of weights into the ith unit.

Momentum and Speed of Convergence

The basic backprop method described above is not known for its fast speed of convergence. Note that though we could simply increase the learning rate, this tends to introduce instability into the learning rule causing wild oscillations in the learned weights. It is possible to speed up the basic method in a number of ways. The simplest is to add a momentum term to the change of weights. The basic idea is to make the new change of weights large if it is in the direction of the previous changes of weights while if it is in a different direction make it smaller. Thus we use $\Delta w_{ij}(t+1) = (1-\alpha).\delta_j.o_i + \alpha \Delta w_{ij}(t)$, where the α determines the influence of the momentum. Clearly the momentum parameter α must be between 0 and 1. The second term is sometimes known as the "flat spot avoidance" term since the momentum has the additonal property that it helps to slide the learning rule over local minima (see below).

Local Minima

Error descent is bedeviled with local minima. You may read that local minima are not much problem to ANNs, in that a network's weights will typically converge to solutions which, even if they are not globally optimal, are good enough. There is as yet little analytical evidence to support this belief. An heuristic often quoted is to ensure that the initial (random) weights are such that the average input to each neuron is approximately unity (or just below it). This suggests randomising the initial weights of neuron j around the value $\frac{1}{\sqrt{N}}$, where N is the number of weights into the jth neuron. A second heuristic is to introduce a little random noise into the network either on the inputs or with respect to the weight changes. Such noise is typically decreased during the course of the simulation. This acts like an annealing schedule.

Weight Decay and Generalisation

While we wish to see as good a performance as possible on the training set, we are even more interested in the network's performance on the test set since this is a measure of how well the network generalises. There is a trade-off between accuracy on the training set and accuracy on the test set. Also generalisation is important not only because we wish a network to perform on new data which it has not seen during learning but also because we are liable to have data which is noisy, distorted or incomplete.

If a neural network has a large number of weights (each weight represents a degree of freedom), we may be in danger of overfitting the network to the training data which will lead to poor performance on the test data. To avoid this danger we may either remove connections explicitly or we may give each weight a tendancy to decay towards zero. The simplest method is $w_{ij}^{new} = (1-\epsilon)w_{ij}^{old}$ after each update of the weights. This does have the disadvantage that it discourages the use of large weights in that a single large weight may be decayed more than a lot of small weights. More complex decay routines can be found which will encourage small weights to disappear.

The Number of Hidden Neurons

The number of hidden nodes has a particularly large effect on the generalisation capability of the network: networks with too many weights (too many degrees of freedom) will tend to memorise the data; networks with too few will be unable to perform the task allocated to it. Therefore many algorithms have been derived to create neural networks with a smaller number of hidden neurons.

4.4.3 Radial Basis Functions

In a typical radial basis function (RBF) network, the input layer is simply a receptor for the input data. The crucial feature of the RBF network is the function calculation which is performed in the hidden layer. This function performs a *non-linear* transformation from the input space to the hidden-layer space. The hidden neurons' functions form a basis for the input vectors and the output neurons merely calculate a linear (weighted) combination of the hidden neurons' outputs.

An often-used set of basis functions is the set of Gaussian functions whose mean and standard deviation may be determined in some way by the input data (see below). Therefore, if $\phi(\mathbf{x})$ is the vector of hidden neurons' outputs when the input pattern $\mathbf{x}$ is presented and if there are M hidden neurons, then

$$\phi(\mathbf{x}) = (\phi_1(\mathbf{x}), \phi_2(\mathbf{x}), \ldots, \phi_M(\mathbf{x}))^T$$
$$\text{where } \phi_i(\mathbf{x}) = \exp(-\lambda_i \parallel \mathbf{x} - \mathbf{c}_i \parallel^2)$$

where the centres $\mathbf{c}_i$ of the Gaussians will be determined by the input data. Note that the terms $\parallel \mathbf{x} - \mathbf{c}_i \parallel$ represent the Euclidean distance between the inputs and the ith centre. For the moment we will only consider basis functions with $\lambda_i = 0$. The output of the network is calculated by

$$y = \mathbf{w}.\phi(\mathbf{x}) = \mathbf{w}^T \phi(x) \tag{4.4}$$

where $\mathbf{w}$ is the weight vector from the hidden neurons to the output neuron.

We may train the network now using the simple LMS algorithm in the usual way. Therefore we can either change the weights after presentation of the ith input pattern $\mathbf{x}^i$ by

$$\Delta w_k = -\frac{\partial E^i}{\partial w_k} = e^i . \phi_k(\mathbf{x}^i) \tag{4.5}$$

or, since the operation of the network after the basis function activation has been calculated is wholly linear, we can use fast batch update methods. Reference [22] is particularly strong in discussing this network.

Finding the Centres of the RBFs

If we have little data, we may have no option but to position the centres of our radial basis functions at the data points. However such problems may be ill-posed and lead to poor generalisation. If we have more training data, several solutions are possible:

1. Choose the centres of the basis functions randomly from the available training data.
2. Choose to allocate each point to a particular radial basis function (i.e. such that the greatest component of the hidden layer's activation comes from a particular neuron) according to the *k-means rule*. The k-means, the centres of the neurons are moved so that each remains the average of the inputs allocated to it.
3. We can use a generalisation of the LMS rule:

$$\Delta c_i = -\frac{\partial E}{\partial c_i} \tag{4.6}$$

This unfortunately is not guaranteed to converge (unlike the equivalent weight change rule) since the cost function E is not convex with respect to the centres and so a local minimum is possible.

Hints and Tips: Comparing RBFs with MLPs

The question now arises as to which is best for specific problems, multilayered perceptrons or radial basis networks. Both RBFs and MLPs can be shown to be universal approximators i.e. each can arbitrarily closely model continuous functions. There are however several important differences:

1. The neurons of an MLP generally all calculate the same function of the neurons' activations e.g. all neurons calculate the logistic function of their weighted inputs. In an RBF, the hidden neurons perform a non-linear mapping whereas the output layer is always linear.
2. The non-linearity in MLPs is generally monotonic; in RBFs we use a radially decreasing function.

3. The argument of the MLP neuron's function is the vector product **w.x** of the input and the weights; in an RBF network, the argument is the distance between the input and the centre of the radial basis function, $\| \mathbf{x} - \mathbf{w} \|$.
4. MLPs perform a global calculation whereas RBFs find a sum of local outputs. Therefore MLPs are better at finding answers in regions of the input space where there is little data in the training set. If accurate results are required over the whole training space, we may require many RBFs i.e. many hidden neurons in an RBF network. However because of the local nature of the model, RBFs are less sensitive to the order in which data is presented to them.
5. MLPs must pass the error back in order to change weights progressively. RBFs do not do this and so are much quicker to train.

4.4.4 Support Vector Machines

Multilayer artificial neural networks can be viewed within the framework of statistical learning theory and a large amount of research investigation has recently been devoted to the study of machines for classification and regression based on empirical risk minimization called Support Vector Machines (SVMs). Vapnik [34] provides a detailed treatment of this particular area of research and tutorials are provided by [3, 33]. In brief, SVMs are powerful discriminative classifiers that have been demonstrated to provide state-of-the-art results for many standard benchmark classification and regression problems. The non-linear discriminative power of SVM classifiers comes from employing a non-linear kernel that returns a distance between two input feature vectors within an implicitly defined high-dimensional non-linear feature space. As the kernel implicitly defines a metric in the associated feature space then it has been proposed that a probabilistic generative model, such as a Hidden Markov Model (HMM), can be used to provide a kernel which would return the distance between points in the parameter space of the particular generative model chosen. A useful online resource considering SVMs is available at the URL http://www.kernel-machines.org

4.4.5 Applications

The type of applications to which supervised learning artificial neural networks have been applied is diverse and growing. Examples include classification, regression, prediction, data compression, optimisation, feature detection, control, approximation and pattern completion. Examples of fields to which they have been applied are handwritten digit classification, prediction of financial markets, robot control, extraction of significant features from remote sensing data and data mining.

4.5 Unsupervised Learning

With the networks which we have met so far, we must have a training set on which we already have the answers to the questions which we are going to pose to the network. Yet humans appear to be able to learn (indeed some would say can only learn) without explicit supervision. The aim of unsupervised learning is to mimic this aspect of human capabilities and hence this type of learning tends to use more biologically plausible methods than those using the error descent methods of the last two chapters. The network must self-organise and to do so, it must react to some aspect of the input data – typically either redundancy in the input data or clusters in the data; i.e. there must be some structure in the data to which it can respond. There are two major methods used

1. Hebbian learning which tends to be used when we want "optimal", in some sense, projections or compressions of the data set
2. Competitive learning which tends to be used when we wish to cluster the data.

We shall examine each of these in turn.

4.5.1 Hebbian Learning

Hebbian learning is so-called after Donald Hebb [23] who in 1949 conjectured:

> When an axon of a cell A is near enough to excite a cell B and repeatedly or persistently takes part in firing it, some growth process or metabolic change takes place in one or both cells such that A's efficiency as one of the cells firing B, is increased.

In the sort of feedforward neural networks we have been considering, this would be interpreted as the weight between an input neuron and an output neuron is very much strengthened when the input neuron's activation when passed forward to the output neuron causes the output neuron to fire strongly. We can see that the rule favours the strong: if the weights between inputs and outputs are already large (and so an input will have a strong effect on the outputs) the chances of the weights growing is large.

More formally, consider the simplest feedforward neural network which has a set of input neurons with associated input vector, $\mathbf{x}$, and a set of output neurons with associated output vector, $\mathbf{y}$. Then we have, as before, $y_i = \sum_j w_{ij} x_j$ where now the (Hebbian) learning rule is defined by $\Delta w_{ij} = \eta x_j y_i$. That is the weight between each input and output neuron is increased proportional to the magnitude of the simultaneous firing of these neurons.

Now we can substitute into the learning rule the value of y calculated by the feeding forward of the activity to get $\Delta w_{ij} = \eta x_j \sum_k w_{ik} x_k = \eta \sum_k w_{ik} x_k x_j$. Writing the learning rule in this way emphasises the statistical properties of the learning rule i.e. that the learning rule identifies the correlation between different parts of the input data's vector components. It does however, also show that we have a difficulty with the basic rule as it stands which is that we have a positive feedback rule which has an associated difficulty with lack of stability. If we do not take some preventative measure, the weights will grow without bound. Such preventative measures include

1. Clipping the weights i.e. insisting that there is a maximum, w_{max} and minimum w_{min} within which the weights must remain.
2. Specifically normalising the weights after each update. i.e. $\Delta w_{ij} = \eta x_j y_i$ is followed by $w_{ij} = \frac{w_{ij}+\Delta w_{ij}}{\sqrt{(\sum_k w_{ik}+\Delta w_{ik})^2}}$ which ensures that the weights into each output neuron have length 1.
3. Having a weight decay term within the learing rule to stop it growing too large e.g. $\Delta w_{ij} = \eta x_j y_i - \gamma(w_{ij})$, where the decay function $\gamma(w_{ij})$ represents a monotonically increasing function of the weights.
4. Create a network containing a negative feedback of activation.

The last is the method we will use in the subsequent case study.

The Negative Feedback Architecture

Principal Component Analysis (PCA) provides the best linear compression of a data set. We [12, 13] have over the last few years investigated a negative feedback implementation of PCA defined by (4.7)–(4.9). Let us have an N-dimensional input vector, **x**, and an M-dimensional output vector, **y**, with W_{ij} being the weight linking the jth input to the ith output. The learning rate, η is a small value which will be annealed to zero over the course of training the network. The activation passing from input to output through the weights is described by (4.7). The activation is then fed back though the weights from the outputs and the error,e calculated for each input dimension. Finally the weights are updating using simple Hebbian learning.

$$y_i = \sum_{j=1}^{N} W_{ij} x_j, \forall i \qquad (4.7)$$

$$e_j = x_j - \sum_{i=1}^{M} W_{ij} y_i \qquad (4.8)$$

$$\Delta W_{ij} = \eta e_j y_i \qquad (4.9)$$

We have subsequently modified this network to perform clustering with topology preservation [14], to perform Factor Analysis [6, 18] and to perform Exploratory Projection Pursuit [15, 17] (EPP).

Exploratory Projection Pursuit [11] is a generic name for the set of techniques designed to identify structure in high dimensional data sets. In such data sets, structure often exists across data field boundaries and one way to reveal such structure is to project the data onto a lower dimensional space and then look for structure in this lower dimensional projection by eye. However we need to determine what constitutes the best subspace onto which the data should be projected.

Now the typical projection of high dimensional data will have a Gaussian distribution [8] and so little structure will be evident. This has led researchers to suggest that what they should be looking for is a projection which gives a distribution as different from a Gaussian as possible. Thus we typically define an index of "interestingness" in terms of how far the resultant projection is from a Gaussian distribution. Since the Gaussian distribution is totally determined by its first two moments, we usually sphere the data (make it zero mean and with covariance matrix equal to the identity matrix) so that we have a level playing field to determine departures from Gaussianity.

Two common measures of deviation from a Gaussian distribution are based on the higher order moments of the distribution. Skewness is based on the normalised third moment of the distribution and basically measures if the distribution is symmetrical. Kurtosis is based on the normalised fourth moment of the distribution and measures the heaviness of the tails of a distribution. A bimodal distribution will often also have a negative kurtosis and therefore kurtosis can signal that a particular distribution shows evidence of clustering. Whilst these measures have their drawbacks as measures of deviation from normality (particularly their sensitivity to outliers), their simplicity makes them ideal for explanatory purposes.

The only difference between the PCA network and this EPP network is that a function of the output activations is calculated and used in the simple Hebbian learning procedure. We have for N dimensional input data and M output neurons

$$s_i = \sum_{j=1}^{N} w_{ij} x_j \tag{4.10}$$

$$e_j = x_j - \sum_{k=1}^{M} w_{kj} s_k \tag{4.11}$$

$$r_i = f\left(\sum_{j=1}^{N} w_{ij} x_j\right) = f(s_i) \tag{4.12}$$

$$\Delta w_{ji} = \eta_t r_i e_j \tag{4.13}$$

$$= \eta_t f\left(\sum_{k=1}^{N} w_{ik} x_k\right) \left\{ x_j - \sum_{l=1}^{M} w_{lj} \sum_{p=1}^{N} w_{lp} x_p \right\} \tag{4.14}$$

where r_i is the value of the function f() at the ith output neuron.

To determine which f() to choose, we decide which feature we wish to emphasise in our projection and use the derivative of the function which would maximise the feature in the projection. For example, clusters in a data set are often characterised by having low kurtosis which is measured by $E(y^4)$, the expected value of the (normalised) fourth moment. Thus we might choose $f(y) = -y^3$ as our function. In practise, robust functions similar to the powers of y are often used and so we would choose in the above $f(y) = \tanh(y)$.

This type of network has been used in exploratory data analysis [29] for remote sensing data, financial data and census data. It has also been applied [19, 20] to the difficult "cocktail-party problem" – the extraction of a single signal from a mixture of signals when few assumptions are made about the signals, the noise or the mixing method.

4.5.2 Competitive Learning

One of the non-biological aspects of the basic Hebbian learning rule is that there is no limit to the amount of resources which may be given to a synapse. This is at odds with real neural growth in that it is believed that there is a limit on the number and efficiency of synapses per neuron. In other words, there comes a point during learning in which if one synapse is to be strengthened, another must be weakened. This is usually modelled as a competition for resources.

In competitive learning, there is a competition between the output neurons to fire. Such output neurons are often called **winner-take-all** units. The aim of competitive learning is to cluster the data. However, as with the Hebbian learning networks, we provide no correct answer (i.e. no labelling information) to the network. It must self-organise on the basis of the structure of the input data.

The basic mechanism of simple competitive learning is to find a winning unit and update its weights to make it more likely to win in future should a similar input be given to the network. We first have a competition between the output neurons and then

$$\Delta w_{ij} = \eta(x_j - w_{ij}), \text{ for the winning neuron i}$$

Note that the change in weights is a function of the *difference* between the weights and the input. This rule will move the weights of the winning neuron directly towards the input. If used over a distribution, the weights will tend to the mean value of the distribution since $\Delta w_{ij} \to 0 \iff w_{ij} \to E(x_j)$, where E(.) indicates the ensemble average.

Probably the three most important variations of competitive learning are

1. Learning Vector Quantisation [26]
2. The ART models [4, 5]
3. The Kohonen feature map [27]

As a case study, we will discuss the last of these.

The Kohonen Feature Map

The interest in feature maps stems directly from their biological importance. A feature map uses the "physical layout" of the output neurons to model some feature of the input space. In particular, if two inputs $\mathbf{x}_1$ and $\mathbf{x}_2$ are close together with respect to some distance measure in the input space, then if they cause output neurons y_a and y_b to fire respectively, y_a and y_b must be close together in some layout of the output neurons. Further we can state that the opposite should hold: if y_a and y_b are close together in the output layer, then those inputs which cause y_a and y_b to fire should be close together in the input space. When these two conditions hold, we have a feature map. Such maps are also called **topology preserving maps**.

Examples of such maps in biology include

the retinotopic map which takes input from the retina (at the eye) and maps
  it onto the visual cortex (back of the brain) in a two dimensional map
the somatosensory map which maps our touch centres on the skin to the somatosensory cortex
the tonotopic map which maps the responses of our ears to the auditory cortex.

Each of these maps is believed to be determined genetically but refined by usage. E.g. the retinotopic map is very different if one eye is excluded from seeing during particular periods of development.

Kohonen's algorithm [27] is exceedingly simple – the network is a simple 2-layer network and competition takes place between the output neurons; however now not only are the weights into the winning neuron updated but also the weights into its neighbours. Kohonen defined a neighbourhood function $f(i, i^*)$ of the winning neuron i^*. The neighbourhood function is a function of the distance between i and i^*. A typical function is the Difference of Gaussians function; thus if unit i is at point $\mathbf{r}_i$ in the output layer then

$$f(i, i^*) = a \exp\left(\frac{-|r_i - r_{i^*}|^2}{2\sigma^2}\right) - b \exp\left(\frac{-|r_i - r_{i^*}|^2}{2\sigma_1^2}\right) \tag{4.15}$$

where r_k is the position in neuron space of the kth centre: if the neuron space is 1 dimensional, $r_k = k$ is a typical choice; if the neuron space is 2 dimensional, $r_k = (x_k, y_k)$, its two dimensional Cartesian coordinates.

Results from an example experiment is shown in Fig. 4.3. The experiment consists of a neural network with two inputs and twenty five outputs. The two inputs at each iteration are drawn from a uniform distribution over the square from -1 to 1 in two directions. The algorithm is

1. Select at random an input point.
2. There is a competition among the output neurons. That neuron whose weights are closest to the input data point wins the competition:

$$\text{winning neuron, } i^* = \arg\min(\|\mathbf{x} - \mathbf{w_i}\|) \tag{4.16}$$

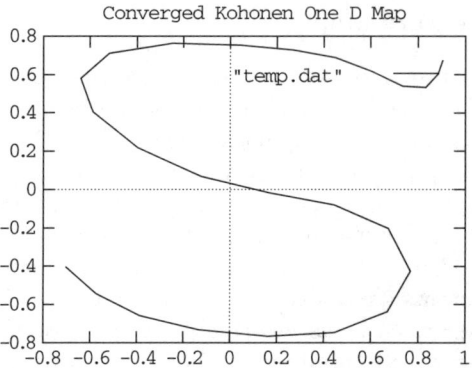

Fig. 4.3. A one dimensional mapping of the two dimensional input space

3. Now update all neurons' weights using

$$\Delta w_{ij} = \alpha(x_j - w_{ij}) * f(i, i^*) \qquad (4.17)$$

where

$$f(i, i^*) = a \exp\left(\frac{-|r_i - r_{i^*}|^2}{2\sigma^2}\right) - b \exp\left(\frac{-|r_i - r_{i^*}|^2}{2\sigma_1^2}\right) \qquad (4.18)$$

4. Go back to the start until some termination criterion has been met.

Kohonen typically keeps the learning rate constant for the first 1000 iterations or so and then slowly decreases it to zero over the remainder of the experiment (we can be talking about 100 000 iterations for self-organising maps).

In [27], a whole chapter is given over to applications of the SOM; examples include text mining, specifically of web pages, a phonetic typewriter, clustering of financial variables, robot control and forecasting.

Hints and Tips for Unsupervised Learning

Unsupervised learning seeks to identify hidden structure of some type or another in a data set. Therefore we must explicitly

1. Identify what type of structure we wish to reveal: if we seek informative projections, we use a variant of Hebbian learning; if we seek clusters, we use a variant of competitive learning. We may use either for data compression: for quantization, use competitive learning; for noise reduction, usually Hebbian learning is better.
2. We tend to start with rather small weights; a weight which is larger than its peers will bias the original learning leading to longer training times though it will be wiped out eventually. Similarly with the centres for competitive learning.

3. It is often best to decrease the learning rate during training. This smoothes out the effect of individual data items and accords well with stochastic gradient descent theory.

4.6 Combining Neural Techniques with Other Technologies

We will not consider, in this chapter, combinations of neural and fuzzy techniques nor combinations of neural and evolutionary techniques since they are described elsewhere in this volume [25, 31]. However, one of the most promising developments of late has been the combination of standard statistical techniques with ANNs.

We discussed above the fact that artificial neural networks have been shown to implement the statistical techniques of Principal Component Analysis and Exploratory Projection Pursuit. There have also been neural implementations of Canonical Correlation Analysis [21, 28], Factor Analysis [7] and so on. It is perhaps, then, surprising that current trends in statistics have not received as much attention as they might within the neural network community. One of these has been the use of ensemble methods in combining predictors [1, 2, 10, 24]. Perhaps the simplest is bagging predictors. The term "bagging" was coined by joining bootstrapping and aggregating- we are going to aggregate predictors and in doing so we are bootstrapping a system. We note that the term "bootstrapping" was derived from the somewhat magical possibilities of "pulling oneself up by one's bootstraps" and the process of aggregating predictors in this way does give a rather magical result – the aggregated predictor is much more powerful than any individual predictor *trained on the same data*. It is no wonder that statisticians have become very convincing advocates of these methods.

4.6.1 Bagging and Bumping

Bootstrapping [1] is a simple and effective way of estimating a statistic of a data set. Let us suppose we have a data set, $D = \{\mathbf{x}_i, i = 1, \ldots, N\}$. The method consists of creating a number of pseudo data sets, D_i, by sampling from D with uniform probability with replacement of each sample. Thus each data point has a probability of $(\frac{N-1}{N})^N \approx 0.368$ of not appearing in each bootstrap sample, D_i. Each predictor is then trained separately on its respective data set and the bootstrap estimate is some aggregation (almost always a simple averaging) of the estimate of the statistic from the individual predictors. Because the predictors are trained on slightly different data sets, they will disagree in some places and this disagreement can be shown to be beneficial in smoothing the combined predictor. In the context of classification, the bootstrap aggregation (bagging) of the classifiers' predictions can be shown to outperform the individual classifiers by quite some margin [2].

Bumping [24] is a rather different technique in that it uses the bootstrapped functions to find the single best function rather than do any averaging over the bootstraps. As with bagging, we draw B bootstrap samples from the training set and fit the predictor or classifier to each. We then choose the model which is best in terms of average minimum error over the whole training set i.e. this estimate includes all members of the training set, D.

Bumping is said to work because the bagging procedure sometimes leaves out data points which have a strong pull on the classifier towards solutions which exhibit poor generalisation. Outliers for example are little use in training an classifier for good generalisation and so a classifier which is trained on data samples minus the outliers is liable to exhibit good generalisation.

Bagging has been applied to predictors in general [1], and artificial neural networks in particular [9], most often to the multilayered perceptron. We [16] have recently carried out a case study comparing bagging, bumping and boosting radial basis networks and multilayered perceptrons in the context of forecasting financial time series. We have shown that

1. Bagging, boosting and bumping have more effect on ensembles of radial basis function networks than they have on the multilayered perceptron. This tends to be because there is greater variability in groups of rbf networks than mlp networks.
2. Boosting is more effective than either bagging or bumping. Indeed, we can design boosting methods based on the nature of rbf nets which are excellent predictors.
3. The method of representation of the data is all-important. Financial time series have often been characterised as random walk models and we find that the best results are achieved by working totally in the space of residuals, $r_t = x_t - x_{t-1}$.

In a related study [32], we have investigated bagging and bumping Self-organising Maps. While the supervised bagging methods have a base on which to hang their classifiers (a piece of training data is known to come from a specific and known class), bagging unsupervised learners has no such ground truth to hang onto. Thus we have a problem illustrated in Fig. 4.4. In the left figure, we show the centres of two SOMs which have been trained on *exactly the same data* i.e. no bagging. We use 1000 two dimensional data points drawn from a mixture of Gaussians and a Self Organising Map with 20 centres. We see that the maps share very little in common. It is very difficult (indeed we believe impossible) to equate these two maps. This problem is a direct consequence of the fact that the dimensionality of the map (1 in this case) is not the same as the dimensionality of the data (2 in this case). Thus the map has more freedom in this space to wind round in any direction which seems appropriate locally. In practice, when the map is trained in an on-line manner, the order in which data happens to be presented to the map will determine where it maps locally. Clearly we can carry out an investigation of the local

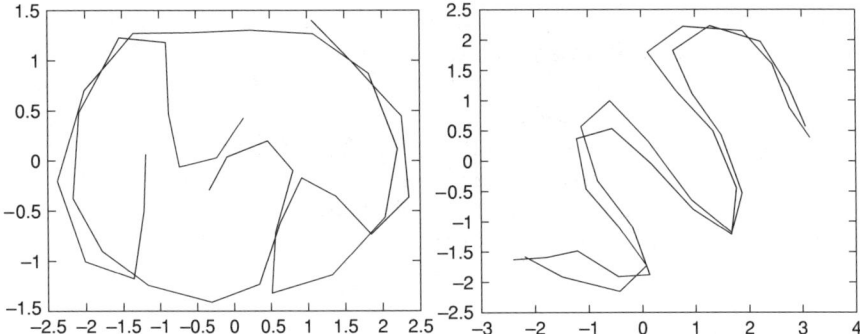

Fig. 4.4. *Left*: Two SOMs trained on the same data; it is difficult/impossible to find a joint map. *Right*: Two SOMs trained on the same data when the centres were initialised to lie on the first Principal Component

dimensionality of the data and create a map of equal dimensionality locally. However

- This is liable to be a very time-consuming process
- There is no guarantee that the resulting map will be as comprehensible as the simple map is and this is one of the main attractions of SOMs as clustering tools.

This problem becomes even more acute when we introduce bagging to the process since different maps will be trained on slightly different data.

We require some way to ground the SOMs so that they tend to learn similar mappings. Note that the grounding cannot be too constricting so that they learn exactly the same mapping which would defeat the purpose of the bagging. In Fig. 4.4 (Right) we show two SOMs trained on the same data when the SOMs' centres were initialised to lie along the first principal component. The first centre is initialised at the projection of the data point with the greatest projection onto this Principal Component, the last centre (centre numbered 20) is initialised to lie at the projection of the data point whose projection is minimal and the other centres were initialised at evenly spaced points along the principal component. We see that we have introduced a great deal of regularity in the mapping found. Notice that while the initial conditions were the same and (in this experiment) the same data points were used, there is still a stochastic element to the trained map in that the order in which data points were presented to the network was arbitrary.

Nevertheless, there is still a degree of disparity between the mappings.

4.7 Conclusion

In this chapter, we have reviewed some of the major types of artificial neural networks. We have suggested that the major differentiation within the artificial neural network field is between those networks which use supervised learning and those networks which use unsupervised learning. We have given some examples of each but any good textbook in the field (and we particularly recommend [22]) will give many more examples.

Finally, we have discussed the use of ensemble methods which uses one of several standard statistical techniques to combine sets of artificial neural networks. The resulting networks are shown to be more powerful than the individual networks and illustrates the use of hybrid techniques in this field.

References

1. L. Breimen. Using adaptive bagging to debias regressions. Technical Report 547, Statistics Dept, University of California at Berkeley, February 1999.
2. L. Breimen. Arcing the edge. Technical report, Statistics Dept, University of California, Berkeley, 486.
3. C. J. C. Burges. A tutorial on support vector machines for pattern recognition. *Knowledge Discovery and Data Mining*, 2(2), 1998.
4. Gail A. Carpenter and Stephen Grossberg. Art 2: Self-organization of stable category recognition codes for analog input patterns. *Applied Optics*, 26:4919–4930, 1987.
5. Gail A. Carpenter and Stephen Grossberg. Art 3:hierarchical search using chemical transmitters in self-organizing pattern recognition architectures. *Neural Networks*, 3:129–152, 1990.
6. D. Charles and C. Fyfe. Modelling multiple cause structure using rectification constraints. *Network: Computation in Neural Systems*, 9:167–182, May 1998.
7. Fyfe C. Koetsier J. MacDonald D. Charles, D. Unsupervised neural networks for the identification of minimum overcomplete basis in visual data. *Neurocomputing*, 2001.
8. Persi Diaconis and David Freedman. Asymptotics of graphical projections. *The Annals of Statistics*, 12(3):793–815, 1984.
9. R. Dybowski and S. Roberts. *Clinical Applications of Artificial Neural Networks*, chapter Confidence Intervals and Prediction Intervals for Feed-Forward Neural Networks, pages 1–30. Cambridge University Press, 2003.
10. J. Friedman, T. Hastie, and R. Tibshirani. Additive logistic regression: a statistical view of boosting, technical report. Technical report, Statistics Dept, Stanford University, 1998.
11. Jerome H Friedman. Exploratory projection pursuit. *Journal of the American Statistical Association*, 82(397):249–266, March 1987.
12. C. Fyfe. Pca properties of interneurons. In *From Neurobiology to Real World Computing, ICANN 93*, pages 183–188, 1993.
13. C. Fyfe. Introducing asymmetry into interneuron learning. *Neural Computation*, 7(6):1167–1181, 1995.
14. C. Fyfe. Radial feature mapping. In *International Conference on Artificial Neural Networks, ICANN95*, Oct. 1995.
15. C. Fyfe. A comparative study of two neural methods of exploratory projection pursuit. *Neural Networks*, 10(2):257–262, 1997.

16. C. Fyfe. Ensembles of supervised artificial neural networks. In B Gabrys and A. Nuernberger, editors, *European Symposium on Intelligent Technologies, Hybrid Systems and their implementation on Smart Adaptive Systems, EUNITE2003*, 2003.
17. C. Fyfe and R. Baddeley. Non-linear data structure extraction using simple hebbian networks. *Biological Cybernetics*, 72(6):533–541, 1995.
18. C. Fyfe and D. Charles. Using noise to form a minimal overcomplete basis. In *Seventh International Conference on Artificial Neural Networks, ICANN99*, pages 708–713, 1999.
19. M. Girolami and C. Fyfe. Blind separation of signals using exploratory projection pursuit. In *Speech and Signal Processing, International Conference on the Engineering Applications of Neural Networks*, 1996.
20. M. Girolami and C. Fyfe. Negentropy and kurtosis as projection pursuit indices provide generalised ica algorithms. In A. Cichocki and A. Back, editors, *NIPS96 Blind Signal Separation Workshop*, 1996. (Invited Contribution).
21. Z. Gou and C. Fyfe. A family of networks which perform canonical correlation analysis. *International Journal of Knowledge-based Intelligent Engineering Systems*, 5(2):76–82, April 2001.
22. Simon Haykin. *Neural Networks- A Comprehensive Foundation*. Macmillan, 1994.
23. D. O. Hebb. *The Organisation of Behaviour*. Wiley, 1949.
24. T. Heskes. Balancing between bagging and bumping. In *Neural Information Processing Sytems,NIPS7*, 1997.
25. N. Kasabov, Z. S. H. Chan, Q. Song, and D. Greer. *Do smart adaptive systems exist*, chapter Neuro-fuzzy systems with evolutionary optimisation, page Chap. 9. Springer, 2003.
26. T. Kohonen. *Self-Organization and Associative Memory*. Springer-Verlag, 1984.
27. Tuevo Kohonen. *Self-Organising Maps*. Springer, 1995.
28. P. L. Lai and C. Fyfe. A neural network implementation of canonical correlation analysis. *Neural Networks*, 12(10):1391–1397, Dec. 1999.
29. D. MacDonald. *Unsupervised Neural Networks for Visualisation of Data*. PhD thesis, University of Paisley, 2001.
30. James McClelland, David E. Rumelhart, and The PDP Research Group. *Parallel Distributed Processing*, volume Volume 1 and 2. MIT Press, 1986.
31. D. Nauck. *Do smart adaptive systems exist*, chapter Neuro-fuzzy Systems for explaining data sets, page Chap. 15. Springer, 2003.
32. L. Petrakieva and C. Fyfe. Bagging and bumping self-organising maps. In B. Gabrys and A. Nuernberger, editors, *European Symposium on Intelligent Technologies, Hybrid Systems and their implementation on Smart Adaptive Systems, EUNITE2003*, 2003.
33. A. J. Smola and B. Scholkopf. A tutorial on support vector regression. Technical Report NC2-TR-1998-030, NeuroCOLT2 Technical Report Series, Oct. 1998.
34. V. Vapnik. *The nature of statistical learning theory*. Springer Verlag, New York, 1995.

5

Machine Learning and Reinforcement Learning

M. van Someren and S. ten Hagen

Informatics Institute, University of Amsterdam
Kruislaan 403, 1098 SJ Amsterdam, The Netherlands
someren,stephanh@science.uva.nl

The application of methods from Machine Learning and Reinforcement Learning for adaptive systems requires a number of decisions. In this chapter we outline the main decisions, summarise basic methods and illustrate these with examples of applications.

5.1 Introduction

Adaptive systems are able to change their behaviour according to their environment. Adaptive systems be viewed as having two functions: (1) a basic, standard function and (2) the optimisation of performance of the standard function with respect to a performance goal. In this chapter we review methods and systems that perform this combination of tasks using methods from Machine Learning and Reinforcement Learning. We focus on the methods on which such systems are based and the conditions that make the methods effective.

Machine Learning provides methods for different forms of adaptivity. We can make the following main distinctions:

- Goal: Generalisation or optimisation.
  The purpose of most methods is to construct a general model of a domain from a sample of observations. This model can be a goal by itself, for inspection by humans or as part of a system that makes classifications or predictions. The model should be accurate and comprehensive as possible, resulting in optimal predictions or classifications. A different type of task is to optimise behaviour with respect to a payoff criterion. In this case the system needs to adjust its behaviour such that the expected value of some criterion will be maximised. This implicitly involves generalisation because behaviour should be optimal not only for the observed situations but in general. An example of a generalisation task is classification of patients by their disease. Patient records are used to construct the general

relation between patient characteristics and diseases. The general model can then be used to predict the disease of new patients. Optimisation would be the choice of an optimal treatment for patients. Treatment can involve multiple parameters (e.g. medication, exercise, diet). The payoff criterion is improvement in condition (and possibly financial or psychological costs). Optimisation problems usually include a generalisation problem but this is often implicit.

- Setting: offline or incremental.
  In incremental learning, data come in one by one and knowledge must be updated continuously. Incremental learning can be achieved by storing data and repeatedly learning offline but this is not an efficient solution in terms of computing time and memory use.
- Stability of domain: Static or dynamic.
  The pattern in the data can be stable or change over time. For dynamic domains, one approach is to adapt, to change the model to follow the changes and the other approach is to include the dynamics in the model.
- Role of prior knowledge: refinement, knowledge intensive learning or empirical learning.
  In the case of refinement the system starts learning with initial knowledge that is entered manually. The learning task is to improve (complete or correct) the initial knowledge on the basis of observations. Refinement is an important form of knowledge intensive learning. Other forms involve prior knowledge, for example about relevance of features or the type of numerical relations. Empirical learning means that no prior knowledge is used for learning.

The first step in applying ML is to classify the problem by the dimensions above because these determine the type of method to be used. Most methods aim at problems that are characterised as generalisation, static domain, offline, empirical. This is what is often called data mining. Applications of ML come in two main types. The most common type, also for data mining, is the application of existing methods to data, to construct a model of the data. The other type of application is the construction of systems to perform a specific learning task. This is needed if learning is part of the function of a system or of no tool is available for the task.

The methodology for data mining has been described with minor variations by various authors, for example [7, 28, 30]. In this chapter we will focus on types of learning tasks with which there is less experience and less established methodology.

5.2 Process Model for Developing Adaptive Systems Using Machine Learning and Reinforcement Learning

In the research literature the emphasis is on methods and their properties. Several authors have described process models for application of Machine

Learning to Data Mining. The best-known model is CRISP [7]. The CRISP process model distinguishes the following steps:

- Problem understanding: Understanding the business problem that should be solved by data mining and formulating the Data Mining problem. In this step the business problem is transformed into a problem stated in terms of the data. (In CRISP terminology this is called "business understanding".)
- Data understanding: Understanding the meaning of the data and the main properties of the data, such as accuracy, completeness, variation and global properties of the data.
- Data preparation: Selecting, converting, cleaning and organising the data to prepare the actual model construction step.
- Modelling: Selecting and applying a learning system to construct a model that solves the Data Mining problem. This may involve experimentation with several methods and it may require additional conversion of the data.
- Evaluation: Assessing the quality of the model that was constructed from both a data analysis and a business perspective: is the model technically sound and is it able to solve the original business problem?
- Deployment: If the model is adequate then it is deployed: used as part of a computer system (e.g. to perform predictions) or brought into a form in which it can be presented to the owners of the original business problem.

The CRISP model is well documented (see http://www.crisp-dm.org/) and used in industrial practice. The method is primarily aimed at data analysis and not at the development of adaptive systems. There are several differences between development of adaptive system and Data Mining. Most Data Mining projects use existing toolkits where adaptive systems are usually developed from scratch. The development of adaptive systems therefore needs a combination of methodological principles from Data Mining and from Software Engineering. Another difference is that in Data Mining often the data already exist. Adaptive systems usually acquire their own data. In this section we outline a process model that is based on this idea. It consists of the following steps:

1. Problem understanding,
2. Characterising the tasks,
3. Designing the data acquisition component,
4. Designing the performance component,
5. Designing the adaptation component,
6. Implementing the adaptive system,
7. Evaluating the adaptive system,
8. Deployment.

In this section we discuss these steps in more depth. In Sects. 5.3.3 and 5.4 we illustrate the process model with examples.

5.2.1 Problem Understanding

Problem understanding amounts first of all to defining the tasks for the adaptive system: the performance task and the adaptation task. For both of these the input, output and function need to be specified. An important point is the type of feedback that is input for the adaptation task. Consider the following example. Suppose that we want to build a system that learns to recognise diseases, as a simple case of an adaptive system.

```
Performance task:
    input:  description of patient
    output: disease

Adaptation task:
    input:  descriptions of patients with diseases
    output: prediction model (for performance task).
```

Performance and adaptation may be interleaved, for example in a scenario in which the system first makes a prediction (performance) and then is given the actual disease from which it adjusts its prediction model (adaptation). Although this may seem a straightforward issue, adaptation problems can sometimes be characterised in different forms. For example, the problem of learning to select an optimal treatment for a patient can be viewed as a single adaptation problem or as a compound task that consists of prediction and maximisation. This gives two different task decompositions:

Single task:

```
Performance task:
    input:  description of patient
    output: best treatment

Adaptation task:
    input:  descriptions of patients -> best treatments
    output: prediction model.
```

Compound task:

```
Performance task 1:
    input:  description of patient + treatment
    output: effect

Performance task 2:
    input:  description of patient + treatment + effect
    output: best treatment

Adaptation task:
```

```
input:  descriptions of patients + treatment -> effect
output: prediction model.
```

The first adaptation subtask may be able to use only data about optimal treatments. If the effect of treatments varies gradually then the second adaptation subtask will produce better results. The decomposition must be such that data for the adaptation task are available. If data are to be collected online then this must be practically feasible.

5.2.2 Characterising the Adaptation Task

In this step the adaptation task is characterised in terms of the features given in Sect. 5.1 above. This provides the basis for the methods to be applied. Often an adaptive system will be integrated in a larger system. Analysis of this context is also performed here.

5.2.3 Designing the Data Acquisition Component

This is specific to the system context. Relevant for the methods is the amount of data and if adaptation needs to be done online or offline (after collecting input for the adaptation process in a buffer).

Data Cleaning

Then, in the data cleansing step the data has to be assessed and approved to be of the required quality for the actual learning. Often, data is not collected with the purpose of using it for a learning technique. In industry, data is often collected for controlling a process. Data formats may change over time, sampling frequency may differ between attributes and other data quality defects may occur. Consequently, the data suffers from missing values and contains records with contradictory or complementary information. Before being suitable for application of a learning technique, some of these defects have to be repaired. *Missing values*, for instance, may lead to removal from the data set, or to assigning a "don't know" value. Moreover, *data enrichment* may be possible: adding information from specific sources to the data set.

Finally, in the feature selection step, the contribution of individual attributes in the data to the task to achieve (classification, prediction or characterization) is assessed, leading to the removal of non-contributing attributes from the data set.

Feature Selection

Deciding which features are relevant is of key importance for effective adaptive systems. This is because redundant features may produce spurious (parts of) models when the amount of data is limited. Eliminating features that are almost certain to have no effect will improve the speed at which the system will adapt correctly.

Feature Transformation

For the choice of a method, the type of data is important. Most problems involve data that can be represented as feature vectors or values on a fixed number of variables. The main scale types are: numerical, ordinal and nominal (also called categorical or symbolic). There are methods for transforming the scale or the scale type. For example, a numerical variable (effect of a treatment) can be represented as a set of ordered labels (e.g. complete recovery, partial recovery, partial recovery and additional new symptoms, no recovery, getting worse). It can also be represented as a binary choice (improved/not improved) or as a numerical scale. Scales can be reduced by merging values. It is also possible to construct a scale from categorical labels. For example, if the country of birth of patients is known then the values of this feature can be scaled by their association with the predicted variable resulting in an ordered (or even numerical variable). The main principles that govern the transformations are: (1) remove redundant information and (2) satisfy requirements of the method that will be applied. If transformations can be reliably made from prior knowledge about the domain, this is preferred.

Feature Construction

Relations between features can also be exploited. If two features are strongly correlated in the data, they can be (normalised and) added to create a single new feature. Features with a "nested" structure (e.g. male/female and pregnant/not-pregnant) can be reformulated as a single feature (male/female-pregnant/female-not-pregnant). If features have many values then they can be mapped to a less fine scale. Feature selection and feature construction can be done from prior knowledge or from initial analysis of the data.

5.2.4 Designing the Performance Component

The performance component can be tightly integrated with the adaptation component. For example, in a system that learns to recognise objects in images or problems from error messages

- Classification: given a collection of object descriptions with classes, the target model should reproduce these classifications and also give plausible classifications for new objects. Classification can be just recognition but it can also be selection of an action from candidate actions.
- Prediction: given a collection of descriptions of situations or events with a continuous variable, the target model should reproduce the values from the descriptions and also give plausible values for new descriptions. The term prediction is used for prediction into the future but in general any estimate of an unknown quantity can be seen as prediction of the outcome of a future measurement.

- Characterization: given a number of classified objects, situations, events, find descriptions that characterise the different classes

5.2.5 Designing the Adaptation Component

The design of the adaptation component depends on the task characteristics, the data and the performance component. It involves the selection or design of a method and representation. There is a wide variety of methods and for a subset of these tools are available. For others research prototypes can be found on the web. In Sect. 5.3 we summarise a few methods. We refer the reader to the textbooks and overviews of Machine Learning and Data Mining.

5.2.6 Implementing the Adaptive System

An adaptive system can be constructed from an existing learning system or tool or it can be developed using only standard software development resources. In offline adaptation it is often better to use available software that produces models in the form of executable code.

5.2.7 Evaluating the Adaptive System

Technical Evaluation

The most important aspect of evaluation of learning is performance. You want to know how well the result of a certain learning method models the domain of the data. There are several ways to answer this question. The extent to which the model reproduces the learning data is not always a good measure because the model is optimised for these data and may work less well on data from the same domain that were not seen before. This is called "overfitting". One way to evaluate the result is to learn on part of the data and count the classification errors or measure the prediction errors (for a continuous variable). This can be done as follows:

- Train-and-test:
  split data in train and test set, train the classifier on the train data, and estimate its accuracy by running it on the test data. If the amount of data is small relative to the strength and complexity of effects then this procedure will underestimate the accuracy because both learning and testing will be unstable.
- Cross validation:
  Divide the data in M sub-samples, For each sub-sample i, train the classifier on the all other sub-samples, and test on sub-sample i. The estimated error rate is the average error rates for the m sub-samples. The classifier to exploit is trained from the entire data set.

- Resampling:
  Draw from the data set a new data set with replacement, the sample being of the same size as the data set. Train the classifier on the drawn set, and estimate the error on the unused parts. Averaging the performance on a large number of experiments gives the final estimate.

Application Evaluation

Another dimension is if the adaptation is adequate from an application viewpoint. What is the risk of serious errors? May the conditions for which the system was developed change and what will be the consequences?

5.2.8 Deploying the Adaptive System

In this stage the system is delivered and integrated in its environment. Besides implementation work this will need planning and involvement of the user environment because the system may need time to adapt and improve performance, which is not what users expect.

5.3 Some Examples of Methods

In this section we summarise a few popular methods, with an emphasis on Reinforcement Learning.

5.3.1 Naive Bayes

The Naive Bayesian classifier is based on the notion of conditional probabilities. According to basic probability theory, if we consider two outcomes A and B

$$P(A\&B) = P(A|B)P(B) = P(B|A)P(A) \tag{5.1}$$

If we take the second and third term and divide by $P(B)$ we get:

$$P(A|B) = \frac{P(B|A)P(A)}{P(B)} \tag{5.2}$$

We now interpret A as "user likes document" and B as a property of the document then we can use this formula to calculate $P(A|B)$ from $P(B|A)$, $P(A)$ and $P(B)$. $P(A)$ can be estimated from observations. We have no single usable property of documents, so instead we use many properties: the presence of a word in the document. The Naive Bayesian classifier is based on the assumption that the probabilities $P(\text{word}|\text{interesting})$ for different words are independent of each other. In that case we can use the following variation of formula (5.1):

$$P(\text{interesting}|W_1 \cap W_2 \cap W_3 \cap \cdots \cap W_n) ; \quad (5.3)$$
$$= \frac{P(W_1|\text{interesting})P(W_2|\text{interesting}) \cdots P(W_n|\text{interesting})}{P(W_1)P(W_2) \cdots P(W_n)}$$

These probabilities can be estimated from data. When more data are collected these estimates will become more accurate.

Hints and Tips

The Naive Bayesian classifier is easy to implement. There are a number of subtleties that must be taken into account. When the amount of data about a word is small then the estimates are unreliable.

5.3.2 Top-Down Induction of Decision Trees

A classic family of methods that learn models in the form of trees. Associated with the nodes in the trees are data variables and the branches correspond to values (or value intervals for numerical variables). Leaves are associated with classes and objects are classified by traversing the tree. The tree is constructed by selecting the best predicting variable, splitting the data by the selected variable, and repeating this recursively for each subset. Another problem is that the process divides the data into ever smaller subsets. At some point the amount of data is too small to decide about extending the tree. Some criteria for stopping the process is evaluation against a separate dataset, statistical significance of tree extensions.

Detailed descriptions can be found in the literature e.g. [19] or [30] and a number of commercial tools are available that usually include procedures for evaluating trees and for displaying and exporting trees.

5.3.3 Reinforcement Learning

Reinforcement Learning (RL) combines machine learning with optimising control. Where machine learning mainly focuses on modeling data, in RL the past experiences form the data that is used to improve a control policy. So the use of RL results in an adaptive system that improves over time. In retrospect we can say that it has been investigated for half a century now (see historical notes in [23]). Research was inspired by trial and error learning in psychology and animal learning and it found its way into artificial neural networks [2]. In the late eighties/early nineties the close relation between RL and the already matured field of Dynamic Programming (DP) became clear. This boosted the reached in RL and allowed for expressing existing heuristics in the unifying terminology of DP. A more solid theoretic underpinning was the result and soon proofs of convergence for reinforcement learning algorithms followed [8, 13, 22]. These proofs applied to only a restricted problem domain and since then research focused more on extending the applicability of RL to other, more practical, domains.

An RL applications should at least have the following elements:

- A set of states
- A set of actions
- A dynamic system
- A scalar evaluation
- An objective

Here the state is the variable that can change over time and that has to be controlled. The dynamics of the system describes how the state changes, given the past states and actions. There should be a conditional dependency between the actions and the state change, so that the actions can be used to control the system. In automatic control the actions are selected as a function of at least the current state. In RL terminology the function that computes that action is referred to as the *policy*. The scalar evaluation is the reinforcement. This can be regarded as a reward or punishment of the current state and action. The objective indicates the criterion based on which the policy is being improved. So the criterion indicated whether the reinforcement should be interpreted as reward of punishment.

Since the system is dynamic an action now can have consequences for future reinforcements, therefore the objective of RL is usually expressed as the sum or the average of the future reinforcements. RL solves this problem, the *temporal credit assignment* problem, by propagating evaluations back in time to all past actions that could be responsible for these evaluation.

RL in Theory

A special class of problems are those that can be formulated as a Markov Decision Problem (MDP). Here the states s are elements from a finite set $\mathcal{S}$, and an action a can be chosen from a finite set $\mathcal{A}(s)$. The state transition probability is given for all state/action/next-state combinations: $P^a_{ss'} = Pr(s'|s,a)$. This specifies the probability of s' being the next state when in state s action a is taken. The expected reinforcement received at a certain moment only depends on current state/action/next-state combination, so it is a stationary process. The reinforcement that is received at time step k is indicated by r_k. Depending on whether r indicates a reward or punishment the objective is to minimize or maximize the (discounted) sum of future reinforcements. The value function maps the (discounted) sum of future reinforcements to the state space. When s is the state at time step k the value is given by:

$$V^\pi(s) = \sum_{i=k}^{\infty} \gamma^{i-k} r_i \quad (5.4)$$

Here $\gamma \in [0,1]$ is the discount factor that makes sure that the sum is finite. The policy π determines for all states which action to take, and in (5.4) it indicates what actions will be taken in future states. Because the state transition $P^a_{ss'}$ is conditionally independent of the past (the Markov property), the future only depends on the current state and all actions taken in future. In other words, the values $V^\pi(s)$ for all states do exist and can be computed when $P^a_{ss'}$, π and all reinforcements are known. The objective is now to find that policy for which the value for all states is minimal or maximal. A classical solution to the MDP is Dynamic Programming (DP) [3, 5], in which the state transition probabilities $P^a_{ss'}$ are used to compute the optimal value function. Once the optimal value function is given the optimal policy can be found by selecting for each state that action that has the highest expected value for the next time step. In RL the value function is not computed off-line. Instead the values for each state are estimated on-line by controlling the system using a policy. The estimation is based on the *temporal difference* [22]. The idea is that the current value should equal the reinforcement received plus the value of the next state.

According to (5.4):

$$V^\pi(s_k) = r_k + \gamma \sum_{i=k+1}^{\infty} \gamma^{i-k-1} r_i = r_k + \gamma V^\pi(s_{k+1}) \qquad (5.5)$$

The same should hold for the approximated value function $\hat{V}$. If it does not, the estimated value of the current state is adjusted based on the estimated value of the next state plus the reinforcement received.

$$\hat{V}(s_k) \leftarrow (1-\alpha)\hat{V}(s_k) + \alpha(r_k + \hat{V}(s_{k+1})) \qquad (5.6)$$

Here $\alpha \in [0,1]$ is the learning rate. Initially the values for each state are random, but after sufficient iterations using (5.6) the values will approach V^π, after which the improved policy can be derived using the same method as in DP. This does require the use the state transition. The only benefit of using RL is that the value function approximation uses only those state transitions that actually occur instead of all possible state transitions. An RL approach that does not use the model of the state transition is Q-learning [27], where values are assigned to all possible state and action combinations. The training is based on the temporal difference, but now it is sufficient to select for each state that action with the highest of lowest value to obtain the improved policy.

The main difference between RL and DP is that RL only updates for state transitions that actually take place. State transitions that never take place are not taken into account and do not contribute the the value function. This means that if all actions are selected according to a deterministic policy π, the value function can not be used to improve the policy. This is because the consequences of alternative actions are unknown. To estimate the effect of alternative actions a stochastic *exploration* component is added to the deterministic policy π. The exploration is responsible for the data acquisition part. If an exploration strategy is used for which all state action combinations are guaranteed to be tried often enough, then the RL algorithms converge to the optimal value function and optimal policy.

RL in Practice

RL in theory can be guaranteed to converge to the optimal solution, given the right conditions. Unfortunately in most practical applications these conditions are not met. This does not mean that RL is useless. It just means that for practical application some parts of the RL approach have to be modified to make it work. This may have the consequence that the optimal result cannot be guaranteed, but it still may result in a significant improvement of the performance. RL addresses a more complex task than those of regular ML methods. In terms of Sect. 5.1 RL can be characterised as follows. Goal is optimisation and only in some variants also generalisation, setting is incremental, domains are static (although RL can be adapted to dynamic domains) and learning is empirical. When using RL one has to ask the following additional questions:

- Is learning required?
  If a good control policy can be designed manually, there is no need to improve it using RL.
- Is the state transition function known?
  If the state transition function is known training can be performed in simulation and after training the policy can be transferred to the real system. If the state transition function is not known simulations are not possible and also only Q-learning methods can be applied. There are also RL methods that estimate the state transition functions and use this to speed up the learning process.
- Is there a finite and small enough set of states?
  In theory RL stores for each state the value and decides for each state the action. In case of a continuous state space this is no longer possible. It is possible to discretize the continuous state space, but usually this requires some knowledge about the problem domain. Alternative a function approximator like a neural network can be used to represent the value function and policy.
- Is enough data available?
  RL may require many epochs of training, which is no problem if the system is simulated on a computer. However, when learning is applied to a real system it may be that the system itself is too slow. In that case generalization over the state and actions space or the reuse of data may be required to learn fast enough.

- How are reinforcements distributed over states?
  Learning will be very slow when, for example, the reinforcement only has a positive value for the goal state and zero everywhere else. The policy can only be improved when the task can be completed by the learning policy. This may not be the case for a random initialized policy. There are also problems where the reinforcement is available at any time step. Here the danger is that the policy focuses on short term goals and may not find the optimal solution.

5.4 Example Applications of Machine Learning for Adaptive Systems

In this section we briefly review some examples of applications of ML and RL for adaptive systems.

5.4.1 Learning to Fly

Adaptive systems can be trained to mimic human performance. This type of learning was named "behavioural cloning" by Michie [17, 20]. In this example, the goal is to construct a system that controls an airplane flying a predefined manoeuver. The example is adapted from [20].

The characteristics of this adaptive task are: goal is *generalisation* (the target system must act correctly also under conditions different from those during training), setting is *offline* (training is done offline, not during flying), stability is *static* (what is a correct action does not change over time) and learning is *empirical* (no prior knowledge is entered into the system). Problem analysis was simple in this case because the goal was to select actions for an airplane controller. Yet there are some details that need to be resolved.

Data Understanding and Data Acquisition

Because this application involves a simulation rather than a real airplane it is natural to use the features that can be recorded from the simulator. Twenty features were incorporated. One difficulty of this problem is the acquisition of data. This was done using a simulator of which measurements and actions were logged with time stamps. Another difficulty is the use of continuous or discrete models. Some of the actions are clearly discrete and some are continuous. Adaptation is easier for discrete models. In this case, a discrete approach was chosen. The actions were discretised (e.g. continuous throttle positions were divided into discrete positions). Similarly, time was discretised to obtain a series of situation-action pairs instead of a continuous stream. The grain size is an important choice for discretising time. This was experimentally optimised.

Designing the Adaptive System and Evaluation

A difficulty is that some pilot actions were anticipatory rather than reactive. Some actions were taken not as a reaction to some measurable change but because the pilot anticipated new conditions. No action was taken to deal with this problem. Top-down induction of decision trees was selected as method because this produces relatively simple controllers that can be reviewed by experts and that was sufficient for constructing an adequate controller. The model was constructed offline by a standard tool for decision tree learning (C 4.5, now succeeded by C 5.0; see [19] for a detailed description of C 4.5). This tool can generate a decision tree and a procedure for applying this tree (in C) that can be run by itself or integrated into a system that connects with the measurements and controls. This was found to work very well. Experiments showed that the synthesized code cannot be distinguished from human pilots, for these manoeuvres. The system was not deployed but acted as a demonstration of the technology that was followed up by a project to explore this further.

5.4.2 Learning to Recognise Addresses

A Dutch company that is active in customer-relationship management has a system for discovering duplicate addresses. This uses an address recogniser. The company considers extending its activities to other countries and wants to adapt its software systems. This required adapting the address recogniser, more specifically, the grammar, to the address formats of other countries. Consider the following examples. Although the form varies, the Dutch addresses "Frans Halsstraat 5-2", "Frans Halsstr. 5-2" and "Frans Halsstraaat 5-II" all refer to the same address. There is substantial variety in the morphology of addresses but the underlying structure is approximately the same. Most addresses consist of a country (optional), a city or village, a street or square and a name. Both this underlying structure and much of the variety is quite constant over different languages but yet there are many small differences. For example, in Dutch addresses some numbers are parts of names. For example, there are a number of squares called "Plein 40-45" (plein = square) and this number refers to the years of the second World War and not to the number of the house as would be the cased in "Frans Halsstraat 40-45" which would be interpreted as numbers 40 to 45 in the Frans Halsstraat. Similarly, in Italian Roman numbers are very frequent in the name of streets while in Dutch there are used to indicate floor in a house with apartments on different floors.

Adapting a grammar to a new language can be characterised as generalisation, offline, static and knowledge intensive learning. A system was developed [26] to perform this task. Specifically, available were an initial grammar for Dutch addresses, a lexicon for Italian addresses and a set of unlabelled Italian addresses. Its task was to adapt the grammar to Italian addresses. The grammar allowed multiple analyses of an address. All analyses were rated and the highest rating was used to recognise the components of addresses. The grammar adaptation system used a hill climbing search strategy and operators that add and delete terms in the grammar rules and modify the ratings of grammar rules. Grammars were evaluated by the number of addresses that they can recognise. The results show that using a dataset of 459 addresses, the system constructed a grammar that correctly recognised 88% of the addresses. To evaluate this, a grammar was constructed manually by a linguist. The accuracy of the automatically adapted grammar is equal to the accuracy of a grammar that was constructed manually by a human language engineer. The grammar constructed by the system was somewhat more complex than the "human" grammar. Presumably this can be avoided by adding a premium for simplicity to the evaluation of candidate grammars.

Evaluation and Deployment

To assess the potential of the approach another experiment was performed in which a grammar for Dutch addresses was adapted to Italian addresses. The results were the same. Again a grammar with a high accuracy was obtained from a relatively small set of addresses.

5.4.3 An Adaptive Web Browser

The World Wide Web contains about 800 million webpages and that number is increasing by the day. Finding the right information becomes harder, finding interesting sites might become impossible. Try searching with AltaVista for websites about "machine learning". You will probably find over 4 million hits! Even if you only try 1% of these and if you would take 1 minute to visit one site, it would take you about 8.5 months of continuous surfing to visit them all.

In this section we summarise the development of an early recommender system, Syskill & Webert [18]. Here we focus on the problem of assisting a person to find information that satisfies long-term, recurring goals (such as finding information on machine applications in medicine) rather than short-term goals (such as finding a paper by a particular author). The "interestingness" of a webpage is defined as the relevance of the page with respect to the user's long-term information goals. Feedback on the interestingness of a set of previously visited sites can be used to learn a profile that would predict the interestingness of unseen sites. In this section, it will be shown that revising profiles results in more accurate classifications, particularly with small training sets. For a general overview of the idea, see Fig. 5.1.

The user classifies visited sites into "interesting" and "not interesting". This website collection is used to create a user profile, which is used by the adaptive web browser to annotate the links on a new website. The results are shown to the user as in Fig. 5.2. The main problem to be addressed in this adaptive system is how to give advice to the user about the interestingness

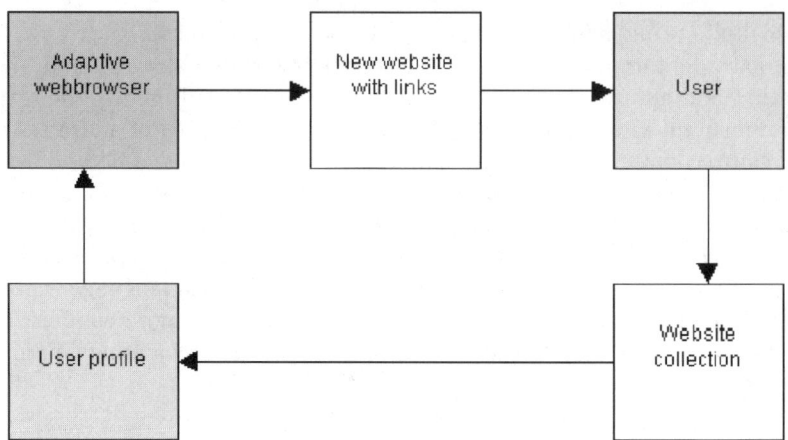

Fig. 5.1. Overview of the adaptation process

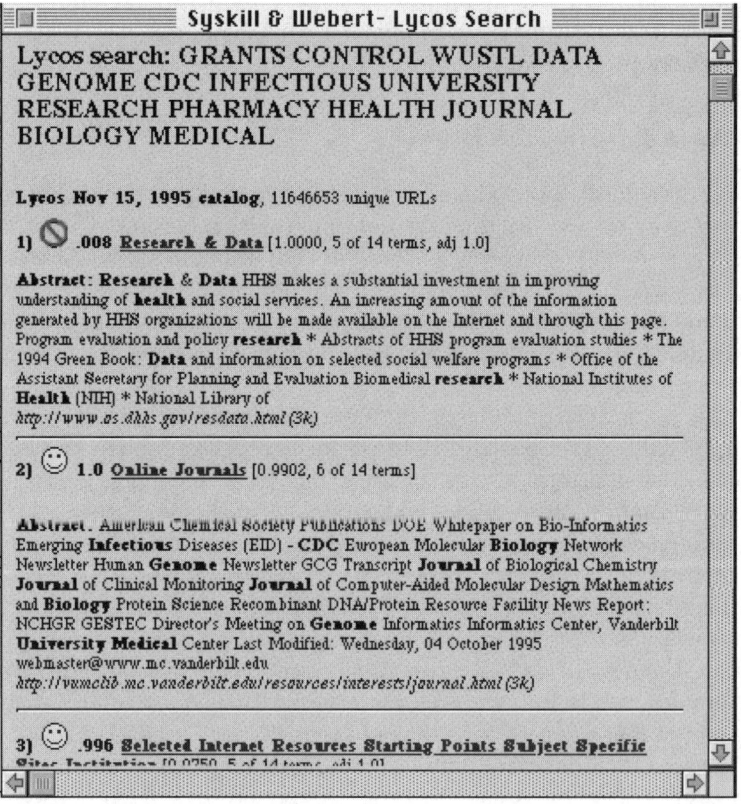

Fig. 5.2. The adaptive web browser shows its advice to the user

of a website, and on what data this advice should be based. The only data available is a list of websites visited by the user, and a corresponding list of ratings. Using this data, the system must be able to predict whether a webpage is interesting or not. And if it's wrong, it has to learn from its mistakes.

Data Preparation

The available data consists of HTML files that the user visited in the past. These files contain HTML markup tags and actual text. For this purpose, the HTML markup can be deleted because it plays only a minor role for understanding the content and its meaning. Everything between HTML tags ($<$ and $>$) is therefore omitted. Very short words are often very frequent and not too discriminative between documents and therefore words with a length smaller than three are ignored. In addition about 600 frequent longer words (taken from a "stoplist") were also omitted.

Selecting single words has its drawbacks: words like "computer science" and "science fiction" will result in counting twice the word "science", although the meaning of this word depends on its context. To be able to compare words to each other in a later stage, all words are transformed to lower case. Cleaning the HTML page shown in Data Design like this, we obtain a list of words. Further processing is possible. For example, the words "application" and "applications" appear both. One could argue that they refer to the same concept and that merging them will give better estimates of probabilities because the data for both words can be pooled together.

When the user uses the adaptive web browser, the learning method will predict which links on the current page are interesting, by reading ahead those links and analyzing their content. If the user follows some of the links, he can enter his opinion about the suggested links. If he agrees, the browser will record the prediction has succeeded, but if the user doesn't, the prediction has failed. The success or failure of the prediction can be can be used to improve the predictions. In this way, the browser learns from the user what type of sites should be recommended and what type of sites should be avoided.

5.4.4 Method Selection

Several methods can be used to predict the interestingness of websites. Neural networks and decision trees are candidates, but here we use the *Naive Bayesian classifier* because the number of features is very large and because the resulting model can well be interpreted and checked. Experiments with several methods show that compared to other methods such as neural networks and decision trees, the Bayesian classifier performs pretty well. In about 73% of the cases, the user agreed with the websites the adaptive browser suggested. A feed-forward backpropagation neural network performed just as well, but decision trees performed worse. Additional features that can enhance the performance include predefining user profiles. Users themselves can provide words that they consider good indicators for websites of their interest. In this way, the browser uses prior knowledge about the domain without having to learn it. Also, using lexical knowledge can enhance performance further. Using knowledge about relations between words can improve the quality of the most informative words. Lexical knowledge can improve the quality of the list by removing some non-informative words.

Evaluation and Deployment

Experiments were performed with users on a small set of documents to evaluate if the method worked. These produced satisfactory results. The recommender is integrated into a webpage by adding a frame that gives links to recommended documents and asks the user to evaluate documents. If the user does this, this information is used to update the probabilities. The system described here is a simple content-based recommender. Other systems use similarities between users. Some extensions to the basic approach that are currently studied are the use of additional information for constructing user profiles, systems that combine the use of similarities in content between documents with similarities between users, the use of more advanced language processing and applications to other types of information such as images and video.

5.4.5 Mobile Robot Control

In [24] RL is applied to a mobile robot that has to learn to follow a line. Learning is based on the data gathered by the robot, so it can be viewed as a self improving adaptive system. The application illustrates process control applications where the model of the dynamics is not available or expensive to obtain. The sample rate is in order of seconds or minutes and the state and action space is continuous. This means that only little data is available and that the standard RL approaches cannot be applied. In case of the robot the model is known, so insight into the correct policy is available in advance. When the robot is facing the right direction and is in close proximity to the line, then the robot can move forward and steer only a little. If the robot is far away from the line the robot should first make a 90 degree turn and approach the line. Once it is close enough it can turn to follow the line.

Design of the Adaptive System

One RL approach able to deal with continuous state and actions spaces is based on Linear Quadratic Regulation (LQR) [6, 12, 15, 29]. Here the state and actions are represented by vectors and the state transition function is given by a linear mapping from current state and action to the next state. The reinforcements represent cost and are computed as quadratic functions of state and actions. The value function or Q-function for such problem is again a quadratic functions. In case of a Q-function the improved linear policy can be derived directly from the estimated parameters of the Q-function.

For the estimation of these parameters it is possible to create an input vector containing all quadratic combination of elements of the state and action vector. The Q-function is formed by a linear mapping from input vector a scalar value. Now the temporal difference can be rewritten such that a linear least squares estimating can be given of the parameters of the Q-function. The dynamics of many systems, like the mobile robot, do not have a linear state transition function. The LQR approach can not be applied, because it would lead to very unreliable results. In [24] a feed forward neural network is used instead of the linear mapping that forms the quadratic Q-function. The network has one linear output unit and the hidden units have hyperbolic tangent activations functions. If all the weights of the hidden units are very small the network forms a linear mapping and the represented Q-function is quadratic. This is how the network is initialized. The network is trained based on the same error as in the least squares estimation. During training certain weights of hidden units can become larger and the Q-function is no longer a quadratic function. Obtaining the improved policy is a bit more complicated than in the LQR case. A two step approach can be used where first the hidden layer is ignored to form a global linear policy that can be corrected in the second step by using the hidden layers. This results in a nonlinear policy.

Evaluation

Using the above described configuration the robot was placed in a large area. A random linear policy plus noise for exploration was used to make the robot move. The robot started at four different initial positions where it ran for 20 seconds to gather the data that was used to train the neural network. The nonlinear policy was extracted from the trained network and tested on the robot. For initial positions close to the line it started turning smoothly in the right direction. For the position far from the line the robot first turned in the direction of the line. Here is had a slight overshoot and turn a bit too much. Closer to the line the robot starts to turn to follow the line again. Only 80 second of running the system was sufficient to learn the correct behavior.

Hints and Tips

- For most problem domains solution methods already exist. Sometimes these solutions can be formulated in RL terminology such that a dedicated RL approach can be developed for the particular application. Formulating the LQR task in RL terms made it possible to deal with the continuous state and actions pace.

- Sometimes problems can be simplified by changing the representation of the state and actions. Instead of the state and action vectors this application used the quadratic input vectors.
- The generalization of the function approximator can speed up the training.

5.4.6 Robot Arm

A control policy for a tracking task of a two linked robot arm is optimized using RL in [16].

The control policy should take into account possible different loads on the arm and ensure stability while minimizing the tracking error. The purpose of [16] is to introduce a stable fuzzy controller that can be tuned using RL. The dynamics of the robot arm is described by a nonlinear function, which makes it hard to obtain an optimal controller. A policy that tracks well for a heavy load may cause instability when the load is reduced. In spite of the difficult nonlinear model, the robot arm is an engineered artifact that is not completely unknown. This allows for a fuzzy controller to compensate for the nonlinearities. RL is only needed for the improvement of the performance. The closed loop behavior can be analyzed and conditions for the tunable parameters can be given. These conditions can be taken into account so that during learning the system remains always stable.

Approach

RL methods to tune fuzzy controllers were developed [4, 14] and used [11]. These approaches use the original architecture from [2], where an actor forms the policy and an critic the approximation of the value function. Instead of a function approximator for the actor a fuzzy controller with tunable parameters is be used. When learning is on-line the risk of an unstable closed loop should be avioded. In [1] this is solved by using two policies, one stabilizing controller and one that is improved using Q-Learning. The second policy is used except when instability may occur.

In [16] fuzzy is used to compensate for the nonlinearities in the system. This is possible because the kinematic model of the robot arm is known. To improve the performance the defuzzified output is weighted with parameters that are updated based on the critic. By choosing the right fuzzy inference model, a closed loop configuration can be obtain for which the parameters can never cause any instability.

Evaluation

The result is a fuzzy controller that provides stability and performs well as tracker.

Hints and Tips

- The dynamics of man made machines are usually not completely unknown. In that case use the prior knowledge available.
- In on-line learning the stability of the closed loop is very important. If possible modify the learning mechanism in such way that instability cannot occur.

5.4.7 Computer Games

In [9] RL is used to make an agent learn to play the computer game Digger.

The player controls a machine that digs tunnels and the objective of the game is to collect all the emeralds while avoiding or shooting monsters. The purpose of [9] is to demonstrate that human guidance can be exploited to speed up learning, specially in the initial phase of learning

The state representation of a game like Digger is rather difficult. The position of the player, monsters and the presents of emeralds should be encoded in the state. Also the monsters can only move in tunnels dug by the player such that the tunnels have to be encode as well. This makes the state space very large and virtually impossible to explore. However, most parts of the state space will be irrelevant. In the original game the machine is controller by a human player, and this is used as guidance in the learning process. The policies of human players are more reasonable to start with than a random learning policies.

Approach

The most famous game application of RL is TD-Gammon [25], which resulted in a program capable of beating the world champion. This game has a finite, but very large, state space and neural network function approximators were used to generalize over this state space. Self-play was used to train the networks and during training sufficient state/action combinations occurred. In Digger the state is more a combination of smaller states, like the positions of each character. This results in a large state space of which most part may be irrelevant, for example the positions of monsters far away from the player. The way to deal with such situations is to only consider relevant parts of the state space. In [15] local approximators were used to model a value function over relevant parts of the state space, those are the states that actually occurred. But the idea of mapping states to values can also be completely abandoned. In [21] state/action combinations are stored and Q-values are associated with these state/action combination. The application was a mobile robot and a human user first guided the robot around obstacles to obtain the initial set of state/action combinations. After that learning changed the value of the associated Q-values. Relational RL [10] also uses the idea of store instances of state/action combinations and relate them to values. This is done through a first order logic decision tree, in which structural descriptions are mapped to real numbers. For Digger this means that predicates like "visibleMonster" and "nearestEmerald" could be used. Learning was performed using different number of guidance traces.

Result

The results was that the game of digger could be played after learning. The result of using guidance was a bit different than expected. The initial learning speed did not increase much, but instead it increased performance a little bit.

Hints and Tips

- Sometimes states cannot be represented as elements from a finite set or vectors. In that case relational reinforcement learning can be applied.
- Instead of learning from scratch, the learner can be initialised with a reasonable policy derived from prior knowledge. It also allows for making sure that the right part of the state space is being explored.

5.5 Conclusion

In this chapter we outlined a method for the development of adaptive systems using methods from Machine Learning and Reinforcement Learning and demonstrated this with a variety of adaptive systems. The main questions are the translation of the problem into a learning problem and the assessment of the feasibility of adaptive systems for the problem at hand.

Acknowledgements

The description of the development of Syskill and Webert was based on an earlier version developed by Floor Verdenius for the Networks of Excellence MLNET and CoIL, funded by the European Commission.

References

1. P.E. An, S. Aslam-Mir, M. Brown and C.J. Harris. A reinforcement learning approach to on-line optimal control. In *Proceedings of the International Conference on Neural Networks*, 1994.
2. A.G. Barto, R.S. Sutton and C.W. Anderson. Neuronlike adaptive elements that can solve difficult learning control problems. *IEEE Transactions on Systems, Man and Cybernetics*, 1983.
3. R. Bellman. *Dynamic Programming*. Princeton University Press, 1957.

4. H.R. Berenji and P. Khedkar. Learning and tuning fuzzy logic controller through reinforcements. *IEEE Trans. on Neural Networks*, 1992.
5. D.P. Bertsekas. *Dynamic Programming: Deterministic and Stochastic Models*. Prentice-Hall, 1987.
6. S.J. Bradtke. Reinforcement learning applied to linear quadratic regulation. In *Advances in Neural Information Processing Systems*, 1993.
7. P. Chapman., J. Clinton, T. Khabaza, T. Reinartz and R. Wirth. The CRISP-DM Process Model. Technical report, Crisp Consortium, 1999, http://www.crisp-dm.org/.
8. P. Dayan. The Convergence of TD(λ) for general λ *Machine Learning*, 8: 341-362, 1992.
9. K. Driessens and Saso Dzeroski. Integrating experimentation and guidance in relational reinforcement learning. In *Proceedings of the Nineteenth International Conference on Machine Learning*, pages 115-122. Morgan Kaufmann Publishers, Inc, 2002.
10. S. Dzeroski, L. De Raedt and K. Driessens. Relational reinforcement learning. *Machine Learning*, 43(1): 7-52, 2001.
11. D. Gu and H. Hu. Reinforcement learning of fuzzy logic controller for quadruped walking robots. In *Proceedings of 15th IFAC World Congress*, 2002.
12. S.H.G. ten Hagen and B.J.A. Kröse. Linear quadratic regulation using reinforcement learning. In F. Verdenius and W. van den Broek, editors, *Proc. of the 8th Belgian-Dutch Conf. on Machine Learning*, pages 39-46, Wageningen, October 1998. BENELEARN-98.
13. T. Jaakkola, M.I. Jordan and S. P. Singh. On the convergence of stochastic iterative dynamic programming algorithms. *Neural Computation*, 1994.
14. L. Jouffe. Fuzzy inference system learning by reinforcement methods. *IEEE Trans. on System, Man and Cybernetic*, 28(3), 1998.
15. T. Landelius. *Reinforcement learning and distributed local model synthesis*. PhD thesis, Linköping University, 1997.
16. C.-K. Lin. A reinforcement learning adaptive fuzzy controller for robots. *Fuzzy Sets and Systems*, 137(3): 339-352, 2003.
17. D. Michie and C. Sammut. Behavioural clones and cognitive skill models. In K. Furukawa, D. Michie and S. Muggleton, editors, *Machine Intelligence 14*. Oxford University Press, Oxford, 1995.
18. M.J. Pazzani and Daniel Billsus. Learning and revising user profiles: The identification of interesting web sites. *Machine Learning*, 27(3): 313-331, 1997.
19. J.R. Quinlan. *C4.5: Programs for Machine Learning*. Morgan Kaufmann, 1993.
20. C. Sammut. Automatic construktion of reaktive control systems using symbolic machine learning. *The Knowledge Engineering Review*, 11:27-42, 1996.
21. W.D. Smart and L.P. Kaelbling. Effective reinforcement learning for mobile robots. In *Proceedings of the International Conference on Robotics and Automation*, 2002.
22. R.S. Sutton. Learning to predict by the methods of temporal differences. *Machine Learning*, 1988.
23. R.S. Sutton and A.G. Barto. *Reinforcement Learning: An Introduction*. MIT Press, 1998.
24. S. ten Hagen and B. Kröse. Neural Q-learning. *Neural Computing and Applications*, 12(2): 81-88, November 2003.
25. G. Tesauro. Temporal difference learning in TD-gammon. In *Communications of the ACM*, 1995.

26. T. van der Boogaard and M. van Someren. Grammar induction in the domain of postal addresses. In *Proceedings Belgian-Netherlands Machine Learning Workshop Benelearn*, Brussels, 2004. University of Brussels.
27. Ch. J.C.H. Watkins and P. Dayan. Technical note: Q learning. *Machine Learning*, 1992.
28. S.M. Weiss and N. Indurkhya. *Predictive Data-Mining. A Practical Guide*. Morgan Kaufmann Publishers, San Francisco, California, 1997.
29. P.J. Werbos. Consitency of HDP applied to a simple reinforcement learning problem. *Neural networks*, 1990.
30. I.H. Witten and E. Frank. *Data-Mining: Practical Machine Learning Tools and Techniques with Java Implementations*. Morgan Kaufmann Publishers, San Francisco, 2000.

6

Fuzzy Expert Systems

J.M. Garibaldi

Automated **S**cheduling, Optimis**A**tion and **P**lanning (ASAP) Research Group
School of Computer Science and IT, University of Nottingham,
Jubilee Campus, Wollaton Road, Nottingham, UK, NG8 1BB
jmg@cs.nott.ac.uk

In this chapter, the steps necessary to develop a fuzzy expert system (FES) from the initial model design through to final system evaluation will be presented. The current state-of-the-art of fuzzy modelling can be summed up informally as "anything goes". What this actually means is that the developer of the fuzzy model is faced with many steps in the process each with many options from which selections must be made. In general, there is no specific or prescriptive method that can be used to make these choices, there are simply heuristics ("rules-of-thumb") which may be employed to help guide the process. Each of the steps will be described in detail, a summary of the main options available will be provided and the available heuristics to guide selection will be reviewed.

The steps will be illustrated by describing two cases studies: one will be a mock example of a fuzzy expert system for financial forecasting and the other will be a real example of a fuzzy expert system for a medical application. The expert system framework considered here is restricted to rule-based systems. While there are other frameworks that have been proposed for processing information utilising fuzzy methodologies, these are generally less popular in the context of fuzzy expert systems.

As a note on terminology, the term *model* is used to refer to the abstract conception of the process being studied and hence *fuzzy model* is the notional representation of the process in terms of fuzzy variables, rules and methods that together define the input-output mapping relationship. In contrast, the term *system* (as in fuzzy expert system) is used to refer to the embodiment, realisation or implementation of the theoretical model in some software language or package. A single model may be realised in different forms, for example, via differing software languages or differing hardware platforms. Thus it should be realised that there is a subtle, but important, distinction between the evaluation of a fuzzy model of expertise and the evaluation of (one or more of) its corresponding fuzzy expert systems. A model may be evaluated as accurately capturing or representing the domain problem under consideration, whereas its realisation as software might contain bug(s) that cause undesired artefacts in the output. This topic will be further explored in Sect. 6.7.

It will generally be assumed that the reader is familiar with fuzzy theory, methods and terminology – this chapter is *not* intended to be an introductory tutorial, rather it is a guide to currently accepted best practice for building a fuzzy expert system. For a simple introductory tutorial the reader is referred to Cox [8]; for a comprehensive coverage of fuzzy methods see, for example, Klir and Yuan [22], Ruspini et al [34] or Kasabov [21].

The central question for this chapter is "what are smart adaptive fuzzy expert systems?" In current state-of-the-art it is *not* possible to automatically adapt a system created in one application area to address a novel application area. Automatic tuning or optimisation techniques may be applied to (in some sense) *adapt* a given fuzzy expert system to particular data (see Sect. 6.8). Real "smart adaptive" systems are presently more likely to be found in neuro-fuzzy or hybrid systems covered in subsequent chapters. In each new application area, a fuzzy expert system must effectively be "hand-crafted" to achieve the desired performance. Thus the creation of good fuzzy expert systems is an art that requires skill and experience. Hints and tips to assist those new to this area in making appropriate choices at each stage will be provided.

6.1 Introduction

The generic architecture of a fuzzy expert system showing the flow of data through the system is shown in Fig. 6.1 (adapted from Mendel [26]). The general process of constructing such a fuzzy expert system from initial model design to system evaluation is shown in Fig. 6.2. This illustrates the typical process flow as distinct stages for clarity but in reality the process is not usually composed of such separate discrete steps and many of the stages, although present, are blurred into each other.

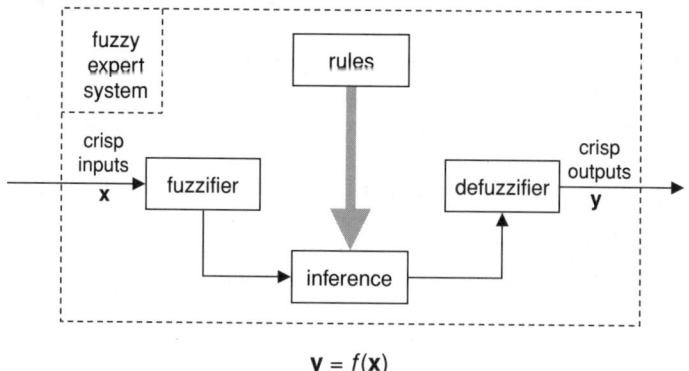

Fig. 6.1. Generic architecture of a fuzzy expert system

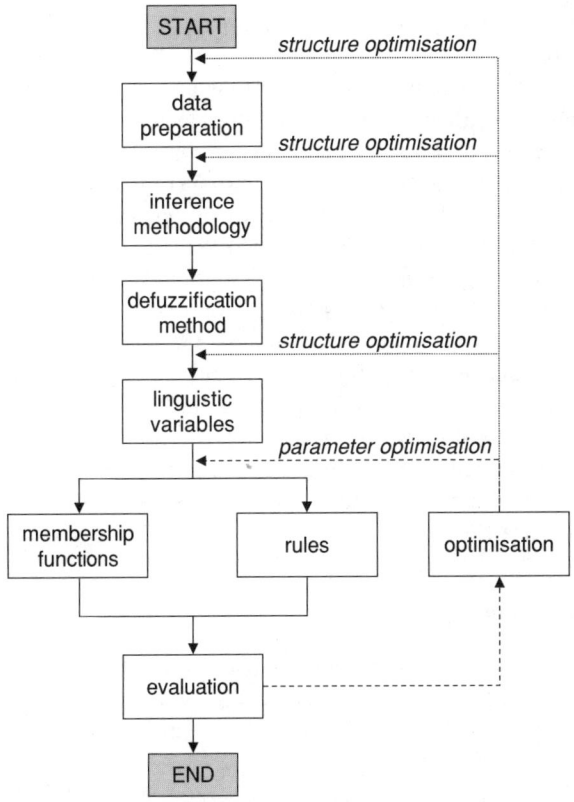

Fig. 6.2. Typical process flow in constructing a fuzzy expert system

Once the problem has been clearly specified (see Chap. 2), the process of constructing the fuzzy expert system can begin. Invariably some degree of data preparation and preprocessing is required, and this stage has been discussed in detail in Chap. 3. The first major choice the designer has to face is whether to use the Mamdani inference method [24] or the Takagi-Sugeno-Kang (TSK) method [37, 39]. The essential difference in these two methodologies is that the result of Mamdani inference is one or more fuzzy sets which must (almost always) then be defuzzified into one or more real numbers, whereas the result of TSK inference is one or more real functions which may be evaluated directly. Thus the choice of inference methodology is linked to the choice of defuzzification method. Once the inference methodology and defuzzification method have been chosen, the process of enumerating the linguistic variables necessary can commence. This should be relatively straightforward if the problem has been well specified and is reasonably well understood. If this is not the case, then the decision to construct a fuzzy expert system may not be appropriate. The next stage of deciding the necessary terms with their defining membership functions and determining the rules to be used is far from trivial however. Indeed, this stage is usually the most difficult and time consuming of the whole process.

After a set of fuzzy membership functions and rules has been established the system may be evaluated, usually by comparison of the obtained output against some *desired* or *known* output using some form of error or distance function. However, it is very rare that the first system constructed will perform at an acceptable level. Usually some form of optimisation or performance tuning of the system will need to be undertaken. Again, there are a multitude of options that a designer may consider for model optimisation. A primary distinction illustrated in Fig. 6.2 is the use of either *parameter optimisation* in which (usually) only aspects of the model such as the shape and location of membership functions and the number and form of rules are altered, or *structure optimisation* in which all aspects of the system including items such as the inference methodology, defuzzification method, or number of linguistic variables may be altered. In general, though, there is no clear distinction. Some authors consider rule modification to be structure optimisation, while others parameterise the rules.

6.1.1 Case Studies

As stated earlier, two case studies will be used through the rest of this chapter to provide a grounding for the discussions. The two case studies are now briefly introduced.

Financial Forecasting Fuzzy Expert System

The problem is to design a fuzzy expert system to predict (advise) when to buy or sell shares in an American company based on three sources of information:

1. the past share price of the company itself ($share\_pr$),
2. the Euro/Dollar exchange rate ($xchg\_rate$), and
3. the FTSE (Financial Times Stock Exchange) share index (the UK stock index) ($FTSE$).

Clearly, this is an artificial scenario as the Dow Jones index would almost certainly be used in any real predictor of an American company, but the case study is designed to be illustrative rather than realistic.

Umbilical Acid-Base Fuzzy Expert System

Childbirth is a stressful experience for both mother and infant. Even during normal labour every infant is being regularly deprived of oxygen as maternal contractions, which increase in frequency and duration throughout labour until delivery, restrict blood supply to the placenta. This oxygen deprivation can lead to fetal "distress", permanent brain damage and, in the extreme, fetal death. An assessment of neonatal status may be obtained from analysis of blood in the umbilical cord of an infant immediately after delivery. The umbilical cord vein carries blood from the placenta to the fetus and the two smaller cord arteries return blood from the fetus. The blood from the placenta has been freshly oxygenated, and has a relatively high partial pressure of oxygen (pO_2) and low partial pressure of carbon dioxide (pCO_2). Oxygen in the blood fuels *aerobic* cell metabolism, with carbon dioxide produced as "waste". Thus the blood returning from the fetus has relatively low oxygen and high carbon dioxide content. Some carbon dioxide dissociates to form carbonic acid in the blood, which increases the acidity (lowers the pH). If oxygen supplies are too low, *anaerobic* (without oxygen) metabolism can supplement aerobic metabolism to maintain essential cell function, but this produces lactic acid as "waste". This further acidifies the blood, and can indicate serious problems for the fetus.

A sample of blood is taken from each of the blood vessels in the clamped umbilical cord and a blood gas analysis machine measures the pH, pO_2 and pCO_2. A parameter termed *base deficit of extracellular fluid* (BD_{ecf}) can be derived from the pH and pCO_2 parameters [35]. This can distinguish the cause of a low pH between the distinct physiological conditions of *respiratory acidosis*, due to a short-term accumulation of carbon dioxide, and a *metabolic acidosis*, due to lactic acid from a longer-term oxygen deficiency. An interpretation of the status of health of an infant can be made based on the pH and BD_{ecf} parameters ("he acid-base status") of both arterial and venous blood. However, this is difficult to do and requires considerable expertise.

A fuzzy expert system was developed for the analysis of umbilical cord acid-base status, encapsulating the knowledge of leading obstetricians, neonatologists and physiologists gained over years of acid-base interpretation. The expert system combines knowledge of the errors likely to occur in acid-base measurement, physiological knowledge of plausible results, statistical analysis of a large database of results and clinical experience of acid-base interpretation. It automatically checks for errors in input parameters, identifies the vessel origin (artery or vein) of the results and provides an interpretation in an objective, consistent and intelligent manner. For a full description of the development and evaluation of this fuzzy expert system, see [11, 12, 14, 15].

6.2 Data Preparation and Preprocessing

Data preparation and preprocessing has already been discussed in detail in Chap. 3. Issues such as feature selection, normalisation, outlier removal, etc., are as important to fuzzy expert systems as to any other systems. The reason data preparation and preprocessing is mentioned again here is that it *must* be considered as an integral part of the modelling process, rather than being a fixed procedure carried out prior to modelling. This implies that data preparation methods may be included as components of model structure that are to be optimised in some way.

As an example, consider the financial forecasting FES. It would be unusual and of little obvious value to use the *absolute* values of any of the three primary variables (*share_pr*, *xchg_rate* and *FTSE*) as input to the FES. A more likely choice would be to use difference information such as daily, weekly or monthly differences. So, for example, an FES could reasonably be imagined which took nine input variables (the daily, weekly and monthly differences of *share_pr*, *xchg_rate* and *FTSE*) and produced one output variable *advice*. Alternatively, it might be that all useful information was contained in daily differences so that only three input variables were used. Viewed this way, an optimisation technique might be employed to carry out combinatorial optimisation from a fixed choice of input variable combinations. Alternatively, the period (no. of days) over which the difference is calculated could be included as a tunable parameter within data preparation which could be carried out under control of a parameter optimisation technique.

From the above it can be seen that data preparation is also closely linked to the choice of linguistic variables for the FES. The most important aspect to be considered is that the universe of discourse (range) of each linguistic variable should be fixed. This is the reason that the absolute values of *share_pr*, *xchg_rate* and *FTSE* are unlikely to be used in a financial forecasting FES. If the universe of discourse of each variable is not fixed in advance, then the FES would have to self-adapt to new ranges. This is a current area of active research, but solutions to this problem have not yet been found.

6.2.1 Hints and Tips

- Data preparation and preprocessing is an essential and fundamental component of fuzzy expert system design. It *must* be considered as another component of the process that can be adjusted, altered or optimised as part of creating a successful fuzzy expert system.
- To paraphrase Einstein, data preparation and preprocessing should be as "simple as possible, but no simpler". Always *consider* utilising raw data as input to the FES where possible (although this may not be possible in situations such as in the presence of high noise levels).

6.3 Inference Methodology

6.3.1 Mamdani Inference

In Mamdani inference [24], rules are of the following form:

$$R_i \quad \text{if } x_1 \text{ is } A_{i1} \text{ and } \ldots \text{ and } x_r \text{ is } A_{ir} \text{ then } y \text{ is } C_i$$
$$\text{for } i = 1, 2, \ldots, L \tag{6.1}$$

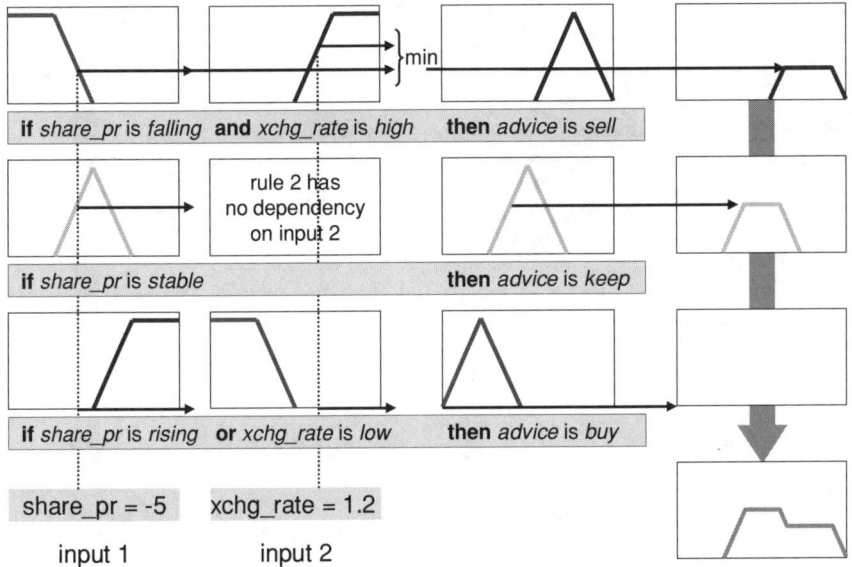

Fig. 6.3. A three-rule Mamdani inference process

where L is the number of rules, x_j ($j = 1, 2, \ldots, r$) are the input variables, y is the output variable, and A_{ij} and C_i are fuzzy sets that are characterised by membership functions $A_{ij}(x_j)$ and $C_i(y)$, respectively. The important thing to note is that the consequence of each rule is characterised by a fuzzy set C_i. An example of Mamdani inference for an illustrative three-rule system is shown in Fig. 6.3. The final output of a Mamdani system is one or more arbitrarily complex fuzzy sets which (usually) need to be defuzzified.

In the context of umbilical acid-base analysis described earlier, the final expert system comprised a set of Mamdani inference rules operating on four input variables – the acidity and base-deficit of the arterial and venous blood (pH_A, BD_A, pH_V and BD_V) – to produce three output variables, severity of *acidemia* (ranging from severely acidemic to alkolotic), *component* of acidemia (ranging from pure metabolic, through mixed, to pure respiratory) and *duration* of acidemia (ranging from chronic to acute).

6.3.2 Takagi-Sugeno-Kang Inference

In Takagi-Sugeno-Kang (TSK) inference [37, 39], rules are of the following form:

$$\begin{aligned}R_i \quad &\text{if } x_1 \text{ is } A_{i1} \text{ and } \ldots \text{ and } x_r \text{ is } A_{ir} \\ &\text{then } y_i = b_{i0} + b_{i1}x_1 + \ldots + b_{ir}x_r \\ &\text{for } i = 1, 2, \ldots, L\end{aligned} \quad (6.2)$$

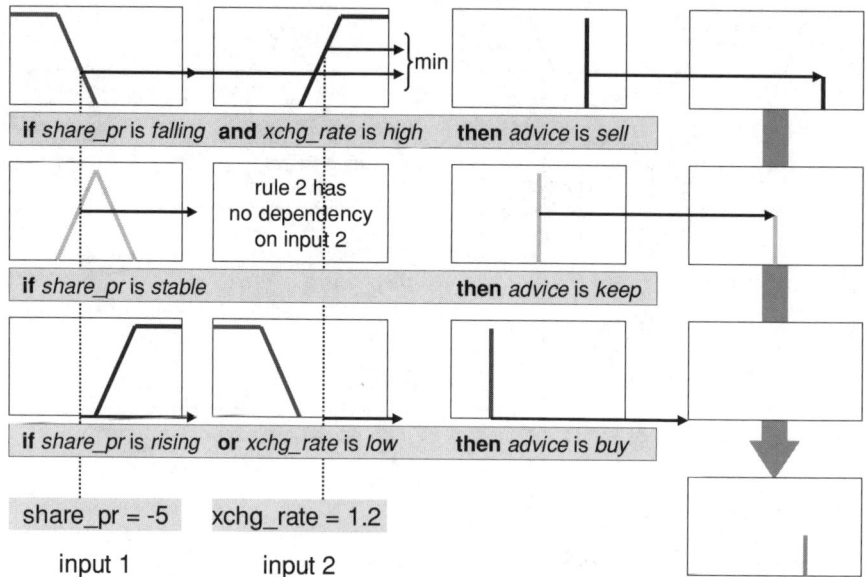

Fig. 6.4. A three-rule zeroth-order TSK inference process

where b_{ij} ($j = 1, 2, \ldots, r$) are real-valued parameters. The important thing to note here is that the consequence of each rule is now characterised by a linear function of the original input variables. The final output the inference process is calculated by:

$$y = \frac{\sum_{i=1}^{L} \alpha_i (b_{i0} + b_{i1} x_1 + \ldots + b_{ir} x_r)}{\sum_{i=1}^{L} \alpha_i} \qquad (6.3)$$

where α_i is the firing strength of rule R_i. The *order* of a TSK system is the degree of (highest exponent of x in) the consequent function. The most commonly used TSK model is the first-order model. An example of zeroth-order TSK inference for an illustrative three-rule system is shown in Fig. 6.4.

6.3.3 Choice of Inference Method

Which method should be chosen? There is no simple answer to this question; rather, the answer depends on the nature of the expert system being developed. Two more specific questions can be framed: (i) Is any expression of uncertainty or non-numeric information required in the answer? (ii) Is processing speed (e.g. real-time operation) or memory usage crucial? If the answer to the first is "yes", then Mamdani inference is clearly favoured. If the answer to the second is "yes", then TSK inference is clearly favoured. Otherwise, the choice is a matter of preference, although in many cases, TSK based methods may result in fewer rules. As an example, consider the two case-studies presented. In the financial forecasting example, there would appear to be little / no need for uncertainty representation in the output. If the output is above a threshold, the advice will be to buy shares; if the output is below a threshold, the advice will be to sell shares held. Although there is no specific requirement for real-time operation, TSK inference might be utilised as a Mamdani system would just introduce an "uncecessary" step of defuzzification. In the umbilical acid-base expert system, there was a specific requirement both for a representation of uncertainty in the output and for the potential of obtaining a linguistic rather than numeric output (see Sect. 6.6 for information on obtaining linguistic output). There was no real-time constraint and hence Mamdani inference was clearly indicated.

6.3.4 Fuzzy Operators

A great deal has been written about the theoretical properties and practical choices of operators necessary to carry out fuzzy operations of set intersection (AND) and union (OR). It is well established that fuzzy intersections are represented by a class of functions that are called *triangular norms* or *T-norms* and that fuzzy unions are represented by *triangular conorms* or *T-conorms*. A T-norm $\otimes$ is a binary function:

$$\otimes : \mathbb{I}^2 \to \mathbb{I}$$

where $\mathbb{I}$ represents the set of real numbers in the unit interval $[0, 1]$, that satisfies the following conditions: for any $a, b \in \mathbb{I}$,

(i) $1 \otimes a = a$ *(identity)*
(ii) $(a \otimes b) \otimes c = a \otimes (b \otimes c)$ *(associativity)*
(iii) $a \otimes b = b \otimes a$ *(commutivity)*
(iv) $a \otimes b \leq a \otimes c$, if $b \leq c$ *(monotonicity)*

A number of operators and parameterised operator families have been proposed which satisfy these conditions, but by far the most common in practical use are either of:

$a \otimes b = \min(a, b)$ *(standard)*
$a \otimes b = ab$ *(algebraic product)*

A T-conorm $\oplus$ is a similar binary function on $\mathbb{I}^2$ that satisfies the same conditions as a T-norm with the exception of:

(i) $0 \oplus a = a$ *(identity)*

The corresponding T-conorms to the two T-norms specified above are:

$a \oplus b = \max(a, b)$ *(standard)*
$a \oplus b = a + b - ab$ *(algebraic sum)*

In most fuzzy expert systems utilising Mamdani inference, the T-norm operator is used for the intersection operator to combine clauses in the rule antecedent joined with an *and*, and for the implication operator to combine the firing strength of the overall antecedent with the consequent fuzzy set. The T-conorm operator is used for the union operator to combine clauses in the rule antecedent joined with an *or*, and for the combination operator to aggregate the consequents of each rule into a single consequent set. The *min* T-norm and the *max* T-conorm are known as the *standard* operators as they were the original operators proposed by Zadeh [43]. They are the only T-norm and T-conorm that also obey the condition of *idempotence*: that is $a \otimes a = a$ and $a \oplus a = a$, respectively.

Several observations concerning current state-of-the-art of fuzzy operators can be made. Firstly, operators are generally used as a consistent family: i.e. if the standard *min* operator is used for intersection and implication, then the *max* operator should be used for union and aggregation, etc. Secondly, the performance differences that are obtained in real expert systems are often very small and of secondary importance compared to, for example, the membership functions of the terms or the rule set. Within the umbilical acid-base expert system a comparison was made of the performance difference between the standard operator family and the algebraic family. The algebraic family was found to increase performance (in comparison with human experts) by around 1% overall. Thirdly, in the context of the same expert system it was also observed that human experts involved in the model design had a definite subjective preference for the algebraic operator family. This appeared to be due to the fact that use of the standard *min* and *max* operators caused plateaus to appear in the consequent sets, as the consequent set is truncated at the height of the rule firing strength. Having become acquainted with the philosophies and notions of "fuzzy logic", the human experts found such plateaus "unnatural". Of course, this subjective preference is possibly less important than the performance gain mentioned above.

6.3.5 Hints and Tips

- Use Mamdani inference if some expression of uncertainty in the output is required. An associated defuzzification technique is usually employed.
- Use TSK inference when a numeric output with no associated expression of uncertainty is required, or processing speed is essential.
- Fuzzy operators should be chosen as a consistent family: i.e. if *max* is used for the intersection operator, *min* should be used for union; similarly the *algebraic product* and *algebraic sum* should be used together.
- The choice of fuzzy operator family rarely makes a sizeable difference to the overall performance of a fuzzy expert system.

6.4 Linguistic Variables and Membership Functions

Obtaining an appropriate set of linguistic variables, their associated terms, and a set of rules lies at the heart of the creation of any fuzzy model of expertise. Rule determination will be discussed further in the next section.

The choice of linguistic variables is primarily a combination of knowledge elicitation and data preparation. For example, in the umbilical acid-base FES there are six primary variables (arterial and venous pH, pCO_2, and pO_2) available, yet four linguistic variables (arterial and venous pH and BD_{ecf}) are used for interpretation, two of which (BD_A and BD_V) are derived. The reason is simply that knowledge elicitation sessions with domain experts indicated that these four variables are the only ones necessary. In the case of the financial forecasting FES, the question of which linguistic variables should be used is more difficult. In such a situation, the processes of data preparation, linguistic variable and membership function specification, rule determination and model optimisation merge into a single interdependent process.

The question of the number and names of the individual terms in each linguistic variable is also difficult. In this case, there are some practically accepted (but not theoretically justified) heuristics. The terms of a linguistic variable should be:

- justifiable in number
  - the number of terms should be small (≤ 7)
  - commonly, there are an odd number of terms
- the terms should not overlap too much
- terms should overlap at around 0.5
- all terms are normal (max. membership is 1)
- all terms are convex
- the terms should span the universe of discourse

Note that these principles are simply guidelines for initial configurations of terms. If (as is probable) the membership functions of the terms are subject to tuning or optimisation procedures, then some (or many) of the principles may be violated.

There is little guidance that can be given as to the most appropriate shape of membership function for a given application domain. The problem of choosing an appropriate shape is exacerbated by the fact that there is no formal definition or interpretation as to the meaning of membership grades [23]. The main shapes used in practical applications are:

- piecewise linear (triangular or trapezoidal)
- Gaussian: $e^{\frac{-(x-c)^2}{\sigma^2}}$
- sigmoidal: of the form $\frac{1}{1+e^{-a(x-c)}}$
- double-sigmoidal: left and right sigmoidals put together to form a term with a central peak

where c, σ and a are parameters that determine the centre and spread of the Guassian or sigmoidal. Again, during development of the umbilical acid-base FES it was found both that clinicians preferred the appearance of non-piecewise linear membership functions (in fact, a combination of sigmoidals and double-sigmoidals were used) and that these membership functions gave a slight performance gain. Many authors, particularly in the context of fuzzy control, prefer to use piecewise linear functions, probably due to their ease of calculation, whereas other such as Mendel [26] prefer Gaussians because they are more simply analysable (e.g. differentiable) and hence are more suitable for automated tuning methodologies.

In the context of a fuzzy expert system, the source of membership functions would ideally be domain experts. However, again, there is no generally accepted method for eliciting membership functions from experts. An excellent survey of methods for membership function elicitation and learning is given by Krishnapuram [23]. The methods include:

- **Membership functions based on perceptions** – in which human subjects are polled or questioned.
- **Heuristic methods** – guessing!
- **Histogram-based methods** – in which membership functions are somehow matched to histograms of the base variables obtained from real data.
- **Transformation of probability distributions to possibility distributions** – if membership functions are considered numerically equivalent to possibility distributions and probability distributions are available, then methods are available to transform the probability distributions to possibility distributions.
- **Neural-network-based methods** – standard feedforward neural networks can be used to generate membership functions from labelled training data.
- **Clustering-based methods** – Clustering algorithms such as fuzzy c-means [4] can be used.

One further observation will be added. Most elicitation methods concentrate on direct or indirect elicitation of the membership functions themselves – i.e. they operate on the input side of the FES. During knowledge elicitation for the umbilical acid-base FES, it was found that such methods led to poor performance. That is, if an expert indicated membership functions directly, then the overall agreement of the FES with *that expert* was poor. In contrast, it was found that if the expert indicated desirable or undesirable features in the output consequent sets for specific (training) cases, then the input membership functions could be adjusted by a sort of *informal* back-propagation methodology to achieve better agreement with the expert.

6.4.1 Hints and Tips

- The number of linguistic variables is generally considered a component of model *structure*; the location and spread (and sometimes number) of membership functions are generally considered to be *parameters* of the model which should be tuned or optimised in some way.
- The choice of shape of membership function (m.f.) (e.g. piecewise linear, Gaussian or sigmoidal) is likely to be less important than the location of the centre of the m.f. and spread of the m.f.
- If a sizeable data set is available, then the use of automated m.f. learning methods should be considered.
- If data is difficult to obtain or human domain expertise is available, then manual knowledge elicitation methods might be favoured.

6.5 Rule Determination

Many of the comments that have just been made pertaining to linguistic variables and the membership functions of their terms also apply to rule determination. Indeed, in any practical construction of an FES the two processes will be closely interdependent. The linkage may be made more explicit by the strong statement that *membership functions have no meaning without the associated rules which will used to perform inference* – i.e. the rules provide the domain context necessary to give meaning to the linguistic variables and membership functions.

Once again, there is no single rule determination methodology that can be recommended. Some of the available options include:

- Elicitation of rules from experts. Although this is often presented as a self-evident option, in practice it is far from obvious how to undertake this. Unless only one expert is used for rule elicitation, there are bound to be disagreements, but how should such disagreements be overcome? This is currently an open question, which might be addressed by, for example, type-2 fuzzy logic methods [26].
- Conversion of rules from a non-fuzzy expert system. In the case of the umbilical acid-base FES, for example, the rules to construct a crisp expert system were initially elicited from experts. Once this crisp system had been successfully evaluated, it was converted to a fuzzy expert system. This allowed direct performance comparisons to be carried out, which confirmed that the fuzzy system indeed out-performed the crisp system [12].
- Crisp rule induction methods. If data is more plentiful than domain expertise then crisp rule induction methods such as Quinlan's ID3 or C4.5 algorithms [32, 33] can be utilised to produce a set of crisp rules that can consequently be fuzzified. Methods for doing this automatically have been proposed by Jang [17] and Baroglio [3].
- Evolutionary learning of rules (see Chaps. 8 & 9).

- Neuro-fuzzy approaches (see Chaps. 7 & 9) or, for example, the methods of Jang [16] and Wang and Mendel [41].

6.5.1 Hints and Tips

- Consider determination of the membership functions and the rules as two components of a single logical step in the design process.
- Whenever possible, use good sources of domain knowledge to lead or guide rule determination. However, always consider including rules generated from domain experts within tuning and optimisation processes.
- Purely automated rule determination procedures can lead to a large number of rules, and rules that may *conflict* (rules with the same antecedents but differing consequents). Most successful fuzzy expert systems feature considerably less than 100 rules. If use of an automated procedure for rule determination results in more than this, then some form of rule pruning [5, 28] should be considered.

6.6 Defuzzification

Once the fuzzy reasoning has been completed it is usually necessary to present the output of the reasoning in a human understandable form, through a process termed *defuzzification*. There are two principal classes of defuzzification, *arithmetic defuzzification* and *linguistic approximation*. In arithmetic defuzzification a mathematical method is used to extract the single value in the universe of discourse that "best" (in some sense) represents the arbitrarily complex consequent fuzzy set (Fig. 6.5). This approach is typically used in areas of control engineering where some crisp result must be obtained. In linguistic approximation the primary terms in the consequent variable's term set are compared against the actual output set in a variety of combinations until the "best" representation is obtained in natural language. This approach might be used in expert system advisory applications where human users view the output, although its actual usage has been relatively rare in the literature.

6.6.1 Arithmetic Defuzzification

The two most popular methods of arithmetic defuzzification are the *centre-of-gravity* (COG) algorithm and the *mean-of-maxima* algorithm. For the consequent set $A = \mu_1/x_1 + \mu_2/x_2 + \ldots + \mu_N/x_N$, the centre-of-gravity algorithm provides a single value by calculating the imaginary balance point of the shape of the membership:

$$x_g = \frac{\sum_{i=1}^{N}(\mu_i \cdot x_i)}{\sum_{i=1}^{N} \mu_i} \qquad (6.4)$$

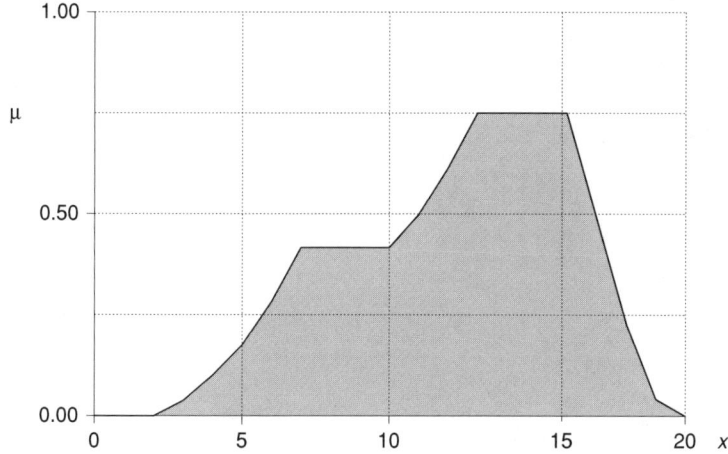

Fig. 6.5. A typical arbitrarily complex fuzzy output set

The mean-of-maxima algorithm finds the point in the universe of discourse with maximum membership grade:

$$x_m = \max_i \mu_i \qquad (6.5)$$

and calculates the mean of all the maxima if more than one maximum is found. An illustration of the output of these two methods is shown in Fig. 6.6. Unfortunately, both these methods have problems. Firstly, they both obviously lose information by trying to represent a complex fuzzy shape as a single scalar number. The COG method is insensitive to the overall height of the fuzzy consequent set, and the mean-of-maxima is prone to discontinuities in output, as only a small change in shape (for instance if there are two similar sized peaks) can cause a sudden large change in output value.

A number of alternative parameters can also be calculated to provide more information on the shape of the output as well as its location. A variety of such parameters are described, and illustrated through the example fuzzy output sets, A, B, C, D, E, and F, as shown in Fig. 6.7.

Membership Grade at the Defuzzification Point

The membership grade of the output set at the centre-of-gravity, μ_g, or mean-of-maxima, μ_m, provides an indication of confidence in the result. For example, in Fig. 6.7, the COG (x_g) of set D is 50, and the membership value at this point (μ_g) is 0.90. Actually, in the same figure, the output fuzzy sets D and E have the same COG ($x_g = 50$), but set D has a higher confidence in the result ($\mu_g = 0.90$) than set E ($\mu_g \approx 0.16$).

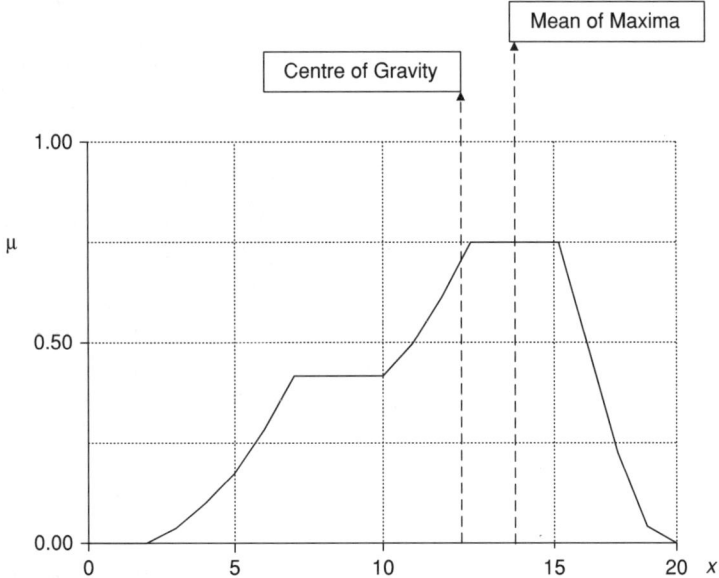

Fig. 6.6. An illustration of the difference between *centre-of-gravity* and *mean-of-maxima* defuzzification

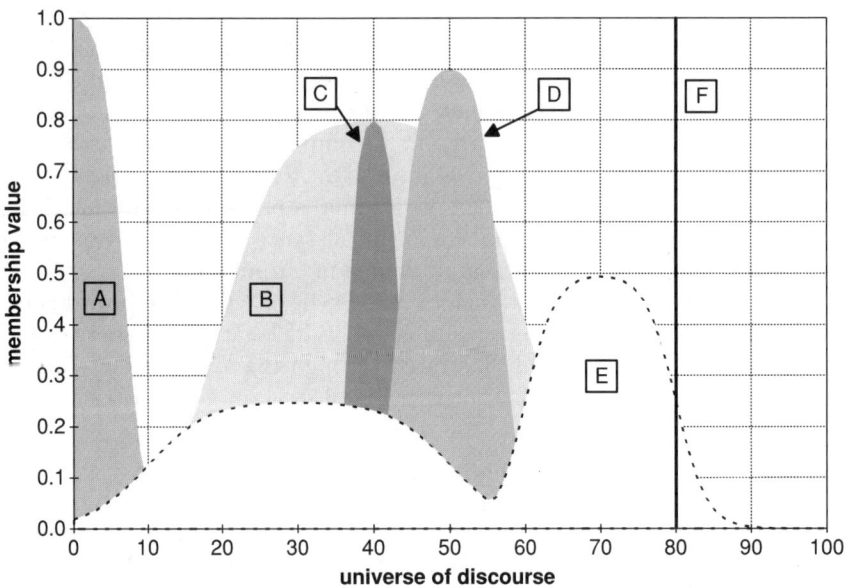

Fig. 6.7. Illustration of a variety of fuzzy output sets, A, B, C, D, E, and F, with different defuzzification parameters

Maximum Membership Grade

The maximum membership grade attained by the consequence fuzzy set, μ_{max}, provides a direct measure of the maximum strength that the antecedents fired a rule. It is especially useful for edge sets such as set A in Fig. 6.7, as the centroid cannot be at the maximum point, so that $\mu_g < \mu_{max}$.

Normalised Area

The area of the output set normalised to its maximum value is given by:

$$\text{area} = \frac{\sum_{i=1}^{N} \mu_i}{N} \tag{6.6}$$

This gives a value of 1.0 for the *unknown* set ($\mu = 1$ across the universe of discourse), a value of 0.0 for the *undefined* set ($\mu = 0$ across the universe of discourse), and would give a minimal value (≈ 0) for a fuzzy singleton. In Fig. 6.7 the output fuzzy sets B and C have the same COG ($x_g = 40$) and the same membership at this point ($\mu_g = 0.80$), but set B has a larger area and hence a larger uncertainty.

Fuzzy Entropy

Yet another measure, termed the *entropy* of a fuzzy set is defined by:

$$S = \frac{\sum_{i=1}^{N}(-\mu_i \log_2(\mu_i) - (1-\mu_i)\log_2(1-\mu_i))}{N} \tag{6.7}$$

This is normalised to its maximum value to give a value between zero and one, and provides an indication of the lack of information contained in the output set in terms of the distance away from the extremes of $\mu = 0.0$ and $\mu = 1.0$. It therefore gives a value of 0.0 for the *unknown* and *undefined* sets, and gives a value of 1.0 for the *indeterminate* set ($\mu = 0.5$ across the universe of discourse). Similarly to the normalised area, it too gives a minimal value for fuzzy singletons.

Summary of Arithmetic Defuzzification

Table 6.1 summarises the various arithmetic defuzzification parameters defined above for sets A, B, C, D, E, and F, in Fig. 6.7, and for the level sets *unknown*, *indeterminate* and *undefined*. It can be seen that for the fuzzy singleton F, which represents a crisp output from a fuzzy system, the values for μ_g and μ_{max} are both high (1.00) and the values for area and entropy are both low (0.00). The same tendencies can be seen for the example sets A, C, and D (only). There are a whole range of possibilities for other measures, such as the span of the fuzzy set (percentage of the set that is non-zero), the percentage of total area within a certain distance of the centroid point, the number of peaks in the fuzzy set, and so on.

Table 6.1. Summary of various arithmetic defuzzification parameters for sets A, B, C, D, E, and F, in Fig. 6.7, with values obtained for level sets *unknown*, *indeterminate* and *undefined*

Set	x_g	μ_g	μmax	Area	Entropy
A	3	0.95	1.00	0.07	0.06
B	40	0.80	0.80	0.32	0.50
C	40	0.80	0.80	0.05	0.08
D	50	0.90	0.90	0.12	0.15
E	50	0.16	0.50	0.19	0.60
F (fuzzy singleton)	80	1.00	1.00	0.00	0.00
unknown (1.0/x)	50	1.00	1.00	1.00	0.00
indeterminate (0.5/x)	50	0.50	0.50	0.50	1.00
undefined (0.0/x)	50	0.00	0.00	0.00	0.00

6.6.2 Linguistic Approximation

In linguistic approximation a similarity measure is used to compute the distance between the actual output set and an arbitrary collection of primary terms, connectives and hedges. For example, a shower control variable with primitive terms such as *cold*, *medium* and *hot*, and allowable hedges of *fairly* and *very*, might produce a composite linguistic term such as *medium and fairly hot*. One such similarity metric is the Euclidean distance between fuzzy sets, given by:

$$\delta = \sqrt{\sum_i (\mu_i - \eta_i)^2} \tag{6.8}$$

where μ_i is the membership of the output set and η_i is the membership grade of the currently considered linguistic approximation – the minimum value of δ will determine the best match. Alternatively, the degree of overlap, γ, of two fuzzy sets, A and B, can be calculated by dividing the area of intersection by the area of the union of the sets:

$$\gamma = \frac{A \cap B}{A \cup B} \tag{6.9}$$

to give a value between zero (for disparate sets) and one (for coincidental sets) – the maximum value of γ will determine the best match.

A search is then initiated to find the best match whilst attempting to limit the complexity of the combination of terms, in order to produce comprehensible output. For example, although the linguistic combination *not extremely cold and fairly medium and medium and fairly hot* might produce a better match than *medium and fairly hot* for Fig. 6.6, the latter term would be preferred due to its relative simplicity.

6.6.3 Hints and Tips

- Defuzzification is a fundamentally important step in Mamdani inference. Although many alternatives for numeric defuzzification have been proposed, there is little theoretical guidance as to which is better in any given situation. It is probably safest to stick to centre-of-gravity defuzzification unless there is a good reason otherwise.
- Linguistic defuzzification has been infrequently used, and is a poorly understood and little researched area.
- Type-2 fuzzy logic [26] might be considered if linguistic output is required.

6.7 Evaluation

Many authors have used the terms *verification*, *validation*, *assessment* and *evaluation* in differing and inconsistent manners in the literature [29, 30]. In this section the following terminology, designed specifically for the European Advanced Informatics in Medicine (AIM) project [10], is adopted:

- *verification* is the process of ensuring that the expert system is functioning according to its specification,
- *validation* is the process of ensuring that the knowledge embedded within the expert system is an accurate representation of the domain, and
- *assessment* is the process of determining the effect that the expert system has in the real-world setting – this can be further split into two further sub-tasks:
  1. *human factors assessment* – determining whether the system is useful to and usable by its target users, and
  2. *performance assessment* – determining whether the system makes a measurable difference (improvement) when deployed.

Evaluation is a global term that refers to the collective processes of *verification*, *validation* and *assessment*.

In an ideal framework of scientific investigation the entire evaluation methodology would be established and fixed a priori. That is to say, the data, experiments, statistical tests and acceptable performance measures to be used in deciding whether an expert system was acceptable would all be decided before construction of the system commenced. This is very rarely the case in practical expert system development and fuzzy expert systems are no exception. It is not possible to cover the process of system evaluation thoroughly here; the reader is referred to [29, 30] for general guidance.

6.7.1 Model v. System Evaluation

As mentioned in the introduction there is a subtle distinction between evaluation of the theoretical fuzzy model (*model evaluation*) and evaluation of a specific implementation of a fuzzy expert system (*system evaluation*). In order to carry out a thorough system evaluation, it is necessary to verify and validate the implementation itself, in terms of examining the code to ensure that there are no coding bugs, examining the environment to ensure, for example, that variables cannot overflow, etc. In short, software engineering evaluation [31, 36] must be carried out in order to claim that an FES has been evaluated. On the other hand, fuzzy model evaluation can be carried out by tasks such as comparison of membership function shape/location with expert opinion, or discussion of rules-sets with experts, etc.

6.7.2 Data Categories

Data can be distinguished into three categories:

- *training data* – training data is used in the development of the model as part of the tuning or optimisation process;
- *testing data* – testing data is used to measure the performance of the system within model development process;
- *evaluation data* – evaluation data should be novel or entirely unused data that is used to evaluate performance after the final model has been fixed.

Often training data and testing data are simply logical partitions of one data set. For example, in two-fold cross validation the data is split into two halves. Firstly, one half is used for training the system and when training has been completed the system is fixed and other half of the data is used for testing. The process is then repeated with the data sets swapped and the performance obtained from the two runs is pooled. Another common training and testing procedure is the "leave-one-out" method, in which all the data except one instance are used for training the system which is then tested on the one remaining datum. This process is repeated, leaving a differing datum out, until the whole data set has been used once for testing. The performance results are then aggregated in some way (usually simply averaged).

Evaluation data should be an entirely separate data set which has not been used in any way in the construction, training (optimisation) or testing of the system. Indeed, it is scientifically desirable for the system designer to not have had any contact with the evaluation data at all, as any viewing or other appreciation of the data could unwittingly influence construction of the system. In the ideal situation, collection of novel evaluation data takes place *after* completion of system construction, but to pre-specified evaluation procedures. This rarely happens in practice.

6.7.3 Performance Measures

In order to evaluate a system, a measure of performance is necessary. Sometimes appropriates measures are obvious. As an example, take the financial forecasting FES. As the system is to advise when to trade shares, the obvious performance measure is the profit (or loss!) made when trading shares using real share prices. Either past data that has not been used for system training and testing could be used for evaluation or, better, real data collected after the system can be proven to have been fixed. Often some variation of mean-squared-error (MSE) between actual output and ideal (or desired) output is utilised.

However, frequently in expert systems (particularly medical expert systems) there is no objective measure of correct performance and other means must be used. The solution in such cases is usually to compare the output of the (fuzzy) expert system against human expert opinion on a range of data. An indirect measure of performance is then created. If the expert opinion and expert system output is categorical (e.g. classification into one of a number of specified disease categories) then the proportion of correct matches can be used. It is better to use a statistical method to correct for chance agreement such as the Kappa statistic [6], which can be used to measure either exact agreements only or can be modified to allow for partial agreement(s) [7]. In the case of the umbilical acid-base FES, a set of 50 difficult cases (see below) was selected for evaluation. Clinical experts and the fuzzy expert system were then asked to rank those cases from worst to best in terms of their indication of the infant's state of health. Spearman rank order correlation, effectively a form of MSE, was then used to measure agreement.

Imagine, as often the case, that the result of an FES is a number on a continuous scale obtained by, for example, centre-of-gravity defuzzification. If this number is representing, say, diagnosis of the presence of a disease, then what threshold should be used to indicate that the disease is indeed present? Often an arbitrary threshold is chosen (e.g. 50 on a scale 0 ... 100). However, a better solution is to use a technique known as Receiver Operating Characteristic (ROC) curves [9, 38] in which the threshold is continuously varied in order to achieve the best agreement. Note that, if ROC analysis is used, a method of generating accurate confidence intervals should be employed [40].

Picking suitable data to be used for evaluation purposes is another tricky area. Ideally, suitable data should cover the range of output possibilities of the fuzzy expert system in a systematic manner. But how can this be known until after evaluation has been carried out? There is no simple answer to this. Probably the best solution, if possible, is to have an independent, acknowledged domain expert pick a suitable set of cases. However, it is often difficult to find such an expert and undesirable, as is technically necessary, to then not use this expert for the evaluation exercise.

6.7.4 Hints and Tips

- Consider the method of evaluation at the earliest possible stage of system construction. Give serious thought to how rigorous and reliable evaluation will be assured.

- Consider incorporating a tunable performance parameter into the final model and use ROC analysis or similar to evaluate the system.

6.8 Model Optimisation

Model optimisation will be covered in several chapters of this book, so details will not be duplicated here. Some points of specific relevance to fuzzy expert system optimisation will be made. Firstly, some form of optimisation is an essential part of developing any fuzzy expert system. It may be used to adapt a fuzzy expert system in a certain domain to changes in the environment, i.e. adaption in the first category. Effectively, it can be viewed as the process of adapting an FES to the specific characteristics of the training and testing data available at the time. Thus the optimisation process could be applied at several stages in the lifetime of an FES in order to update the knowledge embodied in the FES to respond to drifts in the environment. However, in current state-of-the-art such an adapted FES would have to be (manually) re-evaluated. Accepted practice for automatically adapting a fuzzy expert system and evaluating it have not yet been developed.

Secondly, any optimisation technique may be used. Probably the most popular method currently is some form of evolutionary optimisation (see Chap. 13), but other techniques have been used (see below). Thirdly, before undertaking (for example) complex evolutionary optimisation, it is probably worth attempting more simplistic methods such as sampling, hill climbing or Monte Carlo approaches. Using any method, it is unlikely that a true global optimum of performance will be determined. It is much more likely that optimisation will be halted when some predetermined or subjective threshold of performance has been reached, or a time limit has been passed. It might well be the case that implementation of a simple Monte Carlo optimisation will allow adequate performance to be achieved in a similar time taken to program complex evolutionary optimisation.

Finally, the most significant influences on the performance of an FES are likely to be the number and location of the membership functions and the number and form of the rules. The precise shape of membership function (as in triangular, Gaussian, double-sigmoidal, etc., see Sect. 6.4) is unlikely to make a sizeable difference. Finally, *do not forget data preparation*!

Available FES optimisation techniques include:

- Exhaustive Search – The practice of systematically altering all parameters by the smallest interval such that *entire* coverage of the entire search space is obtained. This is unlikely to be feasible for an expert system with realistic dimensions.
- Sampling – The practice of systematically altering parameters by fixed, discrete intervals of sufficient size such that a *reasonable* coverage of the entire search space is obtained. The difference between this and exhaustive

search can be clarified with an example. Suppose a search is carried out to locate the optimal value for the centre of a triangular membership function across an integer universe of discourse from 0 to 100. In exhaustive search, all 101 values (0 to 100) would be considered. In sampling, perhaps only 9 values (e.g. $10, 20, \ldots, 90$) would be considered. Clearly, the optimal value may not be found in this way.

- Hill Climbing – The practice of systematically altering parameters (sometimes independently) in direction of greatest improvement until no further improvement is obtained.
- Monte Carlo – The process of assigning parameter values randomly until an acceptable solution (parameter set) is found. Monte Carlo approaches can be combined with Hill Climbing.
- Simulated Annealing – For a detailed review of Simulated Annealing, see Aarts [1]. Simulated Annealing is most often applied to combinatorial optimisation, but it can also be applied to continuous optimisation of both the structure and parameters of fuzzy models [11].
- Evolutionary Algorithms – See Chaps. 7, 9 & 13 for further details and references.
- Neuro-fuzzy or other hybrid approaches – See Chaps. 7 & 15 for further details and references.
- Other – Other generic optimisation techniques such as, for example, ant colony optimisation might be applied, but little research has been carried out in such areas.

Although different methods are preferred by different authors, there is no general result to state that any one optimisation technique is guaranteed to be better than any other. Indeed, the No Free Lunch Theorem [42] states that all optimisation techniques perform *on average* the same over all optimisation problems. This does not necessarily apply to any given subset of optimisation problems (e.g. specifically to optimisation of fuzzy models) and it does not mean that if one is experienced at applying a particular technique (e.g. evolutionary optimisation) then one should not continue to use that technique. However, in the absence of contrary evidence, it does suggest that there is no theoretical reason to favour e.g. evolutionary optimisation over e.g. simulated annealing optimisation. From a practical point of view, there are strong arguments in favour of using evolutionary algorithms. They are relatively easily applied, general purpose techniques that are well suited to the multi-modal, discontinuous and non-differentiable nature of the "landscape" of the evaluation functions often found.

6.9 Authoring Tools

There are a number of commercially available and freely available authoring tools for creating fuzzy expert systems. The most comprehensive (but, unfortunately, possibly the most expensive) is MATLAB® together with the optional Fuzzy Logic Toolbox. This is a comprehensive package that provides Mamdani and TSK inference, together with many choices of membership function generation and fuzzy operators, all accessible via a command line interface or an advanced graphical user interface with graphical capabilities for constructing and evaluating the output of fuzzy expert systems. MATLAB® also provides a comprehensive programming environment that allows data preprocessing and optimisation techniques to be coded and automated. Other fuzzy methodologies such as ANFIS (Adaptive-Network-Based Fuzzy Inference System) [16] and Fuzzy C-Means clustering [4] are also provided.

Freely available tools include Fuzzy CLIPS. To quote from the Fuzzy CLIPS website [46]:

> FuzzyCLIPS is an extension of the CLIPS (C Language Integrated Production System) expert system shell from NASA. It was developed by the Integrated Reasoning Group of the Institute for Information Technology of the National Research Council of Canada and has been widely distributed for a number of years. It enhances CLIPS by providing a fuzzy reasoning capability that is fully integrated with CLIPS facts and inference engine allowing one to represent and manipulate fuzzy facts and rules. FuzzyCLIPS can deal with exact, fuzzy (or inexact), and combined reasoning, allowing fuzzy and normal terms to be freely mixed in the rules and facts of an expert system. The system uses two basic inexact concepts, fuzziness and uncertainty. It has provided a useful environment for developing fuzzy applications but it does require significant effort to update and maintain as new versions of CLIPS are released.

Note that FuzzyCLIPS is free for educational and research purposes only; commercial uses of the software **require** a commercial licence. Full licence details are available from the FuzzyCLIPS website. Several books such as Cox [8] also supply various fuzzy toolsets, although the capabilities, flexibility and quality of the software varies. Of course, it is also relatively straightforward to code a fuzzy expert system from "first principles" using any programming language of choice, including low-level languages such as C, C++, Java or higher-level languages such as Python [47].

6.10 Future Directions

Leaving aside difficulties in defining "smart", it is probably fair to state that, strictly, classical (non-hybrid) fuzzy expert systems are currently neither smart nor adaptive. While this may sound like a harsh assessment, the reality of the current position is that fuzzy methods that are adaptive and are much closer to being "smart" are to be found in neuro-fuzzy, evolutionary fuzzy or other hybrid methods as covered in Chaps. 7, 8 & 9. It is in these hybrid methods that much leading-edge research is currently focussed.

So, what are the future directions for classical (non-hybrid) fuzzy expert systems? This is a difficult question to address with any clarity or certainty. Zadeh continues to attempt to promote a move to "computing with words" [44]; a paradigm that has yet to be realised. Partly to address this, Zadeh has recently introduced the concept of a protoform [45], the implementation of which Kacprzyk has investigated (See Chap. 16). Other current research into computing with words is examining type-2 fuzzy logic, in which membership functions are "blurred" to represent uncertainty in the membership values associated with any specific value of the base variable. Although first introduced by Zadeh in 1975 [43], type-2 fuzzy logic was largely neglected until particularly Mendel and his group reactivated interest more recently. Research activity and hence publications in the area are now proliferating, for example [13, 18, 19, 20, 25]. For introductory material on type-2 fuzzy logic see Mendel's recent book [26] or Mendel and John's tutorial paper [27].

6.11 Conclusions

The construction of a successful fuzzy expert system is a difficult process that requires considerable expertise from the fuzzy knowledge engineer. It is probably the large number of design choices and tunable parameters of the resultant expert system that are partly responsible for the success of fuzzy expert systems. It is not currently possible to automate the entire process so that a "generic" fuzzy expert system may be designed which will then automatically adapt itself somehow to each new environment it encounters, although frameworks such as Abraham's EvoNF [2] or systems such as Jang's ANFIS [16] can automate many of the steps. Fully automated "smart" adaptation is very much an ongoing research issue.

References

1. E. Aarts and J. Korst. *Simulated Annealing and Boltzmann Machines: A Stochastic Approach to Combinatorial Optimization and Neural Computing.* John Wiley & Sons, New York, 1989.
2. A. Abraham. EvoNF: A framework for optimisation of fuzzy inference systems using neural network learning and evolutionary computation. In *Proceedings of the 17th IEEE International Symposium on Intelligent Control (ISIC'02)*, pages 327–332, IEEE Press, 2002.
3. C. Baroglio, A. Giordana, M. Kaiser, M. Nuttin, and R. Piola. Learning control for industrial robots. *Machine Learning*, 2/3:221–250, 1996.
4. J.C. Bezdek. *Pattern Recognition with Fuzzy Objective Function Algorithms.* Plenum, New York, 1981.

5. J.L. Castro, J.J. Castro-Schez, and J.M. Zurita. Learning maximal structure rules in fuzzy logic for knowledge acquisition in expert systems. *Fuzzy Sets and Systems*, 101:331–342, 1999.
6. J. Cohen. A coefficient of agreement for nominal scales. *Educational Psychological Measurement*, 20:37–46, 1960.
7. J. Cohen. Weighted kappa: Nominal scale agreement with provision for scaled disagreement or partial credit. *Psychological Bulletin*, 70:213–220, 1968.
8. E. Cox. *The Fuzzy Systems Handbook: A Practitioner's Guide to Building, Using and Maintaining Fuzzy Systems*. AP Professional, San Diego, CA, second edition, 1998.
9. J.P. Egan. *Signal Detection Theory and ROC Analysis*. Academic Press, New York, 1975.
10. R. Engelbrecht, A. Rector, and W. Moser. Verification and validation. In E.M.S.J. van Gennip and J.L. Talmon, editors, *Assessment and Evaluation of Information Technologies*, pages 51–66. IOS Press, 1995.
11. J.M. Garibaldi and E.C. Ifeachor. Application of simulated annealing fuzzy model tuning to umbilical cord acid-base interpretation. *IEEE Transactions on Fuzzy Systems*, 7(1):72–84, 1999.
12. J.M. Garibaldi and E.C. Ifeachor. The development of a fuzzy expert system for the analysis of umbilical cord blood. In P. Szczepaniak, P.J.G. Lisboa, and J. Kacprzyk, editors, *Fuzzy Systems in Medicine*, pages 652–668. Springer-Verlag, 2000.
13. J.M. Garibaldi and R.I. John. Choosing membership functions of linguistic terms. In *Proceedings of the 2003 IEEE International Conference on Fuzzy Systems (FUZZ-IEEE 2003)*, pages 578–583, St. Louis, USA, 2003. IEEE, New York.
14. J.M. Garibaldi, J.A. Westgate, and E.C. Ifeachor. The evaluation of an expert system for the analysis of umbilical cord blood. *Artificial Intelligence in Medicine*, 17(2):109–130, 1999.
15. J.M. Garibaldi, J.A. Westgate, E.C. Ifeachor, and K.R. Greene. The development and implementation of an expert system for the analysis of umbilical cord blood. *Artificial Intelligence in Medicine*, 10(2):129–144, 1997.
16. J.-S.R. Jang. ANFIS: Adaptive-network-based fuzzy inference system. *IEEE Transactions on Systems, Man, and Cybernetics*, 23(3):665–685, 1993.
17. J.-S.R. Jang. Structure determination in fuzzy modeling: A fuzzy CART approach. In *Proceedings of IEEE International Conference on Fuzzy Systems (FUZZ-IEEE'94)*, pages 480–485, Orlando, FL, 1994.
18. R.I. John. Embedded interval valued fuzzy sets. *Proceedings of FUZZ-IEEE 2002*, pages 1316–1321, 2002.
19. R.I. John, P.R. Innocent, and M.R. Barnes. Neuro-fuzzy clustering of radiographic tibia image data using type-2 fuzzy sets. *Information Sciences*, 125/1-4:203–220, 2000.
20. N.N. Karnik and J.M. Mendel. Operations on type-2 fuzzy sets. *Fuzzy Sets and Systems*, 122:327–348, 2001.
21. N.K. Kasabov. *Foundations of Neural Networks, Fuzzy Systems and Knowledge Engineering*. MIT Press, Cambridge, Massachussets, 1996.
22. G.J. Klir and B. Yuan. *Fuzzy Sets and Fuzzy Logic: Theory and Applications*. Prentice-Hall, Upper Saddle River, NJ, 1995.

23. R. Krishnapuram. Membership function elicitation and learning. In E.H. Ruspini, P.P. Bonissone, and W. Pedrycz, editors, *Handbook of Fuzzy Computation*, pages B3.2:1–B3.2:11. Institute of Physics, Bristol, UK, 1998.
24. E.H. Mamdani and S. Assilian. An experiment in linguistic synthesis with a fuzzy logic controller. *International Journal of Man-Machine Studies*, 7:1–13, 1975.
25. J.M. Mendel. On the importance of interval sets in type-2 fuzzy logic systems. In *Proceedings of Joint 9th IFSA World Congress and 20th NAFIPS International Conference*, pages 1647–1652, Vancouver, Canada, 2001. IEEE, New York.
26. J.M. Mendel. *Uncertain Rule-Based Fuzzy Logic Systems: Introduction and New Directions*. Prentice Hall PTR, 2001.
27. J.M. Mendel and John R.I. Type-2 fuzzy sets made simple. *IEEE Transactions on Fuzzy Systems*, 10(2):117–127, 2002.
28. D. Nauck and R. Kruse. A neuro-fuzzy approach to obtain interpretable fuzzy systems for function approximation. In *Proceedings of the IEEE International Conference on Fuzzy Systems (FUZZ-IEEE'98)*, pages 1106–1111, Anchorage, AK, 1998.
29. R.M. O'Keefe, O. Balci, and E.P. Smith. Validating expert system performance. *IEEE Expert*, 2(4):81–90, 1987.
30. T.J. O'Leary, M. Goul, K.E. Moffitt, and A.E. Radwan. Validating expert systems. *IEEE Expert*, 5(3):51–58, 1990.
31. R.S. Pressman. *Software Engineering: A Practitioner's Approach*. McGraw Hill, Maidenhead, UK, fifth edition, 2000.
32. J.R. Quinlan. Induction of decision trees. *Machine Learning*, 1:81–106, 1986.
33. J.R. Quinlan. *C 4.5: Programs for Machine Learning*. Morgan Kaufmann, San Mateo, CA, 1993.
34. E.H. Ruspini, P.P. Bonissone, and W. Pedrycz, editors. *Handbook of Fuzzy Computation*. Institute of Physics, Bristol, UK, 1998.
35. O. Siggaard-Andersen. *The Acid-Base Status of the Blood (4th edn)*. Munksgaard, Copenhagen, 1976.
36. I. Sommerville. *Software Engineering*. Addison-Wesley, Harlow, UK, sixth edition, 2001.
37. M. Sugeno and G.T. Kang. Structure identification of fuzzy model. *Fuzzy Sets and Systems*, 28:15–33, 1988.
38. J.A. Swets and R.M. Pickett. *Evaluation of Diagnostic Systems: Methods from Signal Detection Theory*. Academic Press, New York, 1982.
39. T. Takagi and M. Sugeno. Fuzzy identification of systems and its applications to modeling and control. *IEEE Transactions on Systems, Man, and Cybernetics*, 15(1):116–132, 1985.
40. J.B. Tilbury, P.W.J. Van-Eetvelt, J.M. Garibaldi, J.S.H. Curnow, and E.C. Ifeachor. Reciever operator characteristic analysis for intelligent medical systems – a new approach for finding confidence intervals. *IEEE Transactions on Biomedical Engineering*, 47(7):952–963, 2000.
41. L.-X. Wang and J.M. Mendel. Generating fuzzy rules by learning from examples. *IEEE Transactions on Systems, Man, and Cybernetics*, 22(6):1414–1427, 1992.
42. D.H. Wolpert and W.G. Macready. No free lunch theorems for optimization. *IEEE Transactions on Evolutionary Computation*, 1(1):67–82, 1997.
43. L.A. Zadeh. The concept of a linguistic variable and its application to approximate reasoning – I,II,III. *Information Sciences*, 8;8;9:199–249;301–357;43–80, 1975.

44. L.A. Zadeh. From computing with numbers to computing with words – from manipulation of measurements to manipulation of perceptions. *IEEE Transactions on Circuits and Systems*, 45(1):105–119, 1999.
45. L.A. Zadeh. A prototype-centered approach to adding deduction capability to search engines – the concept of protoform. *Proceedings of NAFIPS 2002*, pages 523–525, 2002.
46. National Research Council Canada, Institute for Information Technology, Fuzzy CLIPS Website, http://www.iit.nrc.ca/IR_public/fuzzy.
47. Python Website, http://www.python.org.

7

Learning Algorithms for Neuro-Fuzzy Systems

D.D. Nauck

Computational Intelligence Group
Intelligent Systems Lab
BT Exact
Adastral Park, Ipswich IP5 3RE
United Kingdom
detlef.nauck@bt.com

In this chapter we look at techniques for learning fuzzy systems from data. These approaches are usually called neuro-fuzzy systems, because many of the available learning algorithms used in that area are inspired by techniques known from artificial neural networks. Neuro-fuzzy system are generally accepted as hybrid approaches although they rarely combine a neural network with a fuzzy system as the name would suggest. However, they combine techniques of both areas and therefore the term justified.

Methods for learning fuzzy rules are important tools for analysing and explaining data. Fuzzy data analysis – meaning here the application of fuzzy methods to the analysis of crisp data – can lead to simple, inexpensive and user-friendly solutions. The rule based structure and the ability of fuzzy systems to compress data and thus reducing complexity leads to interpretable solutions which is important for business applications. While many approaches to learning fuzzy systems from data exist, fuzzy solutions can also accommodate prior expert knowledge in form of simple to understand fuzzy rules – thus closing the gap between purely knowledge-based and purely data-driven methods. This chapter reviews several basic learning methods to derive fuzzy rules in a data mining or intelligent data analysis context. We present one algorithm in more detail because it is particularly designed for generating fuzzy rules in a simple and efficient way which makes it very useful for including it into applications.

7.1 Introduction

Modern businesses gather vast amounts of data daily. For example, data about their customers, use of their products and use of their resources. The computerisation of all aspects of our daily live and the ever-growing use of the Internet make it ever easier to collect and store data. Nowadays customers expect that businesses cater for their individual needs. In order to personalise services, intelligent data analysis (IDA) [3] and adaptive (learning) systems are required. Simple linear statistical analysis as it is mainly used in today's businesses cannot model complex dynamic dependencies that are hidden in the collected data. IDA goes one step further than today's data mining approaches and also considers the suitability of the created solutions in terms like usability, comprehension, simplicity and cost. The intelligence in IDA comes from the expert knowledge that can be integrated in the analysis process, the knowledge-based methods used for analysis and the new knowledge created and communicated by the analysis process.

Whatever strategy businesses pursue today cost reduction is invariably at the heart of it. In order to succeed they must know the performance of their processes and find means to optimise them. IDA provides means to find combine process knowledge with the collected data. Learning systems based on IDA methods can continuously optimise processes and also provide new knowledge about business processes. IDA is therefore an important aspect in modern knowledge management and business intelligence.

In addition to statistical methods, today we also have modern intelligent algorithms based on computational intelligence and machine learning. Computational intelligent methods like neuro-fuzzy systems and probabilistic networks or AI methods like decision trees or inductive logic programming provide new, intelligent ways for analysing data. Research in data analysis more and more focuses on methods that allow both the inclusion of available knowledge and the extraction of new, comprehensible knowledge about the analysed data.

Methods that learn fuzzy rules from data are a good example for this type of research activity. A fuzzy system consists of a collections of fuzzy rules which use fuzzy sets [70] to specify relations between variables. Fuzzy rules are also called linguistic rules, because fuzzy sets can be conveniently used for describing linguistic expressions like, for example, small, medium or large.

We interpret fuzzy systems as convenient models to linguistically represent (non-linear) mappings [71]. The designer of a fuzzy system specifies characteristic points of an assumed underlying function by encoding them in the form of simple fuzzy rules. This function is unknown except for those characteristic points. The fuzzy sets that are used to linguistically describe those points express the degree of indistinguishability of points that are close to each other [26, 33]. For parts of the function where characteristic points are not known, but where training data is available, fuzzy rules can be conveniently mined from the data by a variety of learning algorithms.

The advantages of applying a fuzzy system are the simplicity and the linguistic interpretation of the approach. This allows for the inexpensive and fast development and maintenance of solutions and thus enables us to solve problems in application areas where rigorous formal analysis would be too expensive and time-consuming.

In this chapter we review several approaches to learn fuzzy rules from data in Sect. 7.2 and explain one particular algorithm (NEFCLASS) in more detail in Sect. 7.3. We close the chapter with an example of applying this learning algorithm to a data set.

7.2 Learning Fuzzy Rules from Data

The structure of a fuzzy system is given by its rules and by the granularity of the data space, i.e. the number of fuzzy sets used to partition each variable. The parameters of a fuzzy system are the shapes and locations of the membership functions.

One benefit of fuzzy systems is that the rule base can be created from expert knowledge. However, in many applications expert knowledge is only partially available or not at all. In these cases it must be possible to create a rule base from scratch that performs well on a given set of data.

Assume we want to determine whether there is a fuzzy system that yields an error value below a given threshold ε for a particular data set. To do this we can enumerate rule bases until we find such a fuzzy system or until all possible rule bases are checked.

In order to restrict the number of possible rule bases, we examine a simplified scenario where we consider Mamdani-type fuzzy systems that use q triangular fuzzy sets for each variable. We assume that for each variable $x \in [l, u] \subset \mathbb{R}$, $l < u$, holds. A membership function $\mu_{a,b,c}$ is given by three parameters $a < b < c$. Instead of selecting parameters $a, b, c \in \mathbb{R}$, which would result in an infinite number of possible membership functions, we sample the variables such that there are $m + 2$ samples $l = x_0 < x_1 < \ldots < x_m < x_{m+1} = u$ for each variable x. We assume $m \geq q$. The jth fuzzy set ($j \in \{1, \ldots, q\}$) of x is given by $\mu_{x_{k_{j-1}}, x_{k_j}, x_{k_{j+1}}}$, $k_j \in \{1, \ldots, m\}$, $k_{j-1} < k_j < k_{j+1}$. We define $k_0 = 0$ and $k_{q+1} = m + 1$. Thus we obtain for each variable x a fuzzy partition, where the degrees of membership add up to one for each value of x. There are $\binom{m}{q}$ possible fuzzy partitions for each variable.

A fuzzy set configuration is given by the fuzzy partitions for all variables. If there are n variables, then we can choose between $\binom{m}{q}^n$ fuzzy set configurations. For each configuration there are $(q + 1)^n$ possible rules, because a rule can either include a variable by selecting one of its q fuzzy sets or the variable is not used by the rule.

A rule base can be any subset of all possible rules, i.e. there are $2^{((q+1)^n)}$ possible fuzzy rule bases for each fuzzy set configuration. Altogether, there are

$$\binom{m}{q}^n 2^{((q+1)^n)}$$

possible fuzzy rule bases.

These considerations show that finding an appropriate fuzzy system by simply enumerating fuzzy rule bases becomes intractable even for moderate values of n, q and m. Thus, there is a need for data driven heuristics to create fuzzy systems.

Both structure and parameters can be derived from training data. Before the parameters of a fuzzy system can be optimised in a training process at least an initial structure (rule base) must be determined. Later, the structure can still be modified in the course of the training process.

If the granularity of the system (i.e. its fuzzy sets) and the rules are to be determined at the same time, then unsupervised learning like cluster analysis can be used. These approaches have drawbacks when the resulting rule base must be interpretable. Readability of the solution can be guaranteed more easily, if the granularity of the data space is defined in advance and the data space is structured by pre-defined fuzzy partitions for all variables. Supervised rule learning algorithms are often based on a structured data space.

We can also use optimisation techniques know as evolutionary computation to determine both structure and parameter of a fuzzy system at the same time. By suitable constraints we can also ensure readability. However, evolutionary techniques can be computational expensive and time-consuming.

In the following sections we look at different possibilities for learning fuzzy rules in more detail.

7.2.1 Cluster-Oriented and Hyperbox-Oriented Fuzzy Rule Learning

Cluster-oriented methods try to group the training data into clusters and use them to create rules. Fuzzy cluster analysis [5, 7] can be used for this task by searching for spherical or hyperellipsoidal clusters. The clusters are multidimensional (discrete) fuzzy sets which overlap. An overview on several fuzzy clustering algorithms can be found, for example, in [16].

Each fuzzy cluster can be transformed into a fuzzy rule, by projecting the degrees of membership of the training data to the single dimensions. Thus for each cluster and each variable a histogram is obtained that must be approximated either by connecting the degrees of membership by a line, by a convex fuzzy set, or – more preferably – by a parameterised membership function that should be both normal and convex and fits the projected degrees of memberships as well as possible [28, 63]. This approach can result in forms of membership functions which are difficult to interpret (see Fig. 7.1)

This procedure also causes a loss of information because the Cartesian product of the induced membership functions does not reproduce a fuzzy cluster exactly (see Fig. 7.2). This loss of information is strongest in the case of arbitrarily oriented hyperellipsoids. To make this problem easier to handle, it is possible to search for axes–parallel hyperellipsoids only [29].

The fuzzy rule base obtained by projecting the clusters is usually not easy to interpret, because the fuzzy sets are induced individually for each rule

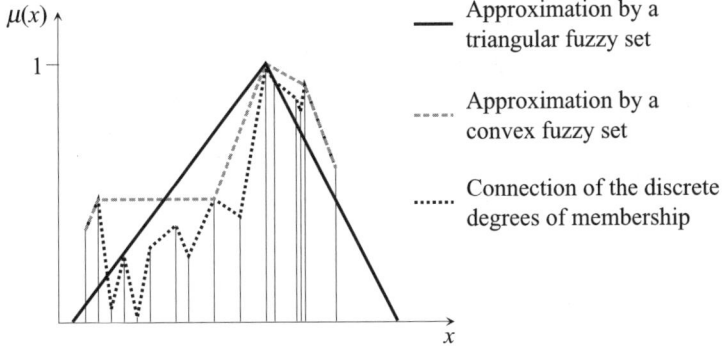

Fig. 7.1. Creation of a fuzzy set from projected degrees of membership

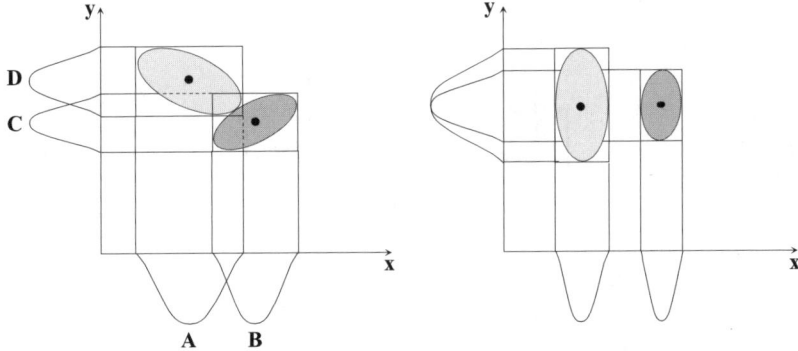

Fig. 7.2. If clusters in the form of hyperellipsoids are projected to obtain fuzzy rules, a loss of information occurs and unusual fuzzy partitions can be obtained

(Fig. 7.2). For each feature there will be as many different fuzzy sets as there are clusters. Some of these fuzzy sets may be similar, yet they are usually not identical. For a good interpretation it is necessary to have a fuzzy partition of few fuzzy sets where each clearly represents a linguistic.

The loss of information that occurs in projecting fuzzy clusters can be avoided, if the clusters are hyperboxes and parameterised multidimensional membership functions are used to represent a cluster. As in fuzzy cluster analysis the clusters (hyperboxes) are multidimensional overlapping fuzzy sets. The degree of membership is usually computed in such a way that the projections of the hyperboxes on the individual variables are triangular or trapezoidal membership functions (Fig. 7.3).

Hyperbox-oriented fuzzy rule learning is usually supervised. For each pattern of the training set that is not covered by a hyperbox, a new hyperbox is created and the output of the training pattern (class information or output value) is attached to this new hyperbox. If a pattern is incorrectly covered by a hyperbox with different output, this hyperbox is shrunk. If a pattern is

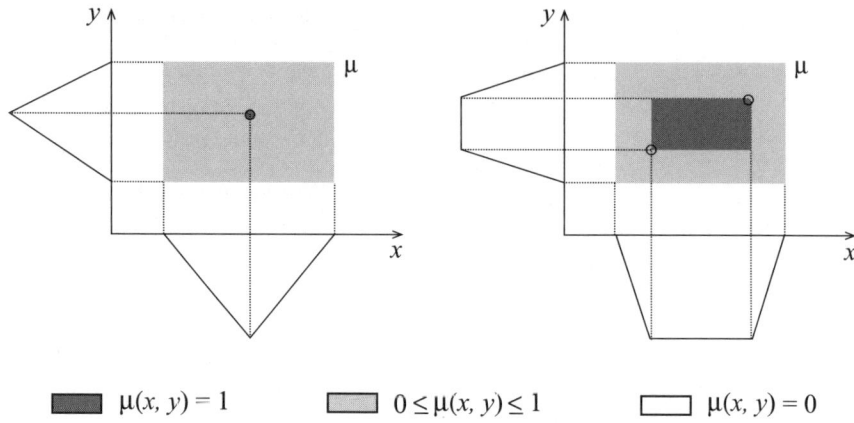

Fig. 7.3. Hyperboxes as multidimensional fuzzy sets

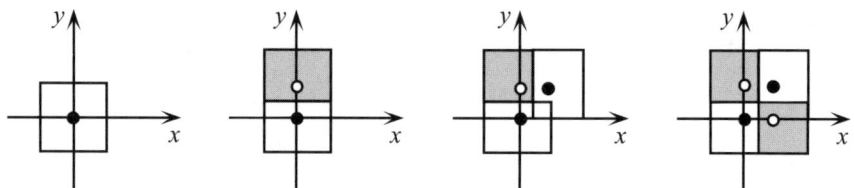

Fig. 7.4. The XOR problem solved by creating four hyperboxes

correctly covered by a hyperbox it is enlarged to increase the degree of membership for the pattern. Figure 7.4 demonstrates this approach for the XOR problem.

Like in fuzzy cluster analysis fuzzy rules can be created by projecting the hyperboxes (see Fig. 7.5). Thus the learning algorithm creates a rule base and membership functions at the same time. However, each fuzzy rule creates its own fuzzy sets for each variable and this usually results in rule bases that cannot be interpreted very well.

Hyperbox-oriented fuzzy rule learning is computationally less demanding than fuzzy cluster analysis and can create solutions for benchmark problems in pattern recognition or function approximation very fast [2, 4, 61, 62, 64]. If there are no contradictions in the training patterns and if there is only one output variable, then hyperbox-oriented learning algorithms can create solutions with no errors on the training data. In the worst case this leads to a situation, where each training pattern is covered by its individual hyperbox.

Grid clustering [25, 27] is a method that can be viewed as a combination of several fuzzy rule learning methods. Each domain is partitioned by fuzzy sets. Then an unsupervised learning algorithm modifies the fuzzy sets to improve the partitioning of the data space. It is required that for each pattern the degrees of membership add up to 1. This approach can be viewed as a cluster

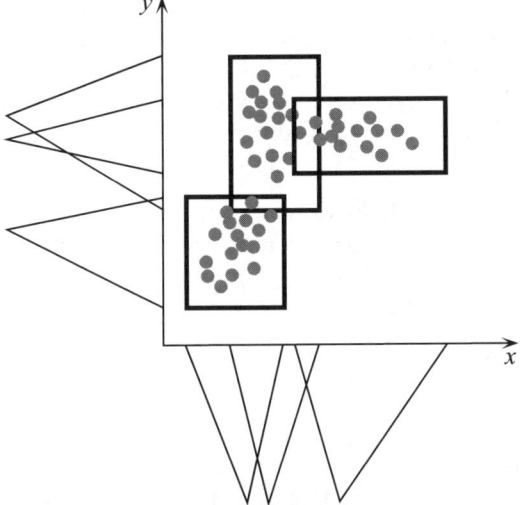

Fig. 7.5. Searching for hyperboxes to create fuzzy rules and fuzzy sets

analysis, where the clusters are hyperboxes which are aligned on a grid. It can therefore also be interpreted as hyperbox-oriented. Because the data space is structured by predefined fuzzy sets it can also be viewed as a structure-oriented learning method as discussed in Sect. 7.3. The clusters are given by membership functions and not vice versa. Therefore the rule base obtained by grid clustering can be well interpreted, because the fuzzy rules do not use individual fuzzy sets.

It is also possible to use neural networks to create fuzzy rule bases. RBF networks can be used to obtain a (Sugeno-type) fuzzy rule base. An RBF network uses multi-dimensional radial basis functions in the nodes of its hidden layer. Each of these functions can be interpreted as a fuzzy cluster. If the RBF network is trained by a gradient descent procedure it adjusts the location of the radial basis functions and – depending on the type of network – also their size and orientation. A fuzzy rule base is determined after training by projecting the radial basis functions.

Pedrycz and Card suggested a way to linguistically interpret Kohonen's self-organizing feature maps (SOM) [31] in order to create a fuzzy rule base [58]. A feature map is used to perform a cluster analysis on the training data and thus to reduce the data set to a set of prototypes represented by neurons. The prototypes are then used to create a fuzzy rule base by a structure-oriented approach as they are discussed in Sect. 7.3.

Kosko suggested a cluster-oriented approach to fuzzy rule learning that is based on his FAM model (Fuzzy Associative Memory) [32]. Kosko uses a form of adaptive vector quantisation that is not topology preserving as in the case of a SOM. In addition, Kosko's procedure determines weights for the resulting fuzzy rules. The problems of using weighted rules is discussed in [39].

All the approaches that are discussed above have drawbacks when interpretable fuzzy systems for data analysis must be created. Cluster-oriented and hyperbox-oriented approaches to fuzzy rule learning both have the following problems:

- each fuzzy rule uses individual fuzzy sets,
- fuzzy sets obtained by projection are hard to interpret linguistically,
- they can only be used for metric data, and
- they cannot cope with missing values.

Cluster-oriented approaches also have the following restrictions:

- they are unsupervised and do not optimise an output error or performance measure,
- a suitable number of clusters must be determined by repeating the cluster analysis with an increasing number of clusters until some validity measure assumes a local optimum,
- the algorithms can become computationally very expensive, especially if clusters are arbitrarily rotated hyperellipsoids and
- a loss of information occurs, if fuzzy rules are created by projection.

Grid clustering does not have most of the drawbacks of fuzzy cluster analysis, and it creates interpretable fuzzy rule bases. However, it is an unsupervised method and therefore its application is restricted.

Fuzzy cluster analysis is very well suited in segmentation tasks [13, 16] — especially in areas where linguistic interpretation plays a minor role. In data analysis fuzzy cluster analysis can be helpful during preprocessing. The cluster analysis can reveal how many rules are needed to suitably partition the data space and can give some insights into the data. It is possible to map the fuzzy sets obtained by clustering to previously defined fuzzy sets [30, 43] and thus obtain an initial rule base that can be tuned further by training the membership functions.

Because the main objective of the data analysis approaches discussed in this chapter is to create interpretable fuzzy rule bases, cluster-oriented and hyperbox-oriented approaches to fuzzy rule learning are considered to be less useful in this context.

7.2.2 Structure-Oriented Fuzzy Rule Learning

Structure-oriented approaches can be seen as special cases of hyperbox approaches that do not search for clusters in the data space, but select hyperboxes from a grid structure. By providing (initial) fuzzy sets for each variable the data space is structured by overlapping hyperboxes (compare Fig. 7.6). This way of learning fuzzy rules was suggested by Wang and Mendel [66, 67].

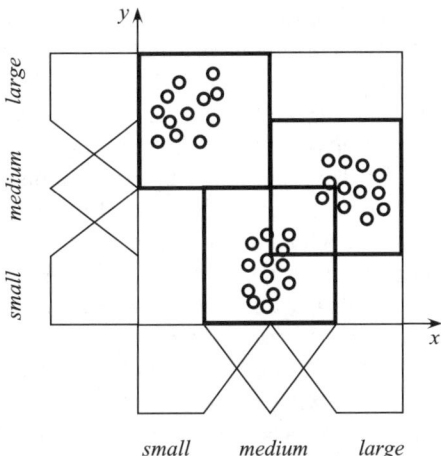

Fig. 7.6. Structure-oriented approaches use initially defined fuzzy sets to structure the data space by overlapping hyperboxes, which represent fuzzy rules

To apply the Wang & Mendel algorithm all variables are partitioned by fuzzy sets. For this purpose equidistant overlapping triangular or trapezoidal membership functions are usually used. By this means the feature space is partitioned by overlapping multidimensional fuzzy sets whose support is a hyperbox. Rules are created by selecting those hyperboxes that contain data. Wang and Mendel designed their algorithm to create fuzzy systems for function approximation. In order to mediate between different output values for the same combination of input values, they used weighted rules. In [67] a proof can be found that this algorithm can create fuzzy rule bases that can approximate any real continuous function over a compact set to an arbitrary accuracy.

Higgins & Goodman [15] suggested a variation of the Wang & Mendel algorithm in order to create fuzzy partitions during rule creation by refining the existing partitions. The algorithm begins with only one membership function for each variable, such that the whole feature space is covered by one large hyperbox. Subsequently, new membership functions are inserted at points of maximum error by refining the fuzzy partitions of all variables. Then the old rules are discarded and a new set of rules is generated based on the new fuzzy partitions. This procedure is iterated until a maximum number of fuzzy sets is created or the error decreases below some threshold. This algorithm was created in order to compensate for a drawback of the Wang & Mendel algorithm, which has problems modelling extreme values of the function to be approximated. However, the Higgins&Goodman algorithm tends to fit outliers because it concentrates on areas with large error.

Fuzzy decision trees are another approach to structure-oriented fuzzy rule learning. Induction of decision trees [19, 20] is a very popular approach in data analysis to generate classification or regression models. Decision trees are based on discriminative learning algorithms working by means of recursive partitioning. The data space is partitioned in a data-driven manner and the partition is represented as a tree. A decision tree can be transformed into a rule base, by following each path from the root node to a leaf node. Well-known algorithms are, for example, ID3 [19] for symbolic domains and C 4.5 [60] that can also incorporate numeric variables. Both approaches are applied to classification problems. CART [10] is an approach that can create decision trees for classification and regression.

Algorithms for building decision trees at the same time try to optimise the performance of the tree and to create a tree as small as possible. This is done by selecting attributes to be included into the tree according to some information theoretical measure like, for example, information gain [60]. A comparative study on different selection measures can be found in [8].

Fuzzy decision trees [7, 9, 17, 22, 69] extend the idea of decision tree learning to the domain of fuzzy systems. Instead of propagating a pattern through the tree based on crisp tests on attribute values, fuzzy tests are used. Each variable of the considered problem must be previously partitioned by fuzzy sets. A test of an attribute in a fuzzy decision tree means determining the degree of membership of the attribute value to a fuzzy set.

Fuzzy decision trees can be viewed as a structure-oriented rule learning procedure with concurrent structure optimisation. By selecting attributes with high information content first, it may turn out that not all variables are needed to solve the learning problem. However, this approach is heuristic and there is no guarantee that the induced tree is optimal in some sense – either in structure (size) or in performance.

The advantage of fuzzy decision tree learning is that not all variables must be included in the rule base at once, as is the case for other structure-oriented or cluster-oriented rule learning approaches which becomes difficult for high-dimensional learning problems. By restricting the height of the tree small rule bases can be enforced even for high-dimensional problems, if a possible loss of performance can be tolerated.

Because the induction of a decision tree is based on heuristics, it can happen that rule learning procedures that include all variables at once produce better results with the drawback of a large rule base. But if a pruning algorithm is applied to such a large rule base it is often possible to reduce the rule base and to retain the performance. In the following we will therefore concentrate on structure-oriented learning algorithms that use all variables in the beginning. After training, the created fuzzy systems can be optimised by pruning methods [41].

Compared to fuzzy clustering or hyperbox-oriented approaches, structure-oriented approaches to fuzzy rule generation have the following advantages:

- they can create rule bases that can easily be interpreted linguistically
- they are very fast and computationally inexpensive,

- they are very easy to implement,
- they can be used if the data contains numeric and non-numeric attributes,
- they can be used if the data contains missing values.

The disadvantages of structure-oriented approaches where fuzzy rules share the fuzzy sets are that decision boundaries between classes are axis-parallel. This can lead to undesired restrictions in classification performance or may require a larger number of rule in order to approximate complex class boundaries [54, 56]. Kuncheva [10] argues that fuzzy classifiers are "look-up tables in disguise". However, this behaviour also depends on the used t-norms and t-conorms. Nürnberger et al. show in [54, 56] that if the minimum is used to evaluate antecedents, then areas of ambiguity can occur where patterns have equal membership to more than class. These areas do not occur, if the product is used instead, for example. Axis-parallel decision boundaries do not necessarily appear, if rule-weights are used. However, rule weights can create problems in interpreting a fuzzy rule base [13] unless in special case where rule weights are used to create an equivalence between a fuzzy system and a naïve Bayes classifier [55].

Structure-oriented approaches to fuzzy rule learning are very suitable in data analysis where the creation of small, interpretable rule bases is usually preferred. In Sect. 7.3 we take a more detailed look at an algorithm to create Mamdani-type fuzzy rule bases for classification problems.

7.2.3 Other Approaches

A common way to apply a learning algorithm to a fuzzy system is to represent it in a special neural-network-like architecture. Then a learning algorithm – such as backpropagation – is used to train the system. There are some problems, however. Neural network learning algorithms are usually based on gradient descent methods. They cannot be applied directly to a fuzzy system, because the functions used in the inference process are usually not differentiable. There are two solutions to this problem:

(a) replace the functions used in the fuzzy system (like min and max) by differentiable functions, or
(b) do not use a gradient-based neural learning algorithm but a better-suited procedure.

Modern neuro-fuzzy systems are often represented as multilayer feedforward neural networks. For an overview see, for example, [11, 12, 16, 21, 35]. Jang's ANFIS model [19], for example, implements a Sugeno-like fuzzy system in a network structure, and applies a mixture of plain backpropagation and least mean squares procedure to train the system. A Sugeno-like fuzzy system uses only differentiable functions, i.e. for ANFIS solution (a) is selected. The GARIC model [1] also chooses solution (a) by using a special "soft minimum" function which is differentiable. The problem with solution (a) is that the models are sometimes not as easy to interpret as for example Mamdani-type fuzzy systems. Other models like NEFCON [36, 57], NEFCLASS [47] and NEFPROX [50] use solution (b) – they are Mamdani-type fuzzy systems and use special learning algorithms.

In addition to multilayer feedforward networks there are also combinations of fuzzy techniques with other neural network architectures, for example self-organising feature maps [6, 65], or fuzzy associative memories [32]. Some approaches refrain from representing a fuzzy system in a network architecture. They just apply a learning procedure to the parameters of the fuzzy system, or explicitly use a neural network to determine them.

7.2.4 Hints and Tips

- When real world data sets must be analysed we often have to deal with different types of variables on different scales, i.e. nominal scales (categorical or symbolic data), ordinal scales, or interval and ratio scales (both metric). Fuzzy systems do no depend on the type of the data they process, i.e. they can work with numerical and non-numerical variables. A neuro-fuzzy algorithm that addresses combinations of numerical and non-numerical data is discussed in [12] and [42].
- In processing real world data we often must deal with missing values. Many learning algorithms cannot cope with this problem and simply delete incomplete patterns from the learning problem. This, however, can lead to a substantial or even unacceptable loss of training data. An approach to learning fuzzy rules if the data contain missing values is described in [41].
- For learning fuzzy controllers usually reinforcement learning [24] – a special form of supervised learning – is used. This type of learning uses reinforcement signals instead of target output values, which are typically unknown for control problem. For an overview on reinforcement learning in fuzzy systems see [16]. Recent advances can be found in [57].

7.3 Learning Mamdani-Type Fuzzy Classifiers

The algorithms presented in this section are extensions to the approach by Wang & Mendel [67] and are used by the neuro-fuzzy approaches NEFCLASS [45, 47, 48] and – in a variation – by NEFPROX [46, 49, 50]. NEFCLASS is used for classification problems and NEFPROX for function approximation. The rule learning algorithms for both approaches refrain from using rule weights and determine the best consequent for a rule by a performance measure or by using an average value. In addition the algorithm tries to reduce the size of the rule base by selecting only a number of rules depending on their performance or on the coverage of the training data. This algorithm is now described in detail.

7.3.1 Rule Learning in NEFCLASS

All variables of the considered problem must be partitioned by fuzzy sets before rule learning can take place. If a domain expert or a user provides these fuzzy sets and labels them appropriately, then these labels can be taken as the "vocabulary" used to describe the problem solution represented by the rule base to be learned. If the fuzzy sets are selected such that they are meaningful to the user, then the interpretability of the rule base depends only on the number of rules and variables. The fuzzy sets can be selected individually for each variable in order to model individual granularities under which the variables are to be observed. Thus a user can hide unwanted information or focus on important areas of the domains of the variables.

Even if a user does not want to individually specify fuzzy sets and prefers to simply use fuzzy partitions of equidistant overlapping membership functions like triangular, trapezoidal or bell-shaped functions, the interpretability of the created rule base will be high. Such fuzzy sets can be conveniently interpreted as fuzzy numbers or fuzzy intervals. Besides, there will be no individual fuzzy sets for each rule as in cluster-oriented or hyperbox-oriented rule learning methods. All created rules share the same fuzzy sets. Thus it is not possible that a linguistic value is represented by different fuzzy sets in a rule base.

The number of membership functions also defines the granularity of the data space. If there are n (input + output) variables, then each hyperbox is a Cartesian product of n fuzzy sets, one from each variable. The number of hyperboxes is equal to the number of all possible fuzzy rules that can be created using the given fuzzy sets.

From an implementation point of view there is no definite requirement to actually create all these hyperboxes (fuzzy rules) and store them in the memory of a learning algorithm. The rules are created on the fly by processing the training data set twice. In the beginning, the rule base is either empty, or contains some rules provided as prior knowledge. In the first cycle all the required antecedents are created. For each point of the data set that is used for rule creation a combination of fuzzy sets is selected. This is done by finding, for each variable, the membership function that yields the highest degree of membership for the current input value. If an antecedent combined from those fuzzy sets does not yet exist in the list of antecedents, it is simply added to it. In the second cycle the best consequent for each antecedent is determined and the rules are completed. The maximum number of rules is bound from above by

$$\min\left\{s, \prod_i^n q_i\right\}$$

where s is the cardinality of the training data set and q_i is the number of fuzzy sets provided for variable x_i. If the training data has a clustered structure and concentrates only in some areas of the data space, then the number of rules will be much smaller then the theoretically possible number of rules. The actual number of rules will normally be bound by criteria defined by the user like "create no more than k rules" or "create so many rules that at least $p\%$ of all training data are covered".

The suitability of the rule base depends on the initial fuzzy partitions. If there are too few fuzzy sets, groups of data that should be represented by different rules might be covered by a single rule only. If there are more fuzzy sets than necessary to distinguish different groups of data, too many rules will be created and the interpretability of the rule base decreases. The example in Fig. 7.6 shows three clusters of data that are represented by the following three rules:

if x is $small$ then y is $large$
if x is $medium$ then y is $small$
if x is $large$ then y is $medium$

In Algorithms 7.1 – 7.3 we present procedures for structure-oriented fuzzy rule learning in classification or function approximation problems. The algorithms are implemented in the neuro-fuzzy approach NEFCLASS. The algorithms use the following notations:

- $\tilde{\mathcal{L}}$: a set of training data (fixed learning problem) with $\left|\tilde{\mathcal{L}}\right| = s$, which represents a classification problem where patterns $\mathbf{p} \in \mathbb{R}^n$ are to be assigned to m classes $C_1, \ldots, C_m$, with $C_i \subseteq \mathbb{R}^n$.
- $(\mathbf{p}, \mathbf{t}) \in \tilde{\mathcal{L}}$: a training pattern consists of an input vector $\mathbf{p} \in \mathbb{R}^n$ and a target vector $\mathbf{t} \in [0, 1]^m$. The target vector represents a possibly vague classification of the input pattern $\mathbf{p}$. The class index of $\mathbf{p}$ is given by the index of the largest component of $\mathbf{t}$: $\text{class}(\mathbf{p}) = \text{argmax}_j\{t_j\}$.
- $R = (A, C)$: a fuzzy classification rule with antecedent $\text{ant}(R) = A$ and consequent $\text{con}(R) = C$, where $A = (\mu_{j_1}^{(1)}, \ldots, \mu_{j_n}^{(n)})$ and C is a class. We use both $R(\mathbf{p})$ and $A(\mathbf{p})$ to denote the degree of fulfilment of rule R (with antecedent A) for pattern $\mathbf{p}$, i.e. $R(\mathbf{p}) = A(\mathbf{p}) = \min\{\mu_{j_1}^{(1)}(p_1), \ldots, \mu_{j_n}^{(n)}(p_n)\}$.
- $\mu_j^{(i)}$: jth fuzzy set of the fuzzy partition of input variable x_i. There are q_i fuzzy sets for variable x_i.
- $\mathbf{c}_A$: a vector with m entries to represent the accumulated degrees of membership to each class for all patterns with $A(\mathbf{p}) > 0$; $\mathbf{c}_A[j]$ is the jth entry of $\mathbf{c}_A$.
- $P_R \in [-1, 1]$: a value representing the performance of rule R:

$$P_R = \frac{1}{s} \sum_{(\mathbf{p},\mathbf{t}) \in \tilde{\mathcal{L}}} (-1)^c R(\mathbf{p}), \text{ with } c = \begin{cases} 0 & \text{if class}(\mathbf{p}) = \text{con}(R), \\ 1 & \text{otherwise.} \end{cases} \quad (7.1)$$

At first, the rule learning algorithm detects all rule antecedents that cover some training data and creates a list of antecedents. In the beginning this list is either empty, or it contains antecedents from rules given as prior knowledge.

7 Learning Algorithms for Neuro-Fuzzy Systems 147

Algorithm 7.1 The NEFCLASS rule learning algorithm

1: **for all** patterns $(\mathbf{p}, \mathbf{t}) \in \tilde{\mathcal{L}}$ **do** (* there are s training patterns *)
2:    **for all** input features x_i **do**
3:       $\mu_{j_i}^{(i)} = \operatorname*{argmax}_{\mu_j^{(i)}, j \in \{1, \ldots, q_i\}} \{\mu_j^{(i)}(p_i)\}$;
4:    **end for**
5:    Create antecedent $A = (\mu_{j_1}^{(1)}, \ldots, \mu_{j_n}^{(n)})$;
6:    **if** ($A \notin$ list of antecedents) **then**
7:       add antecedent A to list of antecedents;
8:    **end if**
9: **end for**
10: **for all** patterns $(\mathbf{p}, \mathbf{t}) \in \tilde{\mathcal{L}}$ **do** (* sum up degrees of fulfilments *)
11:    **for all** $A \in$ list of antecedents **do** (* of antecedents for each class *)
12:       $\mathbf{c}_A[\text{class}(\mathbf{p})] = \mathbf{c}_A[\text{class}(\mathbf{p})] + A(\mathbf{p})$;
13:    **end for**
14: **end for**
15: **for all** A in list of antecedents **do**
16:    $j = \operatorname*{argmax}_{i \in \{1, \ldots, m\}} \{\mathbf{c}_A[i]\}$;
17:    create rule R with antecedent A and consequent C_j;
18:    add R to list of rule base candidates;
19:    $P_R = \dfrac{1}{s}(\mathbf{c}_A[j] - \sum_{i \in \{1, \ldots, m\}, i \neq j} \mathbf{c}_A[i])$; (* performance of rule R *)
20: **end for**
21:
22: **if** (select best rules) **then**
23:    SelectBestRules; (* see Algorithm 7.2 *)
24: **else if** (select best rules per class) **then**
25:    SelectBestRulesPerClass; (* see Algorithm 7.3 *)
26: **end if**

Each time an input training pattern is not already covered by an antecedent from the list, a new antecedent is created and stored (Algorithm 7.1). Next, the algorithm selects an appropriate consequent for each antecedent A and creates a list of rule base candidates. For each antecedent, that specific class is selected that accumulated the largest value in the antecedent's vector $\mathbf{c}_A$. A performance measure $P \in [-1, 1]$ is computed for each rule indicating its unambiguity. For $P = 1$ a rule is general and classifies all training patterns correctly. For $P = -1$ a rule classifies all training patterns incorrectly. For $P = 0$ either misclassifications and correct classifications of a rule are more or less equal, or the rule covers no patterns at all. Only rules with $P > 0$ are considered to be useful.

The last part of the learning procedure is given by Algorithms 7.2 and 7.3. They select the final rule base from the list of rule base candidates computed by Algorithm 7.1. The number of rules is determined by one of the following two criteria:

Algorithm 7.2 Select the best rules for the rule base

SelectBestRules

(* The algorithm determines a rule base by selecting the best rules from *)
(* the list of rule candidates created by Algorithm 7.1. *)

```
 1: k = 0; stop = false;
 2: repeat
 3:    R' = argmax {P_R};
              R
 4:    if fixed rule base size then
 5:       if (k < k_max) then
 6:          add R' to rule base;
 7:          delete R' from list of rule candidates;
 8:          k = k + 1;
 9:       else
10:          stop = true;
11:       end if
12:    else if (all patterns must be covered) then
13:       if (R' covers some still uncovered patterns) then
14:          add R' to rule base;
15:          delete R' from list of rule candidates;
16:          if (all patterns are now covered) then
17:             stop = true;
18:          end if
19:       end if
20:    end if
21: until stop
```

1. The size of the rule base is bound by specified by the user.
2. The size of the rule base is chosen such that each training pattern is covered by at least one rule.

The learning procedure provides two evaluation procedures:

1. "Best rules": the best rules are selected based on the performance measure such that the criterion for the rule base size is fulfilled (Algorithm 7.2). In this case it may happen, that some classes are not represented in the rule base, if the rules for these classes have low performance values.
2. "Best rules per class": for each of the m classes the next best rule is selected alternately until the criterion for the rule base size is fulfilled (Algorithm 7.3). This usually results in the same number of rules for each class. However, this may not be the case, if there are only few rules for some of the classes, or if many rules of some of the classes are needed to fulfil the second rule base size criterion (cover all patterns).

If the rule learning procedure presented in Algorithm 7.1 is to be used to create fuzzy rules for function approximation purposes, the selection of

Algorithm 7.3 Select the best rules per class for the rule base

SelectBestRulesPerClass

(* The algorithm determines a rule base by selecting the best rules for each *)
(* class from the list of rule base candidates created by Algorithm 7.1. *)

1: $k = 0$; stop = false;
2: **repeat**
3:   **for all** classes C **do**
4:     **if** $(\exists R : \text{con}(R) = C)$ **then**
5:       $R' = \underset{R:\ \text{con}(R)=C}{\text{argmax}} \{P_R\}$;
6:       **if** (fixed rule base size) **then**
7:         **if** $(k < k_{\max})$ **then**
8:           add R' to rule base;
9:           delete R' from list of rule candidates;
10:           $k = k + 1$;
11:         **else**
12:           stop = true;
13:         **end if**
14:       **else if** (all patterns must be covered) **then**
15:         **if** (R' covers some still uncovered patterns) **then**
16:           add R' to rule base;
17:           delete R' from list of rule candidates;
18:         **end if**
19:         **if** (all patterns are now covered) **then**
20:           stop = true;
21:         **end if**
22:       **end if**
23:     **end if**
24:   **end for**
25: **until** stop

---

consequents must be adapted. We consider now a training set that contains patterns $(\mathbf{p}, t)$ with $\mathbf{p} \in \mathbb{R}^n$ and $t \in \mathbb{R}$, i.e. we want to approximate an unknown function $f : \mathbb{R}^n \to \mathbb{R}$, $f(x_1, \ldots, x_n) = y$ based on the training data. We have n input variables $x_i \in X_i \subseteq \mathbb{R}$ and one output variable $y \in [y_{\min}, y_{\max}] \subset \mathbb{R}$ with the range $y_r = y_{\max} - y_{\min}$. We consider one output dimension only for the sake of simplicity. The algorithm can be easily extended to create fuzzy rules with multiple output values.

The structure-oriented fuzzy rule learning algorithm presented in this section is very fast, because they only need to processes the training data twice to determine all candidates for a rule base. The selection of the rules to be included into the rule base is guided by a performance measure. The number of rules can be determined automatically such that for each training pattern there is a least one rule with non-zero degree of fulfilment or the number of rules is restricted by some value given by the user. The latter method does not need to process the training data again. Only if the the rule base size is determined automatically, must the training patterns be processed again until so many rules have been selected that all patterns are covered by rules.

If the number of rules is restricted the rule learning algorithm is not very much influenced by outliers. Rules that are only created to cover outliers have a low performance value and will not be selected for the rule base.

The performance of the selected rule base depends on the fuzzy partitions that are provided for the input (and output) variables. To increase the performance the fuzzy sets should be tuned by a suitable algorithm.

7.3.2 Parameter and Structure Tuning

In order to optimise a fuzzy rule base the parameters (fuzzy sets) and structure itself can be tuned. In order to tune fuzzy sets, typically learning algorithms inspired from neural networks are used – hence the name neuro-fuzzy systems. If the rule base is evaluated by differentiable functions and differentiable membership functions are used, for example, Gaussian (bell-shaped) functions, then gradient-descent methods like backpropagation or similar approaches [19, 20] can be applied directly. Otherwise simple heuristics (Fig. 7.7) can be used that simply look if the degree of membership of an antecedent fuzzy set must increased or decreased in order to reduce the error of a selected rule [16, 18].

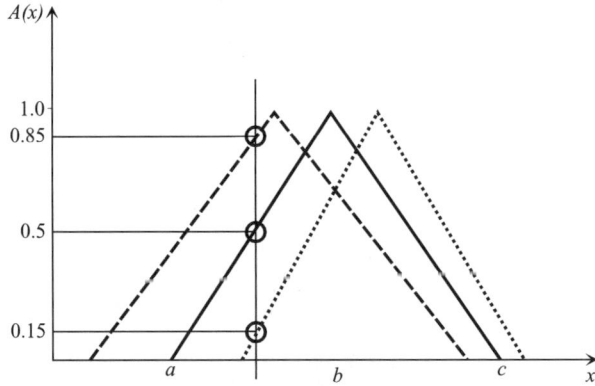

Fig. 7.7. To increase the degree of membership for the current input pattern the original representation of the fuzzy set (*centre*) assumes the representation on the *right*, to decrease the degree of membership, it assumes the representation on the *left*

The individual error of a rule can be computed by

$$\mathcal{E}_r = (\tau_r(1 - \tau_r) + \varepsilon) \, E_{\text{con}(R_r)} , \qquad (7.2)$$

where τ_r is the degree of fulfilment and $E_{\text{con}(R_r)}$ denotes the output error of a rule R_r. This can be, for example, simply the sum of squared errors or a fuzzy error measure [44, 40]. ε is a small positive number, e.g. $\varepsilon = 0.01$, that is used to also train rules with a degree of fulfilment of 0 or 1 to a small extent. Thus we compensate for the absence of adaptable consequent parameters in fuzzy classifiers. This means that a fuzzy cluster that corresponds to a rule can be moved even it it is located in an area of the input space with no data, or if it exactly matches certain outliers.

In order to further tune the structure of a fuzzy system, pruning strategies can be applied. Pruning is well-known from neural networks [14, 53] and decision tree learning [60]. For Fuzzy classifiers like NEFCLASS simple heuristics exist that try to remove variables and rules from the rule base [40, 52].

7.3.3 Hints and Tips

- The rule learning algorithms in this section can be used as an example for implementing a neuro-fuzzy system. They are kept deliberately simple in order to facilitate learning of interpretable fuzzy systems by understanding what is going on in the algorithm.
- Hyperbox-oriented fuzzy rule learning algorithms are similarly simple and also easy to implement [2].
- Free implementations of neuro-fuzzy systems like NEFCLASS [11] or ANFIS [19] can be found on the Internet. Just enter these names into a search engine.
- In order to obtain interpretable solutions fuzzy set learning must be constrained and some modifications to a fuzzy set after a learning step may have to be "repaired" in order to control overlapping [16, 40].

7.4 Creating Neuro-Fuzzy Classifiers

In this section we look at how to create a small well interpretable classifier based on the NEFCLASS algorithm described in the previous section. We use the "Wisconsin Breast Cancer" (WBC) data set[1] to illustrate rule learning and automatic pruning of a rule base. The WBC data set is a breast cancer database that was provided by W.H. Wolberg from the University of Wisconsin Hospitals, Madison [68]. The data set contains 699 cases and 16 of these cases have missing values. Each case is represented by an id number and 9 attributes (x_1: clump thickness, x_2: uniformity of cell size, x_3: uniformity of cell shape, x_4: marginal adhesion, x_5: single epithelial cell size, x_6: bare nuclei, x_7: bland chromatin, x_8: normal nucleoli, x_9: mitoses). All attributes are from the domain $\{1, \ldots, 10\}$. Each case belongs to one of two classes (benign: 458 cases, or malignant: 241 cases).

---

[1] The data set is available at the machine learning repository of the University of Irvine at ftp://ftp.ics.uci.edu/pub/machine-learning-databases.

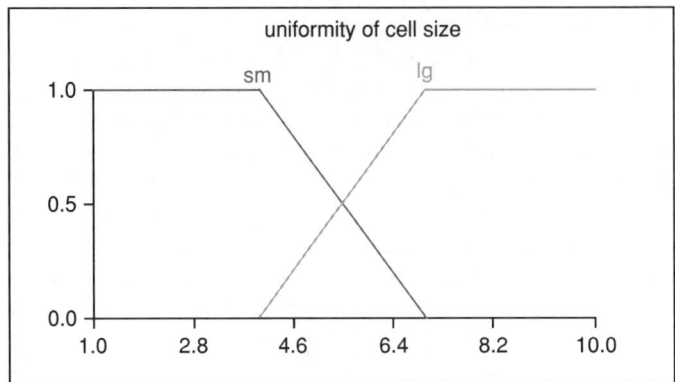

Fig. 7.8. Initial membership functions for the variables of the WBC data set

The goal of this experiment is to obtain a very small classifier. Therefore we use two fuzzy sets (*small* and *large*) per variable. The membership functions are shouldered triangles (Fig. 7.8).

Training is done with 10-fold cross validation. During rule learning the best two rules per class are selected and the membership functions are trained until the error on the validation set reaches a local minimum which the algorithm cannot escape within 30 epochs. The maximum number of training cycles is set to 200 epochs. After training the classifier is automatically pruned [41].

With the described settings we create a fuzzy partition of $2^9 = 512$ overlapping hyperboxes on the data space. This means there are 512 possible fuzzy rules. During the 11 runs of the training process (10 validation runs and a final run to create the classifier) NEFCLASS [11, 18] detects between 127 and 137 rules that actually cover training data. From these rule base candidates the best two rules per class are selected in each run. This means that before pruning a classifier consists of four rules using nine variables.

The classifier that is created based on the complete data set contains the following two rules:

> if   uniformity of cell size (x_2) is *large* and
> uniformity of cell shape (x_3) is *large* and
> bare nuclei is (x_6) *large*
> then class is *malignant*
> if   uniformity of cell size (x_2) is *small* and
> uniformity of cell shape (x_3) is *small* and
> bare nuclei is (x_6) *small*
> then class is *benign*

During cross validation five different rule bases were created, each consisting of two rules. Four rule bases just used one variable (x_3 or x_6). The variables x_3 and x_6 were present in eight and nine rule bases respectively.

x_2 appeared only once in a rule base during cross validation. The training protocol reveals, that it was not pruned from the final rule base, because the error slightly increases and one more misclassification occurs if x_2 is removed.

The mean error that was computed during cross validation is 5.86% (minimum: 2.86%, maximum: 11.43%, standard deviation: 2.95%). The 99% confidence interval for the estimated error is computed to 5.86% ± 2.54%. This provides an estimation for the error on unseen data processed by the final classifier created from the whole data set.

On the training set with all 699 cases the final classifier rules causes 40 misclassifications (5.72%), i.e. 94.28% of the patterns are classified correctly. There are 28 errors and 12 unclassified patterns which are not covered by one of the two rules. The confusion matrix for this result is given in Table 7.1. If one more misclassification can be tolerated, we can also delete variable x_2 from both rules. In this case 41 patterns are misclassified (32 errors and 9 unclassified patterns).

Table 7.1. The confusion matrix of the final classifier obtained by NEFCLASS

	Predicted Class							
	Malignant		Benign		Not Classified		Sum	
malignant	215	(30.76%)	15	(2.15%)	11	(1.57%)	241	(34.48%)
benign	13	(1.86%)	444	(63.52%)	1	(0.14%)	458	(65.52%)
sum	228	(32.62%)	459	(65.67%)	12	(1.72%)	699	(100.00%)

correct: 659 (94.28%), misclassified: 40 (5.72%), error: 70.77.

The linguistic terms *small* and *large* for each variable are represented by membership functions that can be well associated with the terms, even though they intersect at a slightly higher membership degree than 0.5 (Fig. 7.9)

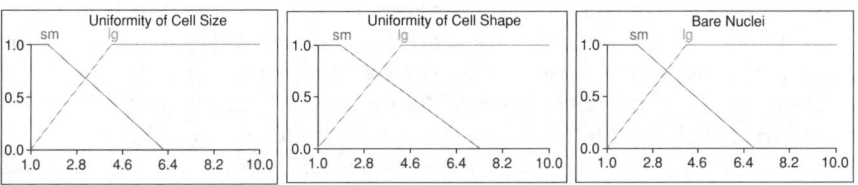

Fig. 7.9. The membership functions for the three variables used by the final classifier

7.4.1 Hints and Tips

- If you have no information about a suitable initial fuzzy partitioning for the variables of your training data, then start with a small number of triangular fuzzy sets. Think about how many fuzzy values you actually need to distinguish. Is it sufficient to distinguish between small and large

values only, or do you need to distinguish between small, medium and large values etc?
- Look at the fuzzy sets after training. If you spot fuzzy sets that overlap very strongly, then reduce the number of fuzzy sets for that particular variable and re-run the learning procedure.
- Restrict the number of rules your algorithm is allowed to learn. Fuzzy systems with a complete rule base do not necessarily perform better than fuzzy systems with small rule bases.
- Keep the number of fuzzy sets and rules small to avoid overgeneralisation and a poor performance on test data.
- Avoid algorithms with rule weights if you want to interpret your rule base linguistically. For a discussion of different effects and interpretations of rule weights see for example [13, 18, 54, 55, 56]
- When we decide to use (neuro-)fuzzy learning we should think about our motivation. For whatever reason we choose a fuzzy system to solve a problem it cannot be because we need an *optimal* solution. Fuzzy systems are used to exploit the tolerance for suboptimal solutions. So it does not make much sense to select a very sophisticated and expensive training procedure to squeeze the last bit of information from the training data. To do this we must usually forsake the standard fuzzy system architectures but, however, we are then confronted with semantical problems instead. The author prefers the view that fuzzy systems are used because they are easy to implement, easy to handle and easy to understand. A learning algorithm to create a fuzzy system from data should also have these features.

7.5 Conclusions

We have reviewed a number of approaches to learning fuzzy rules from data. Cluster and hyperbox-oriented approaches usually have the drawback of creating difficult to interpret rule bases. Structure-oriented approaches use initial fuzzy partitions which make rule learning much easier and faster and result in interpretable fuzzy rules, but are less flexible in learning complex decision boundaries. This chapter does not provide enough space to also discuss the abundance of methods for training membership functions after rule learning. We could only touch on this subject and refer the reader to other publications, for example, [21, 16, 42].

As an example on how to implement a fuzzy rule learning algorithm we have discussed the NEFCLASS approach. NEFCLASS is aimed at deriving a fuzzy classifier from data in a fast and simple way which makes this algorithm very useful for applications. In another contribution to this volume we show one example, where we have used NEFCLASS to derive explanatory fuzzy rules in real time while a user analyses travel pattern of a mobile workforce.

We have chosen fuzzy classifiers for this chapters, because we feel they are the most relevant type of fuzzy system in the context of smart adaptive systems. Obviously, we could only touch the surface of fuzzy classifier design and many other approaches exist. For a more comprehensive discussion of fuzzy classifiers the book by Kuncheva [10] is recommended.

References

1. Hamid R. Berenji and Pratap Khedkar. Learning and tuning fuzzy logic controllers through reinforcements. *IEEE Trans. Neural Networks*, 3:724–740, September 1992.
2. Michael Berthold. Mixed fuzzy rule formation. *Int. J. Approximate Reasoning*, 32:67–84, 2003.
3. Michael Berthold and David J. Hand, editors. *Intelligent Data Analysis: An Introduction*. Springer-Verlag, Berlin, 1999.
4. Michael Berthold and Klaus-Peter Huber. Constructing fuzzy graphs from examples. *Int. J. Intelligent Data Analysis*, 3(1), 1999. Electronic journal (http://www.elsevier.com/locate/ida).
5. James C. Bezdek. *Pattern Recognition with Fuzzy Objective Function Algorithms*. Plenum Press, New York, 1981.
6. James C. Bezdek, Eric Chen-Kuo Tsao, and Nikhil R. Pal. Fuzzy Kohonen clustering networks. In *Proc. IEEE Int. Conf. on Fuzzy Systems 1992*, pages 1035–1043, San Diego, CA, 1992.
7. J.C. Bezdek, J.M. Keller, R. Krishnapuram, and N. Pal. *Fuzzy Models and Algorithms for Pattern Recognition and Image Processing*. The Handbooks on Fuzzy Sets. Kluwer Academic Publishers, Norwell, MA, 1998.
8. Christian Borgelt and Rudolf Kruse. Attributauswahlmaße für die Induktion von Entscheidungsbäumen. In Gholamreza Nakhaeizadeh, editor, *Data Mining. Theoretische Aspekte und Anwendungen*, number 27 in Beiträge zur Wirtschaftsinformatik, pages 77–98. Physica-Verlag, Heidelberg, 1998.
9. Xavier Boyen and Louis Wehenkel. Automatic induction of fuzzy decision trees and its application to power system security assessment. *Fuzzy Sets and Systems*, 102(1):3–19, 1999.
10. L. Breiman, J.H. Friedman, R.A. Olsen, and C.J. Stone. *Classification and Regression Trees*. Wadsworth International, 1984.
11. James J. Buckley and Yoichi Hayashi. Fuzzy neural networks: A survey. *Fuzzy Sets and Systems*, 66:1–13, 1994.
12. James J. Buckley and Yoichi Hayashi. Neural networks for fuzzy systems. *Fuzzy Sets and Systems*, 71:265–276, 1995.
13. A. Grauel, G. Klene, and L.A. Ludwig. Data analysis by fuzzy clustering methods. In A. Grauel, W. Becker, and F. Belli, editors, *Fuzzy-Neuro-Systeme'97 – Computational Intelligence. Proc. 4th Int. Workshop Fuzzy-Neuro-Systeme'97 (FNS'97) in Soest, Germany*, Proceedings in Artificial Intelligence, pages 563–572, Sankt Augustin, 1997. infix.
14. Simon Haykin. *Neural Networks. A Comprehensive Foundation*. Macmillan College Publishing Company, New York, 1994.
15. C. Higgins and R. Goodman. Learning fuzzy rule-based neural networks for control. *Advances in Neural Information Processing Systems*, 5:350–357, 1993.
16. Frank Höppner, Frank Klawonn, Rudolf Kruse, and Thomas Runkler. *Fuzzy Cluster Analysis*. Wiley, Chichester, 1999.

17. H. Ichihashi, T. Shirai, K. Nagasaka, and T. Miyoshi. Neuro-fuzzy ID3: A method of inducing fuzzy decision trees with linear programming for maximizing entropy and an algebraic method for incremental learning. *Fuzzy Sets and Systems*, 81(1):157–167, 1996.
18. H. Ishibuchi and T. Nakashima. Effect of rule weights in fuzzy rule-based systems. In *Proc. IEEE Int. Conf. on Fuzzy Systems 2000*, pages 59–64, San Antonio, 2000.
19. J.S. Roger Jang. ANFIS: Adaptive-network-based fuzzy inference systems. *IEEE Trans. Systems, Man & Cybernetics*, 23:665–685, 1993.
20. J.S. Roger Jang and Eiji Mizutani. Levenberg–marquardt method for anfis learning. In *Proc. Biennial Conference of the North American Fuzzy Information Processing Society NAFIPS'96*, pages 87–91, Berkeley, CA, 1996. IEEE.
21. J.-S.R. Jang, C.T. Sun, and E. Mizutani. *Neuro Fuzzy and Soft Computing*. Prentice Hall, Upper Saddle River, NJ, 1997.
22. Cezary Z. Janikow. Exemplar based learning in fuzzy decision trees. In *Proc. IEEE Int. Conf. on Fuzzy Systems 1996*, pages 1500–1505, New Orleans, 1996.
23. Cezary Z. Janikow. Fuzzy decision trees: Issues and methods. *IEEE Trans. Systems, Man & Cybernetics. Part B: Cybernetics*, 28(1):1–14, 1998.
24. Leslie Pack Kaelbling, Michael H. Littman, and Andrew W. Moore. Reinforcement learning: A survey. *J. Artificial Intelligence Research*, 4:237–285, 1996.
25. Anette Keller and Frank Klawonn. Generating classification rules by grid clustering. In *Proc. Third European Workshop on Fuzzy Decision Analysis and Neural Networks for Management, Planning, and Optimization (EFDAN'98)*, pages 113–121, Dortmund, 1998.
26. Frank Klawonn, Jörg Gebhardt, and Rudolf Kruse. Fuzzy control on the basis of equality relations with an example from idle speed control. *IEEE Trans. Fuzzy Systems*, 3(3):336–350, 1995.
27. Frank Klawonn and Anette Keller. Fuzzy clustering and fuzzy rules. In Milan Mares, Radko Mesiar, Vilem Novak, Jaroslav Ramik, and Andrea Stupnanova, editors, *Proc. Seventh International Fuzzy Systems Association World Congress IFSA'97*, volume I, pages 193–197, Prague, 1997. Academia.
28. Frank Klawonn and Rudolf Kruse. Clustering methods in fuzzy control. In W. Gaul and D. Pfeifer, editors, *From Data to Knowledge: Theoretical and Practical Aspects of Classification, Data Analysis and Knowledge Organization*, pages 195–202. Springer-Verlag, Berlin, 1995.
29. Frank Klawonn and Rudolf Kruse. Constructing a fuzzy controller from data. *Fuzzy Sets and Systems*, 85:177–193, 1997.
30. Frank Klawonn, Detlef Nauck, and Rudolf Kruse. Generating rules from data by fuzzy and neuro-fuzzy methods. In *Proc. Fuzzy-Neuro-Systeme'95*, pages 223–230, Darmstadt, November 1995.
31. Teuvo Kohonen. *Self-Organization and Associative Memory*. Springer-Verlag, Berlin, 1984.
32. Bart Kosko. *Neural Networks and Fuzzy Systems. A Dynamical Systems Approach to Machine Intelligence*. Prentice Hall, Englewood Cliffs, NJ, 1992.
33. Rudolf Kruse, Jörg Gebhardt, and Frank Klawonn. *Foundations of Fuzzy Systems*. Wiley, Chichester, 1994.
34. Ludmilla I. Kuncheva. *Fuzzy Classifier Design*. Springer-Verlag, Heidelberg, 2000.
35. Chin-Teng Lin and C. S. George Lee. *Neural Fuzzy Systems. A Neuro-Fuzzy Synergism to Intelligent Systems*. Prentice Hall, New York, 1996.

36. Detlef Nauck. *Modellierung Neuronaler Fuzzy-Regler.* PhD thesis, Technische Universität Braunschweig, 1994.
37. Detlef Nauck. NEFCLASS. Univ. of Magdeburg, WWW, 1998. http://fuzzy.cs.uni-magdeburg.de/nefclass.
38. Detlef Nauck. Using symbolic data in neuro-fuzzy classification. In *Proc. 18th International Conf. of the North American Fuzzy Information Processing Society (NAFIPS99)*, pages 536–540, New York, NY, 1999. IEEE.
39. Detlef Nauck. Adaptive rule weights in neuro-fuzzy systems. *Neural Computing & Applications*, 9(1):60–70, 2000.
40. Detlef Nauck. *Data Analysis with Neuro-Fuzzy Methods.* Habilitation thesis, Otto-von-Guericke University of Magdeburg, Faculty of Computer Science, Magdeburg, Germany, 2000. Available at http://www.neuro-fuzzy.de/∼nauck.
41. Detlef Nauck. Fuzzy data analysis with NEFCLASS. *Int. J. Approximate Reasoning*, 32:103–130, 2003.
42. Detlef Nauck. Neuro-fuzzy learning with symbolic and numeric data. *Soft Computing*, September 2003. Available as "online first" at http://www.springerlink.com, paper version to appear.
43. Detlef Nauck and Frank Klawonn. Neuro-fuzzy classification initialized by fuzzy clustering. In *Proc. Fourth European Congress on Intelligent Techniques and Soft Computing (EUFIT96)*, pages 1551–1555, Aachen, September 1996. Verlag und Druck Mainz.
44. Detlef Nauck, Frank Klawonn, and Rudolf Kruse. *Foundations of Neuro-Fuzzy Systems.* Wiley, Chichester, 1997.
45. Detlef Nauck and Rudolf Kruse. NEFCLASS – a neuro-fuzzy approach for the classification of data. In K. M. George, Janice H. Carrol, Ed Deaton, Dave Oppenheim, and Jim Hightower, editors, *Applied Computing 1995. Proc. 1995 ACM Symposium on Applied Computing, Nashville, Feb. 26–28*, pages 461–465. ACM Press, New York, February 1995.
46. Detlef Nauck and Rudolf Kruse. Function approximation by NEFPROX. In *Proc. Second European Workshop on Fuzzy Decision Analysis and Neural Networks for Management, Planning, and Optimization (EFDAN'97)*, pages 160–169, Dortmund, June 1997.
47. Detlef Nauck and Rudolf Kruse. A neuro-fuzzy method to learn fuzzy classification rules from data. *Fuzzy Sets and Systems*, 89:277–288, 1997.
48. Detlef Nauck and Rudolf Kruse. NEFCLASS-X – a soft computing tool to build readable fuzzy classifiers. *BT Technology Journal*, 16(3):180–190, 1998.
49. Detlef Nauck and Rudolf Kruse. A neuro-fuzzy approach to obtain interpretable fuzzy systems for function approximation. In *Proc. IEEE Int. Conf. on Fuzzy Systems 1998*, pages 1106–1111, Anchorage, May 1998.
50. Detlef Nauck and Rudolf Kruse. Neuro–fuzzy systems for function approximation. *Fuzzy Sets and Systems*, 101:261–271, 1999.
51. Detlef Nauck and Rudolf Kruse. NEFCLASS-J – a Java-based soft computing tool. In Behnam Azvine, Nader Azarmi, and Detlef Nauck, editors, *Intelligent Systems and Soft Computing: Prospects, Tools and Applications*, number 1804 in Lecture Notes in Artificial Intelligence, pages 143–164. Springer-Verlag, Berlin, 2000.
52. Detlef Nauck, Ulrike Nauck, and Rudolf Kruse. NEFCLASS for JAVA – new learning algorithms. In *Proc. 18th International Conf. of the North American Fuzzy Information Processing Society (NAFIPS99)*, pages 472–476, New York, NY, 1999. IEEE.

53. R. Neuneier and H.-G. Zimmermann. How to train neural networks. In *Tricks of the Trade: How to Make Algorithms Really Work*, LNCS State-of-the-Art-Survey. Springer-Verlag, Berlin, 1998.
54. Andreas Nürnberger, , Aljoscha Klose, and Rudolf Kruse. Discussing cluster shapes of fuzzy classifiers. In *Proc. 18th International Conf. of the North American Fuzzy Information Processing Society (NAFIPS99)*, pages 546–550, New York, 1999. IEEE.
55. Andreas Nürnberger, Christian Borgelt, and Aljoscha Klose. Improving naïve bayes classifiers using neuro-fuzzy learning. In *Proc. 6th International Conference on Neural Information Processing – ICONIP'99*, pages 154–159, Perth, Australia, 1999.
56. Andreas Nürnberger, Aljoscha Klose, and Rudolf Kruse. Analysing borders between partially contradicting fuzzy classification rules. In *Proc. 19th International Conf. of the North American Fuzzy Information Processing Society (NAFIPS2000)*, pages 59–63, Atlanta, 2000.
57. Andreas Nürnberger, Detlef Nauck, and Rudolf Kruse. Neuro-fuzzy control based on the NEFCON-model: Recent developments. *Soft Computing*, 2(4):168–182, 1999.
58. Witold Pedrycz and H.C. Card. Linguistic interpretation of self-organizing maps. In *Proc. IEEE Int. Conf. on Fuzzy Systems 1992*, pages 371–378, San Diego, CA, 1992.
59. J.R. Quinlan. Induction of decision trees. *Machine Learning*, 1:81–106, 1986.
60. J.R. Quinlan. *C4.5: Programs for Machine Learning*. Morgan Kaufman, San Mateo, CA, 1993.
61. P.K. Simpson. Fuzzy min-max neural networks – part 1: Classification. *IEEE Trans. Neural Networks*, 3:776–786, 1992.
62. P.K. Simpson. Fuzzy min-max neural networks – part 2: Clustering. *IEEE Trans. Fuzzy Systems*, 1:32–45, February 1992.
63. M. Sugeno and T. Yasukawa. A fuzzy-logic-based approach to qualitative modeling. *IEEE Trans. Fuzzy Systems*, 1:7–31, 1993.
64. Nadine Tschichold Gürman. *RuleNet – A New Knowledge-Based Artificial Neural Network Model with Application Examples in Robotics*. PhD thesis, ETH Zürich, 1996.
65. Petri Vuorimaa. Fuzzy self-organizing map. *Fuzzy Sets and Systems*, 66:223–231, 1994.
66. Li-Xin Wang and Jerry M. Mendel. Generation rules by learning from examples. In *International Symposium on Intelligent Control*, pages 263–268. IEEE Press, 1991.
67. Li-Xin Wang and Jerry M. Mendel. Generating fuzzy rules by learning from examples. *IEEE Trans. Syst., Man, Cybern.*, 22(6):1414–1427, 1992.
68. W.H. Wolberg and O.L. Mangasarian. Multisurface method of pattern separation for medical diagnosis applied to breast cytology. *Proc. National Academy of Sciences*, 87:9193–9196, December 1990.
69. Yufei Yuan and Michael J. Shaw. Induction of fuzzy decision trees. *Fuzzy Sets and Systems*, 69(2):125–139, 1995.
70. Lotfi A. Zadeh. Fuzzy sets. *Information and Control*, 8:338–353, 1965.
71. Lotfi A. Zadeh. Fuzzy logic and the calculi of fuzzy rules and fuzzy graphs: A precis. *Int. J. Multiple-Valued Logic*, 1:1–38, 1996.

8

Hybrid Intelligent Systems: Evolving Intelligence in Hierarchical Layers

A. Abraham

Natural Computation Lab
Department of Computer Science
Oklahoma State University, USA
ajith.abraham@ieee.org
ajith.softcomputing.net

Hybridization of intelligent systems is a promising research field of modern computational intelligence concerned with the development of the next generation of intelligent systems. A fundamental stimulus to the investigations of hybrid intelligent systems is the awareness in the academic communities that combined approaches might be necessary if the remaining tough problems in artificial intelligence are to be solved. The integration of different learning and adaptation techniques to overcome individual limitations and to achieve synergetic effects through the hybridization or fusion of these techniques has, in recent years, contributed to a large number of new intelligent system designs. Most of these hybridization approaches, however, follow an ad hoc design methodology, justified by success in certain application domains.

This chapter introduces the designing aspects and perspectives of two different hybrid intelligent system architectures involving fuzzy inference system, fuzzy clustering algorithm, neural network learning and evolutionary algorithm. The first architecture introduces a Takagi-Sugeno fuzzy inference system, which is optimized using a combination of neural network learning and evolutionary algorithm. In the second architecture a fuzzy clustering algorithm is used to segregate the data and then a fuzzy inference system is used for function approximation. Both architectures are validated using real world examples. Some conclusions are also provided towards the end.

8.1 Introduction

In recent years, several adaptive hybrid soft computing [27] frameworks have been developed for model expertise, decision support, image and video segmentation techniques, process control, mechatronics, robotics and complicated automation tasks. Many of these approaches use a combination of different knowledge representation schemes, decision making models and learning strategies to solve a computational task. This integration aims at overcoming the limitations of individual techniques through hybridization or the fusion of various techniques. These ideas have led to the emergence of several different kinds of intelligent system architectures [15].

It is well known that intelligent systems, which can provide human-like expertise such as domain knowledge, uncertain reasoning, and adaptation to a noisy and time-varying environment, are important in tackling practical computing problems. Experience has shown that it is crucial, in the design of hybrid systems, to focus primarily on the integration and interaction of different techniques rather than to merge different methods to create ever-new techniques. Techniques already well understood should be applied to solve specific domain problems within the system. Their weaknesses must be addressed by combining them with complementary methods. Nevertheless, developing hybrid intelligent systems is an open-ended concept rather than restricting to a few technologies. That is, it is evolving those relevant techniques together with the important advances in other new computing methods.

We broadly classify the various hybrid intelligent architectures into 4 different categories based on the system's overall architecture namely stand-alone, transformational, hierarchical hybrid and integrated hybrid systems [5, 17]. Fused architectures are the first true form of integrated intelligent systems. They include systems which combine different techniques into one single computational model. They share data structures and knowledge representations. Another approach is to put the various techniques side-by-side and focus on their interaction in a problem-solving task. This method can allow integration of alternative techniques and exploiting their mutuality. The benefits of integrated models include robustness, improved performance and increased problem-solving capabilities. Finally, fully integrated models can provide a full range of capabilities such as adaptation, generalization, noise tolerance and justification. The two architectures presented in this chapter belong to integrated model.

Section 8.2 presents the hybrid framework for the adaptation of fuzzy inference system using a combination of neural network learning and evolutionary computation. An application example is also included in this section. In Sect. 8.3, we present a hybrid combination of fuzzy clustering algorithm and a fuzzy inference system for a Web mining task. Conclusions are given towards the end.

8.2 Adaptation of Fuzzy Inference Systems

A conventional fuzzy inference system makes use of a model of the expert who is in a position to specify the most important properties of the process. Expert knowledge is often the main source for designing fuzzy inference systems. According to the performance measure of the problem environment, the membership functions, the knowledge base and the inference mechanism are to be adapted. Several research works continue to explore the adaptation of fuzzy inference systems [2, 4, 16]. These include the adaptation of membership functions, rule bases and the aggregation operators. They include but are not limited to:

- The self-organizing process controller by Procyk et al. [19] which considered the issue of rule generation and adaptation.
- The gradient descent and its variants which have been applied to fine-tune the parameters of the input and output membership functions [25].
- Pruning the quantity and adapting the shape of input/output membership functions [23].
- Tools to identify the structure of fuzzy models [21].
- In most cases the inference of the fuzzy rules is carried out using the *min* and *max* operators for fuzzy intersection and union. If the T-norm and T-conorm operators are parameterized then the gradient descent technique could be used in a supervised learning environment to fine-tune the fuzzy operators.

The antecedent of the fuzzy rule defines a local fuzzy region, while the consequent describes the behavior within the region via various constituents. The consequent constituent can be a membership function (Mamdani model) or a linear equation (first order Takagi-Sugeno model) [22].

Adaptation of fuzzy inference systems using evolutionary computation techniques has been widely explored [2, 4, 7, 8, 18, 20]. The automatic adaptation of membership functions is popularly known as self-tuning. The genome encodes parameters of trapezoidal, triangle, logistic, hyperbolic-tangent, Gaussian membership functions and so on.

The evolutionary search of fuzzy rules can be carried out using three approaches [10]. In the first (Michigan approach), the fuzzy knowledge base is adapted as a result of the antagonistic roles of competition and cooperation of fuzzy rules. Each genotype represents a single fuzzy rule and the entire population represents a solution. The second method (Pittsburgh approach) evolves a population of knowledge bases rather than individual fuzzy rules. Genetic operators serve to provide a new combination of rules and new rules. The disadvantage is the increased complexity of the search space and the additional computational burden, especially for online learning. The third method (iterative rule learning approach) is similar to the first, with each chromosome representing a single rule, but contrary to the Michigan approach, only the best individual is considered to form part of the solution, the remaining chromosomes in the population are discarded. The evolutionary learning process builds up the complete rule base through an iterative learning process.

In a neuro-fuzzy model [4], there is no guarantee that the neural network-learning algorithm will converge and the tuning of fuzzy inference system be successful (determining the optimal parameter values of the membership functions, fuzzy operators and so on). A distinct feature of evolutionary fuzzy systems is their adaptability to a dynamic environment. Experimental evidence had indicated cases where evolutionary algorithms are inefficient at fine tuning solutions, but better at finding global basins of attraction [2, 6, 14]. The efficiency of evolutionary training can be improved significantly by incorporating a local search procedure into the evolution. Evolutionary algorithms are used to first locate a good region in the space and then a local search procedure is used to find a near optimal solution in this region. It is interesting to consider finding good initial parameter values as locating a good region in the space. Defining that the basin of attraction of a local minimum is composed of all the points, sets of parameter values in this case, which can converge to the local minimum through a local search algorithm, then a global minimum can easily be found by the local search algorithm if the evolutionary algorithm can locate any point, i.e, a set of initial parameter values, in the basin of attraction of the global minimum. Referring to Fig. 8.1, G_1 and G_2 could be considered as the initial parameter values as located by the evolutionary search and W_A and W_B the corresponding final parameter values fine-tuned by the meta-learning technique.

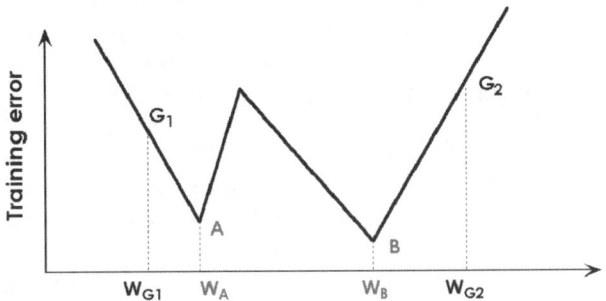

Fig. 8.1. Fine tuning of parameters using hybrid learning

We present the Evolving Neuro Fuzzy (EvoNF) model which optimizes the fuzzy inference system using a meta-heuristic approach combining neural network learning and evolutionary computation. The proposed technique could be considered as a methodology to integrate neural network learning, fuzzy inference systems and evolutionary search procedures [2, 8].

The evolutionary search of membership functions, rule base, fuzzy operators progress on different time scales to adapt the fuzzy inference system according to the problem environment. Figure 8.2 illustrates the general interaction mechanism with the evolutionary search of a fuzzy inference system (Mamdani, Takagi-Sugeno etc.) evolving at the highest level on the slowest time scale. For each evolutionary search of fuzzy operators (for example, best combination of T-norm, T-conorm and defuzzification strategy), the search for the fuzzy rule base progresses at a faster time scale in an environment decided by the fuzzy inference system and the problem. In a similar manner, the evolutionary search of membership functions proceeds at a faster time scale

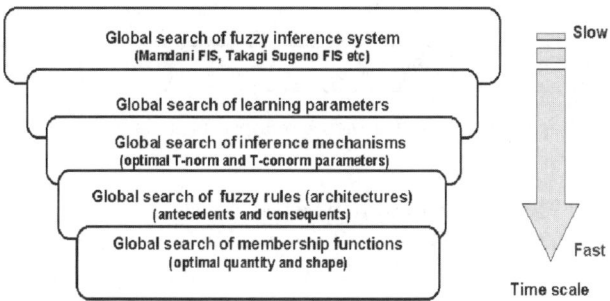

Fig. 8.2. EvoNF general computational framework

(for every rule base) in the environment decided by the fuzzy inference system, fuzzy operators and the problem. Thus, the evolution of the fuzzy inference system evolves at the slowest time scale while the evolution of the quantity and type of membership functions evolves at the fastest rate. The function of the other layers could be derived similarly. The hierarchy of the different adaptation layers (procedures) relies on prior knowledge. For example, if there is more prior knowledge about the knowledge base (*if-then* rules) than the inference mechanism then it is better to implement the knowledge base at a higher level. If a particular fuzzy inference system best suits the problem, the computational task could be reduced by minimizing the search space. The chromosome architecture is depicted in Fig. 8.3.

The architecture and the evolving mechanism could be considered as a general framework for adaptive fuzzy systems, that is a fuzzy model that can change membership functions (quantity and shape), rule base (architecture), fuzzy operators and learning parameters according to different environments without human intervention.

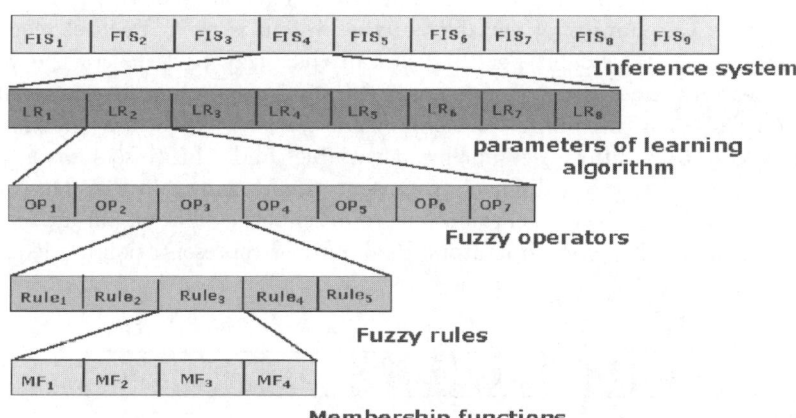

Fig. 8.3. Chromosome structure of EvoNF model

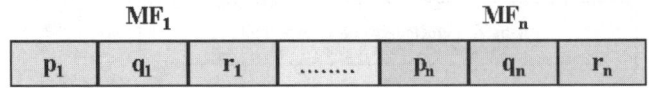

Fig. 8.4. Representation of n membership functions of a bell shape MF

Referring to Fig. 8.3 each layer (from fastest to slowest) of the hierarchical evolutionary search process has to be represented in a chromosome for successful modelling of EvoNF. The detailed functioning and modelling process is as follows.

Layer 1: The simplest way is to encode the number of membership functions per input variable and the parameters of the membership functions. Figure 8.4 depicts the chromosome representation of n *bell* membership functions specified by its parameters p, q and r. The optimal parameters of the membership functions located by the evolutionary algorithm will be further fine tuned by the neural network-learning algorithm. Similar strategy could be used for the output membership functions in the case of a Mamdani fuzzy inference system. Experts may be consulted to estimate the MF shape forming parameters to estimate the search space of the MF parameters.

In our experiments angular coding method proposed by Cordón et al. were used to represent the rule consequent parameters of the Takagi-Sugeno inference system [10].

Layer 2. This layer is responsible for the optimization of the rule base. This includes deciding the total number of rules, representation of the antecedent and consequent parts. Depending on the representation used (Michigan, Pittsburg, iterative learning and so on), the number of rules grow rapidly with an increasing number of variables and fuzzy sets. The simplest way is that each gene represents one rule, and "1" stands for a selected and "0" for a non-selected rule. Figure 8.5 displays such a chromosome structure representation. To represent a single rule a position dependent code with as many elements as the number of variables of the system is used. Each element is a binary string with a bit per fuzzy set in the fuzzy partition of the variable, meaning the absence or presence of the corresponding linguistic label in the rule. For a three input and one output variable, with fuzzy partitions composed of 3, 2, 2 fuzzy sets for input variables and 3 fuzzy sets for output variable, the fuzzy rule will have a representation as shown in Fig. 8.6.

Layer 3. In this layer, a chromosome represents the different parameters of the T-norm and T-conorm operators. Real number representation is adequate

Fig. 8.5. Representation of the entire rule base consisting of m fuzzy rules

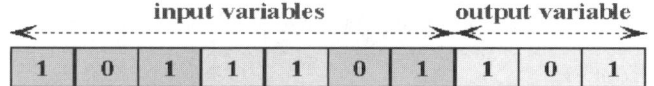

Fig. 8.6. Representation of an individual fuzzy rule

to represent the fuzzy operator parameters. The parameters of the operators could be even fine-tuned using gradient descent techniques.

Layer 4. This layer is responsible for the selection of optimal learning parameters. Performance of the gradient descent algorithm directly depends on the learning rate according to the error surface. Real number representation may be used to represent the learning parameters. The optimal learning parameters decided by the evolutionary algorithm will be used in the learning algorithm to tune the membership functions and the inference mechanism.

Layer 5. This layer basically interacts with the environment and decides which fuzzy inference system (Mamdani type and its variants, Takagi-Sugeno type, Tsukamoto type etc.) will be the optimal according to the environment.

Once the chromosome representation, C, of the entire EvoNF model is done, the evolutionary search procedure could be initiated as follows:

1. Generate an initial population of N numbers of C chromosomes. Evaluate the fitness of each chromosome depending on the problem.
2. Depending on the fitness and using suitable selection methods reproduce a number of children for each individual in the current generation.
3. Apply genetic operators to each child individual generated above and obtain the next generation.
4. Check whether the current model has achieved the required error rate or the specified number of generations has been reached. Go to Step 2.
5. End

8.2.1 Application of EvoNF – Export Behavior Modelling

In this section, we will examine the application of the proposed EvoNF model to approximate the export behavior of multi-national subsidiaries. Several specific subsidiary features identified in international business literature are particularly relevant when seeking to explain Multi-National Company (MNC) subsidiary export behavior. Our purpose is to model the complex export pattern behavior using a Takagi-Sugeno fuzzy inference system in order to determine the actual volume of Multinational Cooperation Subsidiaries (MCS) export output (sales exported) [11]. Malaysia has been pursuing an economic strategy of export-led industrialization. To facilitate this strategy, foreign investment is courted through the creation of attractive incentive packages. These primarily entail taxation allowances and more liberal ownership rights for investments. The quest to attract foreign direct investment (FDI) has proved to be highly successful. The bulk of investment has gone into export-oriented manufacturing industries. For simulations we have used data provided

Table 8.1. Parameter settings of EvoNF framework

Population Size	40
Maximum no of generations	35
FIS	Takagi Sugeno
Rule antecedent MF	2 MF (parameterised Gaussian)/input
Rule consequent parameters	Linear parameters
Gradient descent learning	10 epochs
Ranked based selection	0.50
Elitism	5%
Starting mutation rate	0.50

from a survey of 69 Malaysian MCS. Each corporation subsidiary data set were represented by product manufactured, resources, tax protection, involvement strategy, financial independence and suppliers relationship.

We used the popular grid partitioning method to generate the initial rule base [24]. This partition strategy works well when only few number of inputs are involved since it requires only a small number of MF for each input. We used 90% of the data for training and remaining 10% for testing and validation purposes. The initial populations were randomly created based on the parameters shown in Table 8.1. We used an adaptive mutation operator, which decreases the mutation rate as the algorithm greedily proceeds in the search space 0. The parameters mentioned in Table 8.1 were decided after a few trial and error approaches. Experiments were repeated 3 times and the average performance measures are reported. Figure 8.7 illustrates the meta-learning approach for training and test data combining evolutionary learning and gradient descent technique during the 35 generations.

The 35 generations of meta-learning approach created 76 *if-then* Takagi-Sugeno type fuzzy *if-then* rules compared to 128 rules using the conventional grid-partitioning method. We also used a feed forward neural network with

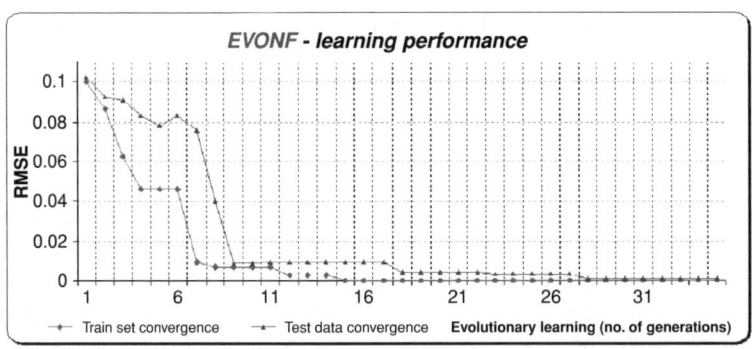

Fig. 8.7. Meta-learning performance (training and test) of EvoNF framework

Table 8.2. Training and test performance of the different intelligent paradigms

	Intelligent Paradigms					
Export Output	EvoNF			Neural Network		
	RMSE		CC	RMSE		CC
	Train	Test		Train	Test	
	0.0013	0.012	0.989	0.0107	0.1261	0.946

12 hidden neurons (single hidden layer) to model the export output for the given input variables. The learning rate and momentum were set at 0.05 and 0.2 respectively and the network was trained for 10,000 epochs using BP. The network parameters were decided after a trial and error approach. The obtained training and test results are depicted in Table 8.2 (RMSE = Root Mean Squared Error, CC = correlation coefficient).

Our analysis on the export behavior of Malaysia's MCS reveals that the developed EvoNF model could learn the chaotic patterns and model the behavior using an optimized Takagi Sugeno FIS. As illustrated in Fig. 8.8 and Table 8.2, EvoNF could approximate the export behavior within the tolerance limits. When compared to a direct neural network approach, EvoNF performed better (in terms of lowest RMSE) and better correlation coefficient.

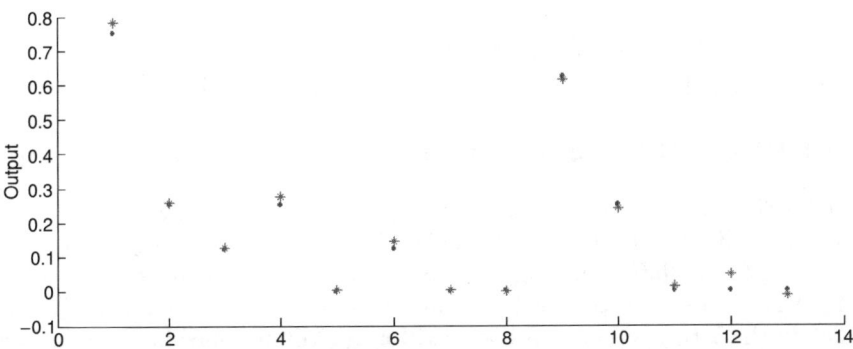

Fig. 8.8. Test results showing the export output (scaled values) for 13 MNC's with respect to the desired values

8.2.2 Hints and Tips to Design EvoNF Architecture

- Incorporate prior knowledge to minimize the number of layers. This will reduce the computational complexity and development time.
- The learning parameter layer could be the first to abolish. By prefixing a low learning rate, we could almost ensure proper meta-learning performance.

- For many function approximation problems, Takagi-Sugeno inference method seems to work very well compared to a Mamdani inference system.
- Computational complexity could be minimized by selecting an appropriate initial rule base which has also minimal number of rules. Instead of the grid partition method, other partition methods (tree partition, scatter partition etc.) could be explored.
- For most function approximation problems Gaussian membership functions seems to work very well. In most cases it is not worth to explore more than 3 membership functions for an input variable.
- Adequate genotype representation of the different layers is very important. Try to use simple representation as much as possible.
- It is efficient to start with a high mutation rate (as high as 0.80) and let the adaptive algorithm pick up the mutation rate according to the generated error.

In the following section we present an extension of the EvoNF architecture discussed in Sect. 8.2. The advanced hybrid framework uses a fuzzy c-means algorithm to segregate the data and a fuzzy inference system for data mining purposes. The parameters of the fuzzy c-means algorithm and the fuzzy inference system are represented in hierarchical layers and are optimized using evolutionary algorithm and gradient decent method.

8.3 Hybrid Fuzzy Clustering and Fuzzy Inference Method for Data Mining

8.3.1 Fuzzy Clustering Algorithm

One of the widely used clustering methods is the fuzzy c-means (FCM) algorithm developed by Bezdek [9]. FCM partitions a collection of n vectors $x_i, i = 1, 2 \ldots, n$ into c fuzzy groups and finds a cluster center in each group such that a cost function of dissimilarity measure is minimized. To accommodate the introduction of fuzzy partitioning, the membership matrix U is allowed to have elements with values between 0 and 1. The FCM objective function takes the form

$$J(U, c_1, \ldots c_c) = \sum_{i=1}^{c} J_i = \sum_{i=1}^{c} \sum_{j=1}^{n} u_{ij}^m d_{ij}^2$$

Where u_{ij}, is a numerical value between [0,1]; c_i is the cluster center of fuzzy group i; $d_{ij} = \|c_i - x_j\|$ is the Euclidian distance between ith cluster center and jth data point; and m is called the exponential weight which influences the degree of fuzziness of the membership (partition) matrix. For every data object the sum of membership values with all the associated clusters should add up to one.

8.3.2 Optimization of Fuzzy Clustering Algorithm

Optimization of usually a number of cluster centers are randomly initialized and the FCM algorithm provides an iterative approach to approximate the minimum of the objective function starting from a given position and leads to any of its local minima. No guarantee ensures that FCM converges to an optimum solution (can be trapped by local extrema in the process of optimizing the clustering criterion). The performance is very sensitive to initialization of the cluster centers. An evolutionary algorithm is used to decide the optimal number of clusters and their cluster centers. The algorithm is initialized by constraining the initial values to be within the space defined by the vectors to be clustered. A very similar approach is used by Hall et al. [12] and [3].

8.3.3 Intelligent Miner (i-Miner) Framework for Web Mining

We propose an integrated framework (*i-Miner*) which optimizes the FCM using an evolutionary algorithm and a Takagi-Sugeno fuzzy inference system using a combination of evolutionary algorithm and neural network learning [1, 3]. The developed framework is used for a Web usage mining problem. Web usage mining attempts to discover useful knowledge from the secondary data obtained from the interactions of the users with the Web. The rapid e-commerce growth has made both business community and customers face a new situation. Due to intense competition on one hand and the customer's option to choose from several alternatives business community has realized the necessity of intelligent marketing strategies and relationship management. Web usage mining has become very critical for effective Web site management, creating adaptive Web sites, business and support services, personalization, network traffic flow analysis and so on. We present how the (*i-Miner*) approach could be used to optimize the concurrent architecture of a fuzzy clustering algorithm (to discover web data clusters) and a fuzzy inference system to analyze the Web site visitor trends. Figure 8.9 illustrates the *i-Miner* framework. A hybrid evolutionary FCM algorithm is used to optimally segregate similar user interests. The clustered data is then used to analyze the trends using a Takagi-Sugeno fuzzy inference system learned using EvoNF approach.

In Web usage mining, data pre-processing involves mundane tasks such as merging multiple server logs into a central location and parsing the log into data fields followed by data cleaning. Graphic file requests, agent/spider crawling etc. could be easily removed by only looking for HTML file requests. Normalization of URL's is often required to make the requests consistent. For example requests for www.okstate.edu and www.okstate.edu/index.html, are all for the same file. All these tasks could be achieved by conventional hard computing techniques which involves text processing, string matching, association rules, simple statistical measures etc. The cleaned and pre-processed raw data from the log files is used by evolutionary FCM algorithm to identify the optimal number of clusters and their centers. The developed clusters of

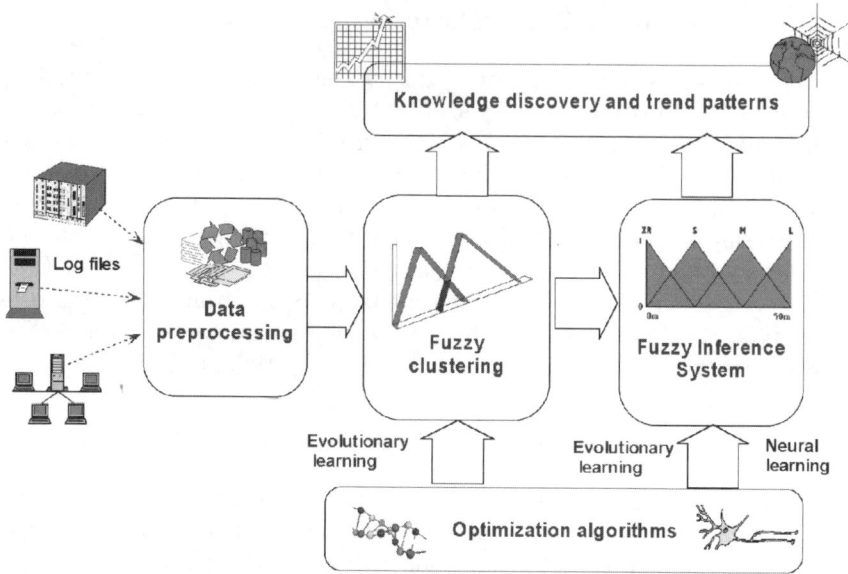

Fig. 8.9. *i-Miner* framework for Web usage mining

data are fed to a Takagi-Sugeno fuzzy inference system to analyze the Web server access trend patterns. The *if-then* rule structures are learned using an iterative learning procedure [10] by an evolutionary algorithm and the rule parameters are fine-tuned using gradient decent algorithm. The hierarchical distribution of *i-Miner* is depicted in Fig. 8.10. The arrow direction indicates the hierarchy of the evolutionary search. In simple words, the optimization of clustering algorithm progresses at a faster time scale at the lowest level in an environment decided by the fuzzy inference system and the problem envi-

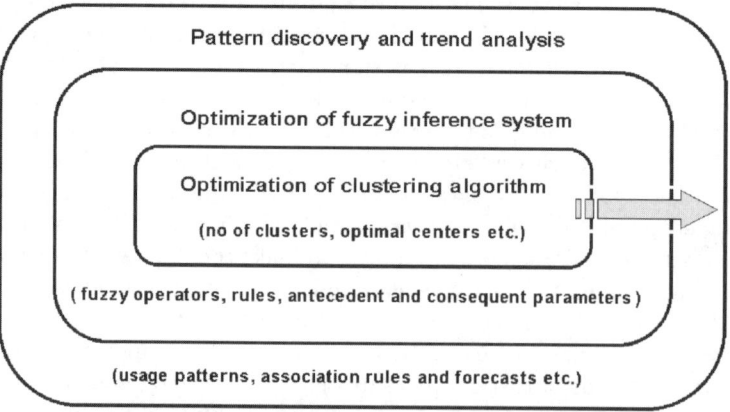

Fig. 8.10. Hierarchical architecture of *i-Miner*

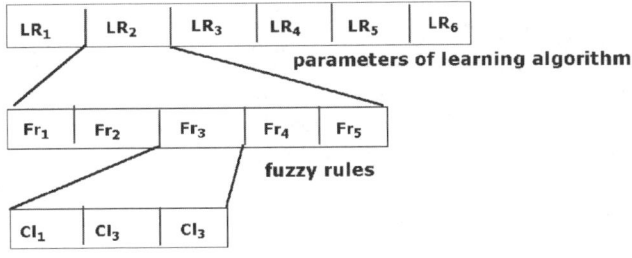

Fig. 8.11. Chromosome structure of the *i-Miner*

ronment. The evolution (optimization) of the fuzzy inference system proceeds at a slower time scale at a higher level with respect to the fuzzy clustering algorithm.

8.3.4 Chromosome Modelling and Representation

Hierarchical evolutionary search process has to be represented in a chromosome for successful modelling of the *i-Miner* framework. A typical chromosome of the *i-Miner* would appear as shown in Fig. 8.11 and the detailed modelling process is as follows.

Layer 1. The optimal number of clusters and initial cluster centers is represented this layer.

Layer 2. This layer is responsible for the optimization of the rule base (same function defined in Fig. 8.3). We used the grid-partitioning algorithm to generate the initial set of rules. An iterative learning method is then adopted to optimize the rules. The existing rules are mutated and new rules are introduced. The fitness of a rule is given by its contribution (strength) to the actual output. To represent a single rule a position dependent code with as many elements as the number of variables of the system is used.

Layer 3. This layer is responsible for the selection of optimal learning parameters. Performance of the gradient descent algorithm directly depends on the learning rate according to the error surface. The optimal learning parameters decided by this layer will be used to tune the parameterized rule antecedents/consequents and the fuzzy operators. In the *i-Miner* approach, rule antecedent/consequent parameters and the fuzzy operators are fine tuned using a gradient descent algorithm to minimize the output error

$$E = \sum_{k=1}^{N}(d_k - x_k)^2 \quad (8.1)$$

where d_k is the kth component of the rth desired output vector and x_k is the kth component of the actual output vector by presenting the rth input vector to the network. The gradients of the rule parameters to be optimized, namely the consequent parameters (P_n) $\frac{\partial E}{\partial P_n}$ for all rules R_n and the premise parameters $\frac{\partial E}{\partial \sigma_i}$ and $\frac{\partial E}{\partial c_i}$ for all fuzzy sets F_i (σ and c represents the MF width and center of a Gaussian MF)are to be computed. As far as rule parameter learning is concerned, the key difference between *i-Miner* and the EvoNF approach is in the way the consequent parameters were determined.

Once the three layers are represented in a chromosome structure C, then the learning procedure could be initiated as defined in Sect. 8.2.

8.3.5 Application of *i-Miner*: Web Usage Mining

The hybrid framework described in Sect. 8.3 was used for Web usage mining [1]. The statistical/text data generated by the log file analyzer from 1 January 2002 to 7 July 2002. Selecting useful data is an important task in the data preprocessing block. After some preliminary analysis, we selected the statistical data comprising of domain byte requests, hourly page requests and daily page requests as focus of the cluster models for finding Web users' usage patterns. It is also important to remove irrelevant and noisy data in order to build a precise model. We also included an additional input "*index number*" to distinguish the time sequence of the data. The most recently accessed data were indexed higher while the least recently accessed data were placed at the bottom. Besides the inputs "*volume of requests*" and " *volume of pages (bytes)*" and "*index number*", we also used the "*cluster information*" provided by the clustering algorithm as an additional input variable. The data was re-indexed based on the cluster information. Our task is to predict the Web traffic volume on a hourly and daily basis. We used the data from 17 February 2002 to 30 June 2002 for training and the data from 1 July 2002 to 6 July 2002 for testing and validation purposes.

The performance is compared with self-organizing maps (alternative for FCM) and several function approximation techniques like neural networks, linear genetic programming and Takagi-Sugeno fuzzy inference system (to predict the trends). The results are graphically illustrated and the practical significance is discussed in detail.

The initial populations were randomly created based on the parameters shown in Table 8.3. Choosing good reproduction operator values is often a challenging task. We used a special mutation operator, which decreases the mutation rate as the algorithm greedily proceeds in the search space [3]. If the allelic value x_i of the ith gene ranges over the domain a_i and b_i the mutated gene x_i' is drawn randomly uniformly from the interval $[a_i, b_i]$.

$$x_i' = \begin{cases} x_i + \Delta(t, b_i - x_i), & if \ \omega = 0 \\ x_i + \Delta(t, x_i - a_i), & if \ \omega = 1 \end{cases} \quad (8.2)$$

where ω represents an unbiased coin flip $p(\omega = 0) = p(\omega = 1) = 0.5$, and

$$\Delta(t, x) = x \left(1 - \gamma^{\left(1 - \frac{t}{t_{\max}}\right)^b}\right) \quad (8.3)$$

8 Evolving Intelligence in Hierarchical Layers

Table 8.3. Parameter settings of *i-Miner*

Population Size	30
Maximum no of generations	35
Fuzzy inference system	Takagi Sugeno
Rule antecedent membership functions	3 membership functions per input variable (parameterized Gaussian)
Rule consequent parameters	linear parameters
Gradient descent learning	10 epochs
Ranked based selection	0.50
Elitism	5%
Starting mutation rate	0.50

and t is the current generation and t_{max} is the maximum number of generations. The function Δ computes a value in the range $[0, x]$ such that the probability of returning a number close to zero increases as the algorithm proceeds with the search. The parameter b determines the impact of time on the probability distribution Δ over $[0, x]$. Large values of b decrease the likelihood of large mutations in a small number of generations. The parameters mentioned in Table 8.3 were decided after a few trial and error approaches (basically by monitoring the algorithm convergence and the output error measures). Experiments were repeated 3 times and the average performance measures are reported. Figures 8.12 and 8.13 illustrates the meta-learning approach combining evolutionary learning and gradient descent technique during the 35 generations.

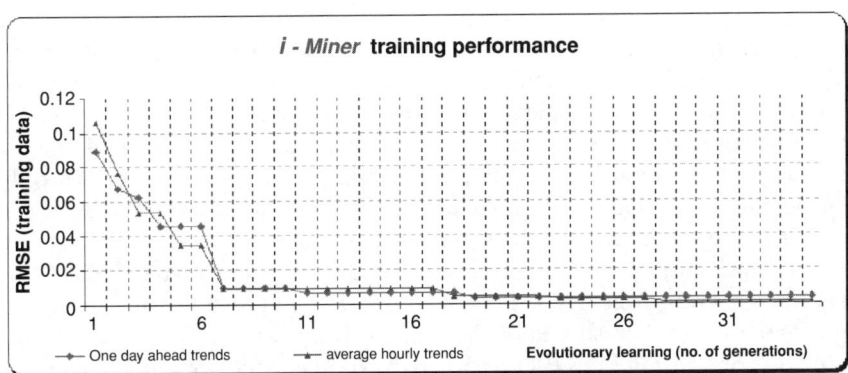

Fig. 8.12. Meta-learning performance (training) of *i-Miner*

Table 8.4 summarizes the performance of the developed *i-Miner* for training and test data. *i-Miner* trend prediction performance is compared with ANFIS [13], Artificial Neural Network (ANN) and Linear Genetic Programming (LGP). The Correlation Coefficient (CC) for the test data set is also

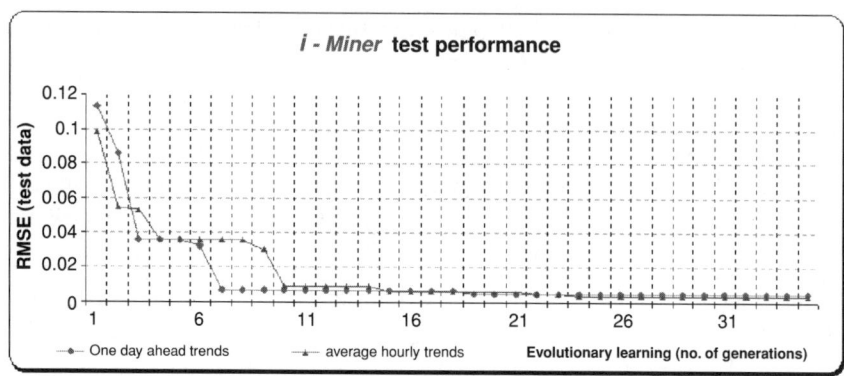

Fig. 8.13. Meta-learning performance (testing) of *i-Miner*

Table 8.4. Performance of the different paradigms

	Period					
	Daily (1 day ahead)			Hourly (1 hour ahead)		
	RMSE			RMSE		
Method	Train	Test	CC	Train	Test	CC
i-Miner	0.0044	0.0053	0.9967	0.0012	0.0041	0.9981
ANFIS	0.0176	0.0402	0.9953	0.0433	0.0433	0.9841
ANN	0.0345	0.0481	0.9292	0.0546	0.0639	0.9493
LGP	0.0543	0.0749	0.9315	0.0654	0.0516	0.9446

given in Table 8.4. The 35 generations of meta-learning approach created 62 *if-then* Takagi-Sugeno type fuzzy rules (daily traffic trends) and 64 rules (hourly traffic trends) compared to the 81 rules reported in [26] using ANFIS [13]. Figures 8.14 and 8.15 illustrate the actual and predicted trends for the test data set. A trend line is also plotted using a least squares fit (6th order polynomial). Empirical results clearly show that the proposed Web usage-mining framework (*i-Miner*) is efficient.

As shown in Fig. 8.16, Evolutionary FCM approach created 7 data clusters for the average hourly traffic according to the input features compared to 9 data clusters for the daily Web traffic (Fig. 8.17). The previous study using Self-organizing Map (SOM) created 7 data clusters for daily Web traffic and 4 data clusters for hourly Web traffic respectively. Evolutionary FCM approach resulted in the formation of additional data clusters.

Several meaningful information could be obtained from the clustered data. Depending on the no of visitors from a particular domain, time and day of access etc. data clusters were formulated. Clusters based on hourly data show the visitor information at certain hour of the day. Some clusters accounted for the visitors during the peak hour and certain weekdays and so on. For example,

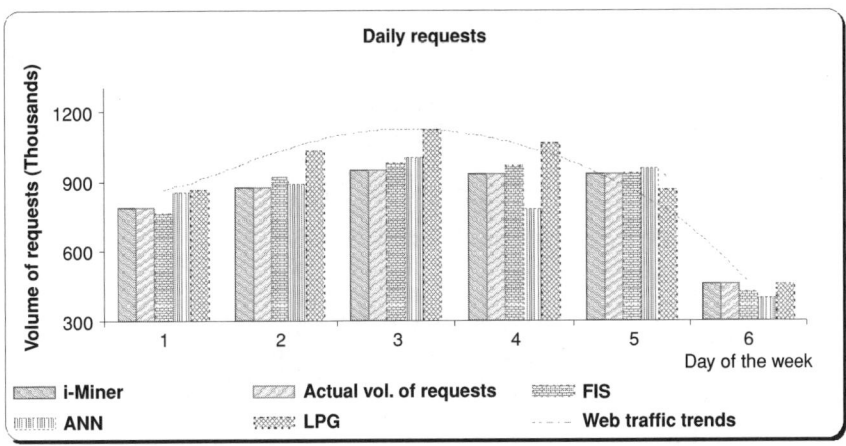

Fig. 8.14. Test results of the daily trends for 6 days

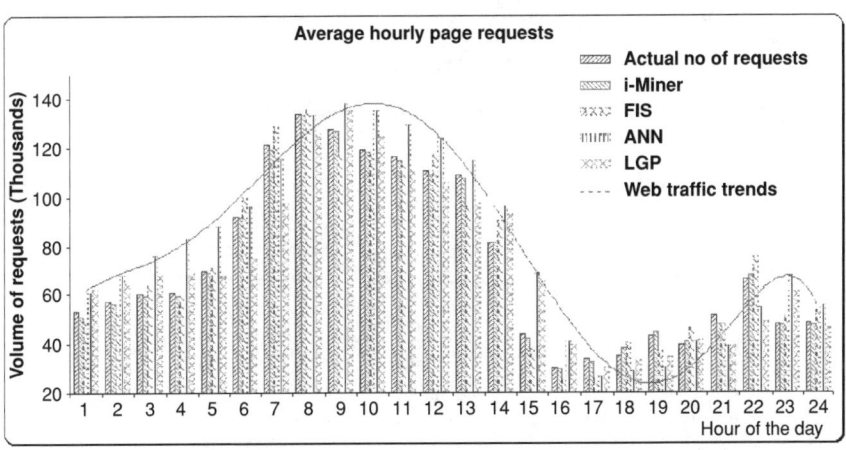

Fig. 8.15. Test results of the average hourly trends for 6 days

Fig. 8.18 depicts the volume of visitors according to domain names from a cluster developed using the evolutionary FCM approach. Detailed discussion on the knowledge discovered from the data clusters is beyond the scope of this chapter.

8.3.6 Hints and Tips to Design *i-Miner*

- Most of the hints and tips given for the design of EvoNF is applicable for *i-Miner*.
- Data pre-processing is an important issue for optimal performance. In most cases, normalization or scaling would suffice.

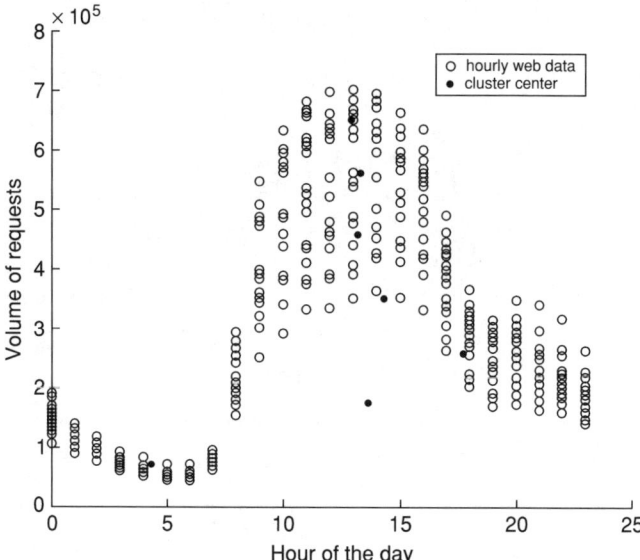

Fig. 8.16. FCM clustering of hourly Web traffic

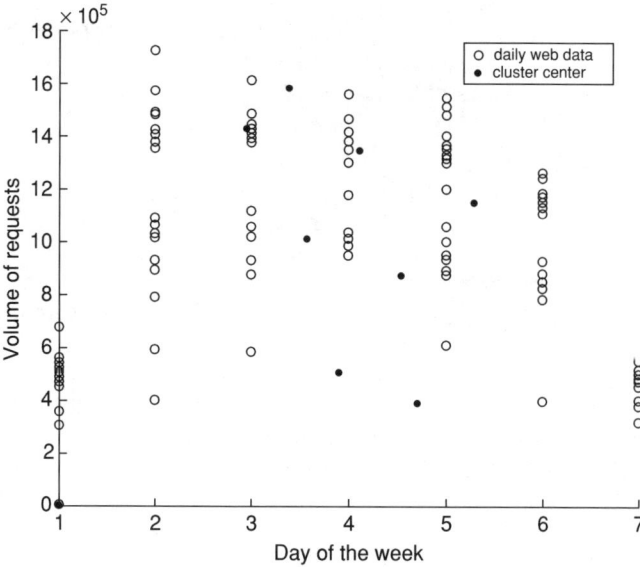

Fig. 8.17. FCM clustering of daily Web traffic

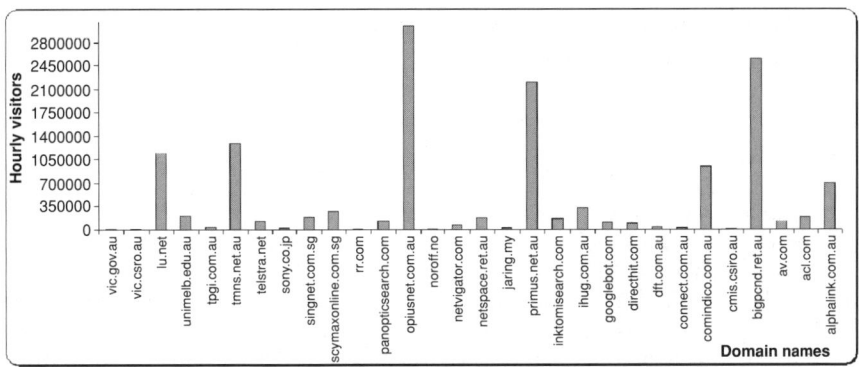

Fig. 8.18. Hourly visitor information according to the domain names from an FCM cluster

8.4 Conclusions

This chapter has presented some of the architectures of hybrid intelligent systems involving fuzzy clustering algorithms, neural network learning, fuzzy inference systems and evolutionary computation. The key idea was to demonstrate the evolution of intelligence in hierarchical layers. The developed hybrid intelligent systems were applied to two real world applications illustrating the importance of such complicated approaches. For the two applications considered, the hybrid models performed better than the individual approaches. We were able to improve the performance (low RMSE and high CC) and at the same time we were able to substantially reduce the number of rules. Hence these approaches might be extremely useful for hardware implementations.

The hybrid intelligent systems has many important practical applications in science, technology, business and commercial. Compared to the individual intelligent constituents hybrid intelligent frameworks are relatively young. As the strengths and weakness of different hybrid architectures are understood, it will be possible to use them more efficiently to solve real world problems. Integration issues range from different techniques and theories of computation to problems of exactly how best to implement hybrid systems. Like most biological systems which can adapt to any environment, adaptable intelligent systems are required to tackle future complex problems involving huge data volume. Most of the existing hybrid soft computing frameworks rely on several user specified network parameters. For the system to be fully adaptable, performance should not be heavily dependant on user-specified parameters.

The real success in modelling the proposed hybrid architectures will directly depend on the genotype representation of the different layers. The population-based collective learning process, self-adaptation, and robustness are some of the key features. Evolutionary algorithms attract considerable computational effort especially for problems involving complexity and huge data volume. Fortunately, evolutionary algorithms work with a population of independent solutions, which makes it easy to distribute the computational load among several processors.

References

1. Abraham A., Business Intelligence from Web Usage Mining, Journal of Information and Knowledge Management (JIKM), World Scientific Publishing Co., Singapore, Volume 2, No. 4, pp. 1-15, 2003.
2. Abraham A., EvoNF: A Framework for Optimization of Fuzzy Inference Systems Using Neural Network Learning and Evolutionary Computation, 2002 IEEE International Symposium on Intelligent Control (ISIC'02), Canada, IEEE Press, pp. 327-332, 2002.
3. Abraham A., i-Miner: A Web Usage Mining Framework Using Hierarchical Intelligent Systems, The IEEE International Conference on Fuzzy Systems FUZZ-IEEE'03, IEEE Press, pp. 1129-1134, 2003.
4. Abraham A., Neuro-Fuzzy Systems: State-of-the-Art Modeling Techniques, Connectionist Models of Neurons, Learning Processes, and Artificial Intelligence, LNCS 2084, Mira J. and Prieto A. (Eds.), Springer-Verlag Germany, pp. 269-276, 2001.
5. Abraham A., Intelligent Systems: Architectures and Perspectives, Recent Advances in Intelligent Paradigms and Applications, Abraham A., Jain L. and Kacprzyk J. (Eds.), Studies in Fuzziness and Soft Computing, Springer Verlag Germany, Chap. 1, pp. 1-35, 2002.
6. Abraham A., Meta-Learning Evolutionary Artificial Neural Networks, Neurocomputing Journal, Elsevier Science, Netherlands, Vol. 56c, pp. 1-38, 2004.
7. Abraham A. and Nath B., Evolutionary Design of Fuzzy Control Systems – An Hybrid Approach, The Sixth International Conference on Control, Automation, Robotics and Vision, (ICARCV 2000), CD-ROM Proceeding, Wang J.L. (Ed.), ISBN 9810434456, Singapore, 2000.
8. Abraham A. and Nath B., Evolutionary Design of Neuro-Fuzzy Systems – A Generic Framework, In Proceedings of The 4-th Japan-Australia Joint Workshop on Intelligent and Evolutionary Systems, Namatame A. et al (Eds.), Japan, pp. 106-113, 2000.
9. Bezdek J.C., Pattern Recognition with Fuzzy Objective Function Algorithms, New York: Plenum Press, 1981.
10. Cordón O., Herrera F., Hoffmann F., and Magdalena L., Genetic Fuzzy Systems: Evolutionary Tuning and Learning of Fuzzy Knowledge Bases, World Scientific Publishing Company, Singapore, 2001.
11. Edwards R., Abraham A. and Petrovic-Lazarevic S., Export Behaviour Modeling Using EvoNF Approach, The International Conference on Computational Science (ICCS 2003), Springer Verlag, Lecture Notes in Computer Science- Volume 2660, Sloot P.M.A. et al (Eds.), pp. 169-178, 2003.
12. Hall L.O., Ozyurt I.B., and Bezdek J.C., Clustering with a Genetically Optimized Approach, IEEE Transactions on Evolutionary Computation, Vol. 3, No. 2, pp. 103-112, 1999.
13. Jang J.S.R., ANFIS: Adaptive-Network-BasedFuzzy Inference System, IEEE Transactions in Systems Man and Cybernetics, Vol. 23, No. 3, pp. 665-685, 1993.

14. Jayalakshmi G.A., Sathiamoorthy S. and Rajaram, An Hybrid Genetic Algorithm – A New Approach to Solve Traveling Salesman Problem, International Journal of Computational Engineering Science, Vol. 2, No. 2, pp. 339-355, 2001.
15. Kandel A. and Langholz G. (Eds.), Hybrid Architectures for Intelligent Systems, CRC Press, 1992.
16. Lotfi A., Learning Fuzzy Inference Systems, PhD Thesis, Department of Electrical and Computer Engineering, University of Queensland, Australia, 1995.
17. Medsker L.R., Hybrid Intelligent Systems, Kluwer Academic Publishers, 1995.
18. Pedrycz W. (Ed.), Fuzzy Evolutionary Computation, Kluwer Academic Publishers, USA, 1997.
19. Procyk T.J. and Mamdani E.H., A Linguistic Self Organising Process Controller, Automatica, Vol. 15, no. 1, pp. 15-30, 1979.
20. Sanchez E., Shibata T. and Zadeh L.A. (Eds.), Genetic Algorithms and Fuzzy Logic Systems: Soft Computing Perspectives, World Scientific Publishing Company, Singapore, 1997.
21. Stepniewski S.W. and Keane A.J., Pruning Back-propagation Neural Networks Using Modern Stochastic Optimization Techniques, Neural Computing & Applications, Vol. 5, pp. 76-98, 1997.
22. Sugeno M. and Tanaka K., Successive Identification of a Fuzzy Model and its Applications to Prediction of a Complex System, Fuzzy Sets Systems, Vol. 42, no. 3, pp. 315-334, 1991.
23. Wang L.X. and Mendel J.M., Backpropagation Fuzzy System as Nonlinear Dynamic System Identifiers, In Proceedings of the First IEEE International conference on Fuzzy Systems, San Diego, USA, pp. 1409-1418, 1992.
24. Wang L.X. and Mendel J.M., Generating Fuzzy Rules by Learning from Examples, IEEE Transactions in Systems Man and Cybernetics, Vol. 22, pp. 1414–1427,1992.
25. Wang L.X., Adaptive Fuzzy Systems and Control, Prentice Hall Inc, USA, 1994.
26. Wang X., Abraham A. and Smith K.A, Soft Computing Paradigms for Web Access Pattern Analysis, Proceedings of the 1st International Conference on Fuzzy Systems and Knowledge Discovery, pp. 631-635, 2002.
27. Zadeh L.A., Roles of Soft Computing and Fuzzy Logic in the Conception, Design and Deployment of Information/Intelligent Systems, Computational Intelligence: Soft Computing and Fuzzy-Neuro Integration with Applications, Kaynak O. et al (Eds.), pp. 1-9, 1998.

9

Evolving Connectionist Systems with Evolutionary Self-Optimisatio

N. Kasabov[1], Z. Chan[1], Q. Song[1], and D. Greer[1]

Knowledge Engineering and Discovery Research Institute
Auckland University of Technology
Private Bag 92006, Auckland 1020, New Zealand
{nkasabov,shchan,qsong,dogreer}@aut.ac.nz

This chapter discusses Evolving Connectionist Systems (ECOS) which are neural network systems that evolve their structure and functionality from incoming data through adaptive, incremental, on-line or off-line, local area learning based on clustering. There are now several models of ECOS, including Evolving Fuzzy Neural Networks (EFuNN), Evolving Classification Systems (ECF), Dynamic Evolving Fuzzy Inference Systems (DENFIS) and Zero Instruction Set Computers (ZISC). These models require self-adjustment of their parameters during their continuous learning in order to react efficiently to changes in the dynamics of the incoming data. In this work, the methods of Evolutionary Computation (EC) are applied for the on-line and off-line optimisation of different ECOS models for different parameters such as learning rate, maximum cluster radius, data normalisation intervals, feature weighting coefficients and feature selection. The methods are illustrated on both time series prediction problems and on classification problems.

9.1 Introduction

Many real-world problems, such as biological data processing, are continuously changing non-linear processes that require fast, adapting non-linear systems capable of following the process dynamics and discovering the rules of these changes. Since the 80's, many Artificial Neural Network (ANN) models such as the popular multilayer perceptrons (MLP) and radial basis network (RBF) have been proposed to capture the process non-linearity that often fail classical linear systems. However, most of these models are designed for batch learning and black-box operation. Only few are capable of performing on-line, incremental learning and/or knowledge discovery along with self-optimising their parameters as they learn incrementally.

Aiming to tackle these weaknesses of common ANN models, a new class of on-line self-evolving networks called Evolving Connectionist Systems (ECOS) was proposed [17, 18, 21, 34]. ECOS are capable of performing the following functions: adaptive learning, incremental learning, lifelong learning, on-line learning, constructivist structural learning that is supported by biological facts, selectivist structural learning and knowledge-based learning. Incremental learning is supported by the creation and the modification of the number and the position of neurons and their connection weights in a local problem space as new data comes. Fuzzy rules can be extracted for knowledge discovery. Derivatives of ECOS, including Evolving Fuzzy Neural Network (EFuNN) [18], Evolving Classification Function (ECF), Evolving Clustering Method for Classification (ECMC) [17], Dynamic Evolving Neural-Fuzzy Inference System (DENFIS) [21] and ZISC [34] have been applied to speech and image recognition, brain signal analysis and modeling dynamic time-series prediction, gene expression clustering and gene regulatory network inference [17, 19].

Evolutionary computation (EC) are robust, global optimisation methods [3, 4, 12, 13, 23, 36] that have been widely applied to various neuro-fuzzy systems for prediction and control [8, 9, 10, 11, 16, 24, 27, 29, 33, 38, 39]. While ECOS models were designed to be self-evolving systems, it is our goal to extend this capability through applying EC to perform various on-line and off-line optimisation of the control parameters. In this work, such example applications are illustrated. EC, in the form of Genetic Algorithms (GA) and Evolutionary Strategies (ES), are applied to the ECOS models of EFuNN, ECF and ECMC. All methods are illustrated on benchmark examples of Mackey-Glass and Iris data [6] to demonstrate the performance enhancement with EC.

The paper is organised as follows. Section 9.2 presents the main principles of ECOS and some of their models – EFuNN, DENFIS, ECF, ECMC, while Sect. 9.3 presents the principle of Evolutionary Computation (EC) and in particular, Evolutionary Strategies (ES) and Genetic Algorithm (GA), which are the two forms of EC employed in this work. Sections 9.4 and 9.5 present the application of ES for on-line and off-line parameter optimisation of EFuNN [7], while Sect. 9.6 presents the application of GA to off-line parameter optimisation of ECF [20, 22]. Section 9.7 applies GA to off-line optimisation of the normalisation ranges of input variables and to feature weighting for the EFuNN and ECMC models [37]. Conclusions and outlook for future research are discussed in Sect. 9.8.

9.2 Principles and Models of Evolving Connectionist Systems (ECOS)

9.2.1 General Principles of ECOS

Evolving connectionist systems (ECOS) are multi-modular, connectionist architectures that facilitate modelling of evolving processes and knowledge discovery. An ECOS may consist of many evolving connectionist modules: it is a neural network that operates continuously in time and adapts its structure and functionality through a continuous interaction with the environment and with other systems according to: (i) a set of parameters that are subject to change during the system operation; (ii) an incoming continuous flow of information with unknown distribution; (iii) a goal (rationale) criteria (also subject to modification) that is applied to optimise the performance of the system over time. ECOS have the following characteristics [17, 18, 21, 34]:

1. They evolve in an open space, not necessarily of fixed dimensions.
2. They are capable of on-line, incremental, fast learning – possibly through one pass of data propagation.
3. They operate in a life-long learning mode.
4. They learn as both individual systems, and as part of an evolutionary population of such systems.
5. They have evolving structures and use constructive learning.
6. They learn locally via local partitioning of the problem space, thus allowing for fast adaptation and process tracing over time.
7. They facilitate different kind of knowledge representation and extraction, mostly – memory-based, statistical and symbolic knowledge.

In this work, we study the application of Evolutionary Computation (EC) to the self-optimisation of the ECOS models of EFuNN, ECF and ECMC. These three ECOS models are briefly described in the following sections.

9.2.2 Evolving Fuzzy Neural Network (EFuNN)

Evolving Fuzzy Neural Networks (EFuNNs) are ECOS models that evolve their nodes (neurons) and connections through supervised incremental learning from input-output data pairs. A simple version of EFuNN is shown in Fig. 9.1 [18]. It has a five-layer structure. The first layer contains the input nodes that represent the input variables; the second layer contains the input fuzzy membership nodes that represent the membership degrees of the input values to each of the defined membership functions; the third layer contains

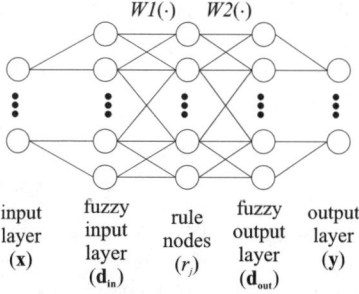

Fig. 9.1. A schematic diagram of EFuNN

the rule nodes that represent cluster centers of samples in the problem space and their associated output function; the fourth layer contains output fuzzy membership nodes that represent the membership degrees to which the output values belong to defined membership functions and finally, the fifth layer contains the output nodes that represent output variables.

EFuNN learns local models from data through clustering of the data into rule nodes and associating a local output function for each cluster. Rule nodes evolve from the input data stream to cluster the data. The first layer connection weights $W1(r_j), j = [1, 2, \ldots, no.\,rule\,nodes]$ represent the co-ordinates of the nodes r_j in the input space and the second layer connection weights $W2(r_j)$ represents the corresponding local models (output functions). Clusters of data are created based on similarity between data samples and existing rule nodes in both the input fuzzy space and the output fuzzy space with the following methods. Samples that have a distance, measured in local normalised fuzzy distance, to an existing cluster center (rule node) r_j of less than a threshold R_{max} in the input space and E_{max} in the output space are allocated to the same cluster and are used to update cluster centre and radius. Let $\mathbf{d_{in}}$ and $\mathbf{d_{out}}$ denote the fuzzified input variable and output variable respectively, Nex_j the number of data associated with r_j, l_1 and l_2 the learning rate for $W1(.)$ and $W2(.)$ respectively. The centres are adjusted according to the following error correction equations:

$$W1\left(r_j^{(t+1)}\right) = W1\left(r_j^{(t)}\right) + \frac{l_1}{Nex_j + 1}\left(\mathbf{d_{in}} - W1\left(r_j^{(t)}\right)\right) \quad (9.1)$$

$$W2\left(r_j^{(t+1)}\right) = W2\left(r_j^{(t)}\right) + \frac{l_2}{Nex_j + 1}\left(\mathbf{d_{out}} - W2\left(r_j^{(t)}\right)\right) \quad (9.2)$$

The learning rates l_1 and l_2 control how fast the connection weights adjust to the dynamics of the data. Samples that do not fit into existing clusters form new clusters as they arrive in time. This procedures form the kernel of EFuNN's incremental learning: that cluster centres are continuously adjusted according to the predefined learning rates according to new data samples, and new clusters are created incrementally.

During learning, EFuNN creates a local output function for each cluster that is represented as the $W2(\cdot)$ connection weights. Each cluster thus represents a local model that can be described by a local fuzzy rule with an antecedent – the cluster area, and a consequent – the output function applied to data in this cluster, for example:

if x_1 is High (0.7) **and** x_2 is Low (0.9) **then**
    y is High (0.8) (* radius of the input cluster 0.3, number of examples in the cluster 13 *)
end if

where High and Low are fuzzy membership functions defined on the range of the variables for x_1, x_2, and y. The number and the type of the membership functions can either be deduced from the data through learning algorithms, or can be predefined based on human knowledge.

EFuNN can be trained either in incremental (on-line) mode, in which data arrive one at a time and local learning is performed upon the arrival of each datum, or in batch (off-line) mode, in which the whole data set is available for global learning. The former follows the local learning procedures described above, whereas the later requires clustering algorithms like k-means clustering and/or Expectation-Maximisation algorithm.

In this work, EC is applied to EFuNN in three main areas: first, for optimising the learning rates l_1 and l_2 in the incremental learning mode (Sect. 9.4); second, for optimising the fuzzy MFs in the batch learning mode (Sect. 9.5) and last, for performing feature weighting and feature selection (Sect. 9.7).

9.2.3 Evolving Classification Function (ECF) and Evolving Clustering Method for Classification (ECMC)

Evolving Classification Function (ECF) and Evolving Clustering Method for Classification (ECMC) (similar to ZISC [4]) are ECOS classifiers that classify a data set into a number of classes in the n-dimensional input space by evolving rule nodes. Each rule node r_j is associated with a class through a label. Its receptive field $R^{(j)}$ covers a part of the n-dimensional space around the rule node. The main difference between ECF (and ZISC) and ECMC is that the former uses spherical receptive field for simplicity while the later uses elliptical receptive field to achieve a smaller number of rule nodes and higher accuracy. A comparison between ECF, ECMC, ZISC and MLP for the classification task of Iris [6] is given in [37].

ECF and the ECMC operations include learning and classification. During learning, data vectors are dealt with one by one with their known classes. Data may be pre-processed by fuzzification with n_{MF} number of Fuzzy membership functions. The learning sequence is described as the following steps:

1. If all training data vectors have been entered into the system, complete the learning phase; otherwise, enter a new input vector from the data set.
2. Find all existing rule nodes with the same class as the input vector's class.
3. If there is not such a rule node, create a new rule node and go back to step (1). The position of the new rule node is the same as the current input vector in the input space, and the radius of its influence field is set to the minimum radius R_{min}.
4. For those rule nodes that are in the same class as the input vector's class, if the input vector lies within node's influence field, update both the node position and influence field. Suppose that the field has a radius of $R^{(j)}$ and the distance between the rule node and the input vector is d; the increased radius is $R^{(j),new} = (R^{(j)} + d)/2$ and the rule node moves to a new position situated on the line connecting the input vector and the rule node.

5. For those rule nodes that are in the same class as the input vector's class, if the input vector lies outside of the node's influence, increase this field if possible. If the new field does not include any input vectors from the data set that belong to a different class, the increment is successful, the rule node changes its position and the field increases. Otherwise, the increment fails and both rule node and its field do not change. If the field increment is successful or the input vector lies within an associated field, go to step (1); otherwise, a new node is created according to step (3) and go to step (1).

The learning procedure takes only one iteration (epoch) to retain all input vectors at their positions in the input/output space. The classification of new input vectors is performed in the following way:

1. The new input vector is entered and the distance between it and all rule nodes is calculated. If the new input vector lies within the field of one or more rule nodes associated with one class, the vector belongs to this class.
2. If the input vector does not lie within any field, the vector will belong to the class associated with the closest rule node. However, if no rules are activated, ECF calculates the distance between the new vector and the M closest rule nodes out of a total of N rule nodes. The average distance is calculated between the new vector and the rule nodes of each class. The vector is assigned the class corresponding to the smallest average distance. This averaging procedure is the so-called M-of-N prediction or classification.

In this work EC is applied to ECF for optimising the main control parameters of ECF, which include R_{min}, R_{max}, M-of-N and n_{MF} (the number of fuzzy membership functions for fuzzifying input data) (Sect. 9.6) and to ECMC for feature weighting (Sect. 9.7).

9.3 Principles of Evolutionary Computation

In this work, Evolutionary Computation (EC) [4, 12, 13] is applied to implement self-optimisation of ECOS. EC represents a general class of global optimisation algorithms that imitate the evolutionary process explained by Darwinian Natural Selection. Its models include Evolutionary Strategies (ES) proposed by Schwefel et al. in the mid 1960s, Genetic Algorithms (GA) proposed by Holland in 1975, and Genetic Programming (GP) by Koza in 1992 [4]. EC searches with multiple search points and requires only the function evaluation of each point. It is therefore robust and convenient for optimising complex, problems that lack derivative information and contain multi-modality, which are the features that defeat classical, derivative-based search methods. However, the large number of function evaluations causes EC to be computationally intensive. Thus although the earliest form of EC has existed since

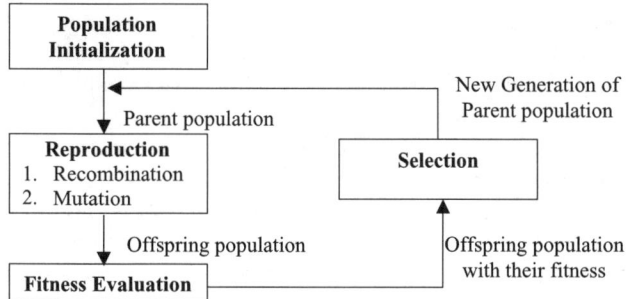

Fig. 9.2. An algorithmic description of EC

the mid 1960s, EC received research attention only after the availability of high-speed computers in the last two decade.

Here we present a brief description of Genetic Algorithm (GA) and Evolutionary Strategies (ES), which are the two forms of EC employed in this work. Both GA and ES adopt the same algorithmic structure shown in Fig. 9.2 that resembles the natural evolutionary process, searching with a set of solutions called the *population*. In canonical GA, each solution (called an *individual*) is expressed in a binary code sequence called *string* or *chromosome* and each bit of the string is called a *gene*, following the biological counterparts. For real number solutions, each individual can be encoded through the standard binary code or Gray's code. In ES, each individual is the real number solution itself. No coding is required. Each iteration, called a *generation*, consists of three phases: reproduction, evaluation and selection.

In the reproduction phase, the current population called the *parent* population is processed by a set of *evolutionary operators* to create a new population called the *offspring* population. The evolutionary operators include two main operators: *mutation* and *recombination*, both imitate the functions of their biological counterparts. Mutation causes independent perturbation to a parent to form an offspring and is used for diversifying the search. It is an asexual operator because it involves only one parent. In GA, mutation flips each binary bit of an offspring string at a small, independent probability p_m (which is typically in the range [0.001, 0.01] [2]). In ES, mutation is the addition of a zero-mean Gaussian random number with covariance Σ to a parent individual to create the offspring. Let $\mathbf{s_{PA}}$ and $\mathbf{s_{OF}}$ denote the parent and offspring vector, they are related through the Gaussian mutation

$$\mathbf{s_{OF}} = \mathbf{s_{PA}} + \mathbf{z} \qquad \mathbf{z} \sim N(0, \Sigma) \qquad (9.3)$$

where $N(\mathbf{a}, \mathbf{B})$ represents a Normal distribution with mean $\mathbf{a}$ and covariance $\mathbf{B}$ and "$\sim$" denotes sampling from the corresponding distribution.

Obviously, the mutation parameters p_m and Σ control the diversity of the search: the larger they are, the more diversified the search becomes. Recombination mixes the features of two or more parents to form an offspring and is used for accelerating convergence. It is a *sexual* operator because it involves contribution from multiple parents in the reproduction process. In GA, the recombination operator is called the *crossover* operator, which exchanges bits at selected bit positions between two parents to generate two offspring (usually one offspring is discarded is reduce statistical dependence in the offspring population). The earliest form of crossover is called k-point crossover, which upon being triggered with the crossover probability p_c, swaps k segments between two parents at k different crossover points on the strings. *Uniform crossover* is latter developed to reduce the induced dependencies between neighboring bits, which upon being triggered with the crossover probability p_c, swaps bits at every bit position with a uniform probability of 0.5. As the performance of GA is less sensitive to the crossover probability p_c than to the mutation rate p_m, p_c is always set to much higher value than p_m, typically in the range [0.5, 0.9]. In ES, recombination performs linear combination of the two parents to produce the offspring. It is not always implemented because ES uses the mutation as the main search operator. The evolutionary operators are critical to the success of the search because they determine the distribution of the new population in the search space.

In the evaluation phase, each member of the offspring population if evaluated for its objective function value, which is its "fitness" for survival. Individuals with favorable objective function values (for example, low objective value in a minimisation problem and high objective value in a maximisation problem) are said to be high in fitness. They are given higher survival probability in the next phase – the selection phase.

In the selection phase, survival of the fittest is simulated. The high fitness individuals are selected from the offspring population or from the joint pool of the offspring and parent population to become the next generation parents. The unselected individuals are abandoned. The selection operator is probabilistic in GA and deterministic in ES. The former uses *Roulette Wheel* selection that assigns each individual a selection probability equal to its fitness divided by the sum of all individuals' fitness. The later uses the so called (μ, λ) that selects the best μ out of λ offspring, or the $(\mu + \lambda)$ that selects the best μ out of the joint pool of μ parents and λ offspring. Many heuristic designs, like the Rank-based selection that assigns to the individuals a survival probability proportional (or exponentially proportional) to their ranking, have also been studied. The selected individuals then become the new generation of parents for reproduction. The entire evolutionary process iterates until some stopping criteria is met. The process is essentially a Markov Chain, i.e. the outcome of one generation depends only on the last. It has been shown [5, 35] that under certain design criteria of the evolutionary operators and selection operator, the average fitness of the population increases and the probability of discovering the global optimum tends towards unity. The search could however, be lengthy.

9.3.1 Integration of Evolving Connectionist System and Evolutionary Computation

Due to EC's robustness and ease of use, it has been applied to many areas of ANN training; in particular, network weights and architecture optimisation [10, 11, 39], regularisation [8], feature selection [14] and feature weighting [15]. It is the goal of this work to extend these applications to ECOS. As ECOS, unlike conventional ANNs, are capable f fast, incremental learning and adapting their network architectures to new data, there is greater flexibility in EC's application area, of which two are identified here:

1. ECOS's fast learning cycle reduces the computation time of fitness evaluation, which in turn speeds up the overall EC search. Thus EC can afford using larger population and longer evolutionary time to achieve consistent convergence.
2. ECOS's capability of incremental learning raises the opportunity of applying on-line EC to optimise the dynamic control parameters.

In this work, such applications of EC to ECOS are reviewed and summarised from the corresponding references. They represent our first effort to develop a completely self-optimised ECOS that require minimal manual parameter adjustments via EC. These applications are:

1. On-line optimisation of the learning rates in EFuNN [7] (Sect. 9.4)
2. Off-line optimisation of the fuzzy membership functions in EFuNN [7] (Sect. 9.5)
3. Off-line control parameter optimisation in ECF [20] (Sect. 9.6)
4. Off-line feature weighting in EFuNN and ECMC and feature selection [37] (Sect. 9.7)

9.4 ES for On-line Optimisation of the Learning Rates in EFuNN

This section presents the first EC application to ECOS discussed in this work: on-line EC is applied to tune the learning rates l_1 and l_2 in (9.1) and (9.2) of EFuNN during incremental training with the objective of minimising the prediction error over the last n_{last} data. The learning rates are sensitive to the dynamics of the data as they control how fast the model adapts to the data; they are therefore time-varying parameters that require on-line adjustment. For rapidly changing data, learning rates should take higher value to obtain faster model response, and vice-versa.

While EC is an intrinsically computation-intensive algorithm for off-line problems, there are also many recent reports on its application to on-line problems such as controller tuning and optimal planning [9, 16, 24, 27, 29, 33, 38]. In these example references, on-line-EC are characterised by small population size (6-50) and high selective pressure (use of elitist selection) and low selection ratio, which aims to minimise computational intensity and to accelerate convergence (at the cost of less exploration of the search space) respectively. Their applications involve optimising only a small number of parameters (2-10), which are often the high-level tuning parameters of a model. For applications involving larger number of parameters, parallel EC can be used to accelerate the process. Due to the fast learning cycle of ECOS and the relatively small data set used in our experiment, our on-line EC suffices in terms of computational speed using small population size and high selection pressure.

The on-line EC is implemented with Evolutionary Strategies (ES). Each individual $\mathbf{s}^{(k)} = (s_1^{(k)}, s_2^{(k)})$ is a 2-vector solution to l_1 and l_2. Since there are only two parameters to optimise, ES requires only a small population and a short evolutionary time. In this case, we use $\mu = 1$ parent and $\lambda = 10$ offspring and a small number of generations of $gen_{max} = 20$. We set the initial values of $\mathbf{s}$ to $(0.2, 0.2)$ in the first run and use the previous best solution as the initial values for subsequent runs. Using a non-randomised initial population encourages a more localised optimisation and hence speeds up convergence. We use simple Gaussian mutation with standard deviation $\sigma = 0.1$ (empirically determined) for both parameters to generate new points. For selection, we use the high selection pressure $(\mu + \lambda)$ scheme to accelerate convergence, which picks the best μ of the joint pool of μ parents and λ offspring to be the next generation parents.

The fitness function (or the optimisation objective function) is the prediction error over the last n_{last} data, generated by using EFuNN model at $(t - n_{last})$ to perform incremental learning and predicting over the last n_{last} data using the EFuNN model at $(t - n_{last})$. The smaller n_{last}, the faster the learning rates adapts and vice-versa. Since the effect of changing the learning rates is usually not expressed immediately but after a longer period, the fitness function can be noisy and inaccurate if n_{last} is too small. In this work we set $n_{last} = 50$. The overall algorithm is as follows:

1. *Population Initialization.* Reset generation counter gen. Initialize a population of μ parents with the previous best solution (l'_1, l'_2)

$$\mathbf{s}_{PA}^{(k)} = (l'_1, l'_2), \mathbf{k} = \{1, 2, \ldots \mu\}$$

2. *Reproduction.* Randomly select one of the μ parents, $\mathbf{s}_{PA}^{(r)}$, to undergo Gaussian mutation to produce a new offspring

$$\mathbf{s}_{OF}^{(i)} = \mathbf{s}_{PA}^{(r)} + \mathbf{z}^{(i)} \quad where \quad \mathbf{z}^{(i)} \sim N(0, \sigma^2 \mathbf{I}), i = \{1, 2, \ldots \lambda\}$$

3. *Fitness Evaluation.* Apply each of the λ offspring to the EFuNN model at $(t - n_{last})$ to perform incremental learning and prediction using data in $[t - n_{last}, t]$. Set the respective prediction error as fitness.

4. *Selection.* Perform $(\mu + \lambda)$ selection

5. *Termination.* Increment gen. Stop if $gen \geq gen_{max}$ or if no fitness improvement has been recorded over 3 generations; otherwise go to step (2).

9.4.1 Testing On-Line EC-EFuNN on Mackey Glass Series

To verify the effectiveness of the proposed on-line ES, we first train EFuNN with the first 500 data of the Mackey-Glass series (using $x(0) = 1.2$ and $\tau = 17$) to obtain a stable model, and then apply on-line ES to optimise the learning rates over the next 500 data. The corresponding prediction error is recorded. Example results are shown in Fig. 9.3 and Fig. 9.4.

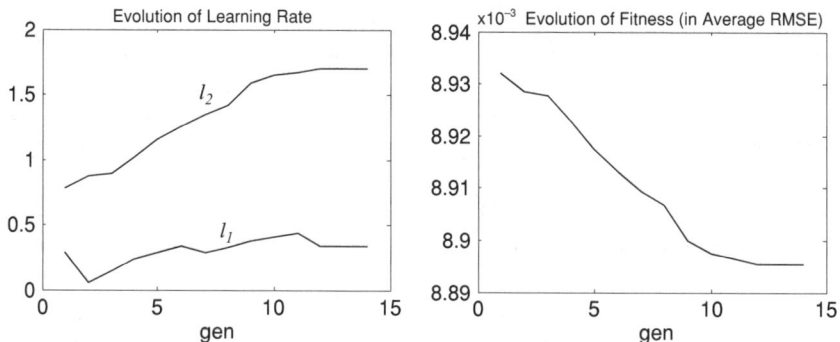

Fig. 9.3. Evolution of the best fitness and the learning rates over 15 generations

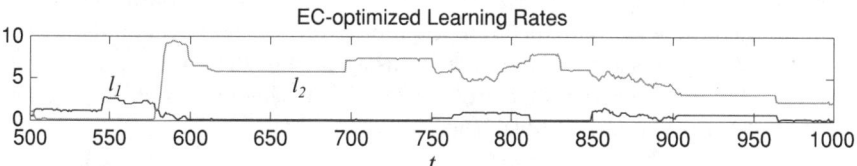

Fig. 9.4. Optimised learning rates l_1 and l_2 over the period $t = [500, 1000]$

Figure 9.3 shows an example of the evolution of the least RMSE (used as the fitness function) and the learning rates. The RMSE decreases unidirectionally, which is characteristics of $(\mu + \lambda)$ selection because the best individual is always kept. The optimal learning rates are achieved quickly after 14 generations.

Figure 9.4 shows the dynamics of the learning rate over the period $t = [500, 1000]$. Both learning rates l_1 and l_2 vary considerably over the entire course with only short stationary moments, showing that they are indeed dynamic parameters. The average RMSE for on-line prediction one step ahead obtained with and without on-line EC are 0.0056 and 0.0068 respectively, showing that on-line EC is effective in enhancing EFuNN's prediction performance during incremental learning.

9.5 ES for Off-line Optimisation of the Fuzzy Membership Functions of EFuNN

Most reported applications of EC are off-line problems, e.g. [8, 32, 39], mainly because of EC's requirement of intense computation for the population search. Here, we apply an off-line ES to optimise the fuzzy input and output membership functions (MFs) at the second and fourth layers respectively during batch learning with the objective of minimising the training error. The approach is similar to that in [33]. Relative to the learning rates, MFs are static control parameters that remain constant over time and hence require less frequent updating. For both the input and output MFs, we use the common triangular function, which is completely defined by the position of the MF center.

Given that there are p input variables and p_{MF} fuzzy quantisation levels for each input variables, m output variables and m_{MF} fuzzy quantisation levels for each output variable, there are $n_c = (p \times p_{MF} + m \times m_{MF})$ centers $\mathbf{c} = [c_1, c_2, \ldots, c_{n_c}]$ to be optimised, which exceeds that (only two) in the on-line case. Thus naturally, the off-line ES requires a larger population size and longer evolution period. This is not problematic as the learning is performed off-line and has lesser time constraint.

The off-line ES represents each individual as an $(p \times p_{MF} + m \times m_{MF})$ real vector solution to the positions of the MFs. We use $\mu = 5$ parent and $\lambda = 20$ offspring and a relatively larger number of generations of $gen_{max} = 40$. Each individual of the initial population is a copy of the evenly distributed MFs within the boundaries of the variables. Standard Gaussian mutation is used for reproduction with $\sigma = 0.1$ (empirically determined). Every offspring is checked for the membership hierarchy constraint, i.e. the value of the higher MF must be larger than that of the lower MF. If the constraint is violated, the individual is resampled until a valid one is found.

1. *Population Initialization.* Reset generation counter gen. Initialize a population of μ parents $[c'_1, c'_2, \ldots, c'_{n_c}]$ with the evenly distributed MFs
$$\mathbf{s}_{PA}^{(k)} = [c'_1, c'_2, \ldots, c'_{n_c}], \quad k = \{1, 2, \ldots, \mu\}$$

2. *Reproduction.* Randomly select one of the μ parents, $\mathbf{s}_{PA}^{(r)}$, to undergo Gaussian mutation to produce a new offspring
$$\mathbf{s}_{OF}^{(i)} = \mathbf{s}_{PA}^{(r)} + \mathbf{z}^{(i)} \text{ and } \mathbf{z}^{(i)} \sim N(0, \sigma^2 \mathbf{I}) \quad for \quad i = \{1, 2, \ldots, \lambda\}$$
Resample if the membership hierarchy constraint is violated.

3. *Fitness Evaluation.* Apply each of the λ offspring to the EFuNN model at $(t - n_{last})$ to perform incremental learning and prediction using data in $[t - n_{last}, t]$. Set the respective prediction error as fitness.

4. *Selection.* Perform $(\mu + \lambda)$ selection

5. *Termination.* Increment gen. Stop if $gen \geq gen_{max}$, otherwise go to step 2.

9.5.1 Testing Off-line EC-EFuNN on Mackey-Glass Series

The proposed off-line EC is tested on the same Mackey-Glass Series described above. We first perform incremental training on EFuNN with the first 1000 data of the Mackey-Glass series to obtain a stable model, and then apply off-line ES during batch learning (over the same 1000 data) to optimise the input and output MFs. Example results are shown in Fig. 9.3

Figure 9.3(a) shows the evolution of the best fitness recorded in each generation. The uni-directional drop in prediction error shows that the optimisation of MFs has a positive impact on improving model performance.

Figure 9.3(b) shows the initial MFs and the EC-optimised MFs and Fig. 9.5(c) shows the frequency distribution of the fourth input variable $\{x_4^{(t)}\}$. Clearly, the off-line ES evolves the MFs towards the high frequency positions, which maximises the precision for fuzzy quantisation and in turn yields higher prediction accuracies. The training RMSEs are 0.1129 and 0.1160 for EFuNN with and without off-line ES respectively, showing that the ES-optimised fuzzy MFs are effective in improving EFuNN's data tracking performance.

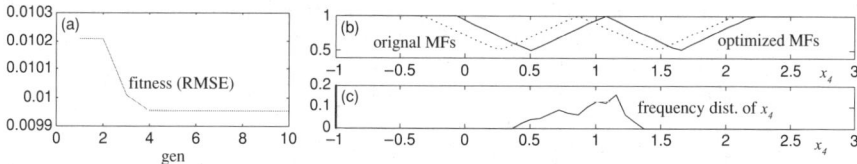

Fig. 9.5. (a) The evolution of the best fitness from the off-line ES. (b) Initial membership functions and EC-optimised membership functions, (c) Frequency distribution of the first input variable

9.6 A for Parameter Optimisation of ECF

ECF can be applied in both on-line and off-line modes. When working in an off-line mode, ECF requires accurate setting of several control parameters to achieve optimal performance. However, as with other ANN models, it is not always clear what the best values for these parameters are. EC provides a robust global optimisation method for choosing values for these parameters. The use of Genetic Algorithm (GA) for the optimisation the parameters of ECF are described in [20] in conjunction with biomedical data. In this section, the same method is illustrated with the Iris data set [6].

A brief description of the operation and the parameters of ECF are found earlier in Sect. 9.2.3. In this work, ECF is applied to optimise the following four parameters:

R_{max}: the maximum radius of the receptive hypersphere of the rule nodes. If, during the training process, a rule node is adjusted such that the radius of its hyper-sphere becomes larger than R_{max} then the rule node is left unadjusted and a new rule node is created.

R_{min}: the minimum radius of the receptive hypersphere of the rule nodes. It becomes the radius of the hypersphere of a new rule node.

n_{MF}: the number of membership functions used to fuzzify each input variable.

M-of-N: if no rules are activated, ECF calculates the distance between the new vector and the M closest rule nodes. The average distance is calculated between the new vector and the rule nodes of each class. The vector is assigned the class corresponding to the smallest average distance.

9.6.1 Testing GA-ECF on the Iris data

Each parameter being optimised by the GA is encoded through standard binary coding into a specific number of bits and is decoded into a predefined range through linear normalisation. A summary of this binary coding information is shown in Table 9.1. Each individual string is the concatenation of a set of binary parameters, yielding a total string-length $l_{tot} = (5+5+3+3) = 16$.

Table 9.1. The number of bits used by GA to represent each parameter and the range into which each bit string is decoded

Parameter	Range	Resolution	Number of Bits
R_{max}	0.01–0.1	2.81e-3	5
R_{min}	0.11–0.8	2.16e-2	5
n_{MF}	1–8	1	3
M-of-N	1–8	1	3

The experiments for GA are run using a population size of 10 for 10 generations (empirically we find that 10 generations are sufficient for GA convergence, see Fig. 9.4 and Fig. 9.5 shown later). For each individual solution, the initial parameters are randomised within the predefined range. Mutation rate p_m is set to $1/l_{tot}$, which is the generally accepted optimal rate for unimodal functions and the lower bound for multi-modal functions, yielding an average of one bit inversion per string [1, 2]. Two-point crossover is used. Rank-based selection is employed and an exponentially higher probability of survival is assigned to high fitness individuals. The fitness function is determined by the classification accuracy. In the control experiments, ECF is performed using the manually optimised parameters, which are $R_{max} = 1$, $R_{min} = 0.01$, $n_{MF} = 1$, M-of-N = 3. The experiments for GA and the control are repeated 50 times and between each run the whole data set is randomly split such that 50% of the data was used for training and 50% for testing. Performance is determined as the percentage of correctly classified test data. The statistics of the experiments are presented in Table 9.2.

The results show that there is a marked improvement in the average accuracy in the GA-optimised network. Also the standard deviation of the

Table 9.2. The results of the GA experiment repeated 50 times and averaged and contrasted with a control experiment

	Average accuracy (%)	Standard Dev.
Evolving Parameters	97.96	1.13
Control	93.68	2.39

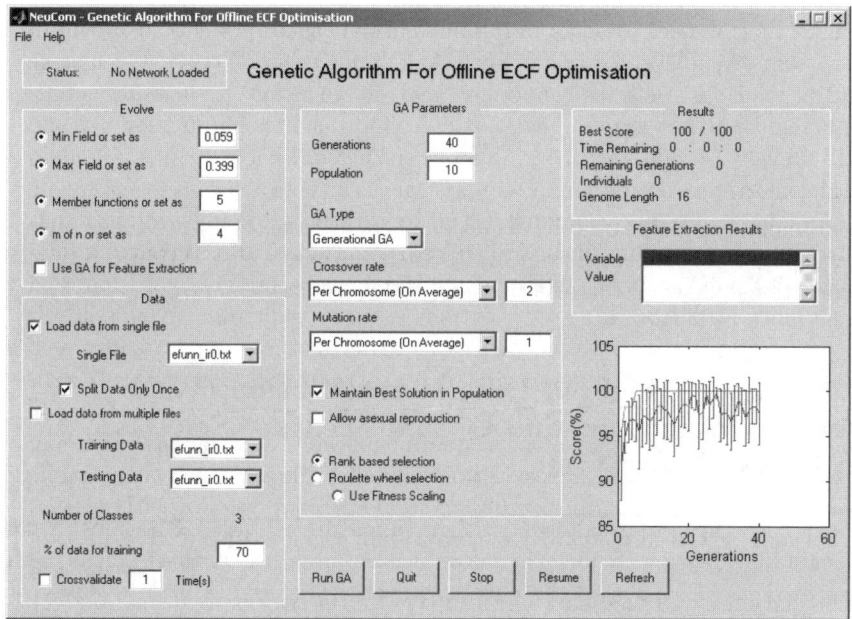

Fig. 9.6. Neucom module that implements GA for the optimisation of ECF

accuracy achieved by the 50 GA experiments is significantly less than that for the control experiments, indicating that there is a greater consistency in the experiments using the GA.

Figure 9.4 shows the Neucom[1] module developed to implement the experiment. In this case, the GA runs for 40 generations. The mean, standard deviation, and the best fitness found at each generation are recorded. Note that within the first ten generations, the average fitness increases most steadily and that the best-fit individual also emerges. For this reason, only 10 generations are used for the GA experiments.

---

[1] Neucom is a Neuro Computing environment developed by the Knowledge Engineering and Discovery Research Institute (www.kedri.info) at Auckland University of Technology. Software is available for download from the Web sites www.kedri.info, or www.theneucom.com

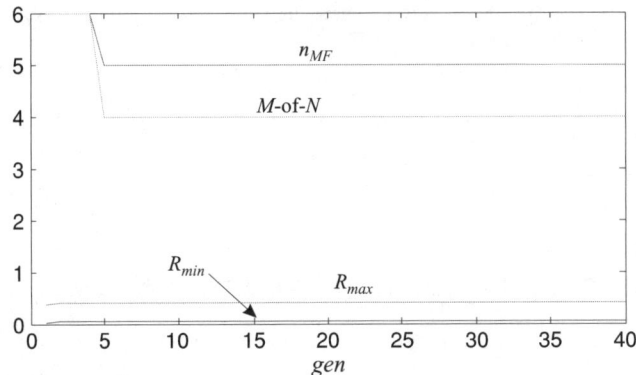

Fig. 9.7. Evolution of the parameters

Finally, Fig. 9.7 shows the evolution of the parameters over 40 generations. Each parameter converges quickly to their optimal value within the first 10 generations, showing the effectiveness of the GA implementation.

9.7 GA for Weighted Data Normalisation (WDN), Feature Weighting, and Optimal Feature Selection in EFuNN and ECMC

Feature weighting is an alternative form of feature selection. It assigns to each variable a weighting coefficient that reduces or amplifies its effect on the model based on its relevance to the output. A variable that has little relevance to the output is given a small weight to suppress its effect, and vice-versa. The purpose of feature weighting is twofold: first, to protect the model from the random and perhaps detrimental influence of irrelevant variables and second, to act as a guide for pruning away irrelevant variables (feature selection) by the size of the weighting coefficients. Feature weighting/selection are generally implemented through three classes of methods: Bayesian methods [26, 28], incremental/sequential methods [25, 31] or stochastic methods [14, 30]. The first two classes are local methods that are fast and yet susceptible to local optima; the last class includes Evolutionary Computation (EC) applications that use computationally intensive population search to search for the global optimum. Here the algorithm proposed for ECOS, called Weighted Data Normalisation (WDN), belongs to such class, using the robust optimisation capability of Genetic Algorithm (GA) to implement feature weighting.

The WDN method described here is introduced in [6]. It optimises the normalisation intervals (range) of the input variables and allocates weights to each of the variables from a data set. The method consists of the following steps:

1. The training data is preprocessed first by a general normalisation method. There are several ways to achieve this: (a) normalising a given data set so that they fall in a certain interval, e.g. $[0,1]$, $[0,255]$ or $[-1,1]$ etc; (b) normalising the data set so that the inputs and targets will have means of zero and standard deviations of 1; (c) normalising the data set so that the deviation of each variable from its mean normalised by its standard deviation. In the WDN, we normalise the data set in the interval $[0, 1]$.
2. The weights of the input variables $[x_1, x_2, \ldots, x_n]$ represented respectively by $[w_1, w_2, \ldots, w_n]$ with initial values of $[1, 1, \ldots, 1]$, form a chromosome for a consecutive GA application. The weight w_i of the variable x_i defines its new normalisation interval $[0, w_i]$.
3. GA is run on a population of connectionist learning modules for different chromosome values over several generations. As a fitness function, the Root Mean Square Error (RMSE) of a trained connectionist module on the training or validation data is used, or alternatively the number of the created rule nodes can be used as a fitness function that needs to be minimised. The GA runs over generations of populations and standard operations are applied such as binary encoding of the genes (weights); Roulette Wheel selection criterion; multi-point crossover operation for crossover.
4. The connectionist model with the least error is selected as the best one, and its chromosome – the vector of weights $[w_1, w_2, \ldots, w_n]$ defines the optimum normalisation range for the input variables.
5. Variables with small weights are removed from the feature set and the steps from above are repeated again to find the optimum and the minimum set of variables for a particular problem and a particular connectionist model.

The above WDN method is illustrated in the next section on two case study ECOS and on two typical problems, namely EFuNN for a time series prediction, and ECMC for classification.

9.7.1 Off-line WDN of EFuNN for Prediction [6]

In the present paper, EFuNN is applied to the time series prediction. Improved learning with the WDN method is demonstrated on the Mackey-Glass (MG) time series prediction task [33]. In the experiments, 1000 data points, from $t = 118$ to 1117, were extracted for predicting the 6-steps-ahead output value. The first half of the data set was taken as the training data and the rest as the testing data.

The following parameters are set in the experiments for the EFuNN model: $R_{max} = 0.15$; $E_{max} = 0.15$ and $n_{MF} = 3$. The following GA parameter values are used: for each input variable, the values from 0.16 to 1 are mapped onto 4 bit string; the number of individuals in a population is 12; mutation rate is 0.001; termination criterion (the maximum epochs of GA operation) is 100 generations; the RMSE on the training data is used as a fitness function. The optimsed weight values, the number of the rule nodes created by EFuNN with

Table 9.3. Comparison between EFuNN without WDN and EFuNN with WDN

	Normalisation Weights	Number of Rule Nodes	Training RMSE	Testing RMSE
EFuNN without WDN	1, 1, 1, 1	87	0.053	0.035
EFuNN with WDN	0.4, 0.8 0.28 0.28	77	0.05	0.031

such weights, the training and testing RMSE and the control experiments are shown in Table 9.3.

With the use of the WDN method, better prediction results are obtained for a significantly less number of rule nodes (clusters) evolved in the EFuNN models. This is because of the better clustering achieved when different variables are weighted accurately acccording to their relevance.

9.7.2 Off-line WDN Optimisation and Feature Extraction for ECMC [6]

In this section, the ECMC with WDN is applied to the Iris data for both classification and feature weighting/selection. All experiments in this section are repeated 50 times with the same parameters and the results are averaged. 50% of the whole data set is randomly selected as the training data and the rest as the testing data. The following parameters are set in the experiments for the ECMC model: $R_{min} = 0.02$; each of the weights for the four normalised input variables is a value from 0.1 to 1, and is mapped into a 6-bit binary string.

The following GA parameters are used: number of individuals in a population 12; mutation rate $p_m = 0.005$; termination criterion (the maximum epochs of GA operation) 50; fitness function is determined by the number of created rule nodes.

The final weight values, the number of rule nodes created by ECMC and the number of classification errors on the testing data, as well as the control experiment are shown in the first two rows of Table 9.4 respectively Results show that the weight of the first variable is much smaller than the

Table 9.4. Comparison between ECMC without WDN and ECMC with WDN

	Normalised Feature Weights	Number of Rule Nodes	Number of Test Errors
4 Inputs without WN	1, 1, 1, 1	9.8	3.8
4 Inputs with WN	0.25, 0.44 0.73 1	7.8	3.1
3 Inputs without WN	1, 1, 1	8.8	3.7
3 Inputs with WN	0.50, 0.92, 1	8.1	2.9
2 Inputs without WN	1,1	7.7	3.2
2 Inputs with WN	1, 0.97	7.4	3.1

weights of the other variables. Now using the weights as a guide to prune away the least relevant input variables, the same experiment is repeated without the first input variable. As shown in the subsequent rows of Table 9.4, this pruning operation slightly reduces test errors. However, if another variable is removed (i.e. the total number of input variables is 2) test error increases. So we conclude that for this particular application the optimum number of input variables is 3.

9.8 Conclusions and Outlook

In this work, we summarised our current efforts on applying EC to build self-optimising ECOS. Two forms of EC, namely GA and ES, have been applied to on-line and off-line parameter optimisation and feature weighting of ECOS and have shown to be effective in enhancing ECOS' performance. The proposed methods could lead to the development of fully autonomous, self-organised and self-optimised systems that learn in a life long learning mode from different sources of information and improve their performance over time regardless of the incoming data distribution and the changes in the data dynamics.

It must however, be emphasised that the listed applications of EC to ECOS are by no means exhaustive. There are still important areas in ECOS unexplored, e.g. clustering of data, aggregation of rule nodes and adjustment of node radii. Presently, new ECOS methods are being developed that are tailored to integrate with EC for optimising the control parameters as well as implementing feature selection/weighting simultaneously.

9.9 Acknowledgements

The research presented in the paper is funded by the New Zealand Foundation for Research, Science and Technology under grant NERF/AUTX02-01. Software and data sets used in this chapter, along with some prototype demo systems, can be found at the Web site of the Knowledge Engineering and Discovery Research Institute, KEDRI – www.kedri.info.

References

1. T. Baeck, D. B. Fogel, and Z. Michalewicz. *Evolutionary Computataion II. Advanced algorithm and operators*, volume 2. Institute of Physics Pub., Bristol, 2000.
2. T. Baeck, D. B. Fogel, and Z. Michalewicz. *Evolutionary Computation I. Basic algorithm and operators*, volume 1. Institute of Physics Publishing, Bristol, 2000.

3. T. Baeck, U. Hammel, and H-P Schwefel. Evolutionary computation: Comments on the history and current state. *IEEE Transactions on Evolutionary Computation*, 1(1):3–17, Apr 1997.
4. Thomas Baeck. *Evolutionary algorithm in theory and practice: evolution strategies, evolutionary programming, genetic algorithms*. Oxford University Press, New York, 1995.
5. H. G. Beyer. *The Theory of Evolution Strategies*. Natural Computing Series. Springer, Heidelberg, 2001.
6. C. Blake and C. Merc. Repository of machine learning databases, 1998.
7. Z. Chan and N. Kasabov. Evolutionary computation for on-line and off-line parameter tuning of evolving fuzzy neural networks. In *International Conference on Neuro Information Processing (submitted)*, Vancouver, 2003.
8. Z. S. H. Chan, H. W. Ngan, A. B. Rad, and T. K. Ho. Alleviating "overfitting" via genetically-regularised neural network. *Electronics Letter*, 2002.
9. W. G. da Silva, P. P. Acarnley, and J. W. Finch. Application of genetic algorithms to the online tuning of electric drive speed controllers. *IEEE Transactions on Industrial Electronics*, 47(1):217–219, Feb 2000.
10. D. B. Fogel. Evolving neural networks. *Biol. Cybern.*, 63:487–493, 1990.
11. D. B. Fogel. Using evolutionary programming to create neural networks that are capable of playing tic-tac-toe. In *International Joint Conference on Neural Networks*, pages 875–880, New York, 1993. IEEE Press.
12. D. E. Goldberg. *Genetic Algorithms in Search, Optimization and machine Learning*. Addison-Wesley, Reading, MA, 1989.
13. J. H. Holland. *Adaptation in natural and artificial systems*. The University of Michigan Press, Ann Arbor, MI, 1975.
14. I. Inza, P. Larranaga, and B. Sierra. *Estimation of Distribution Algorithms. A new tool for Evolutionary Computation*. Kluwer Academic Publishers, 2001.
15. I. Inza, P. Larranaga, and B. Sierra. Estimation of distribution algorithms for feature subset selection in large dimensionality domains. In H. Abbass, R. Sarker, and C. Newton, editors, *Data Mining: A Heuristic Approach*, pages 97–116. IDEA Group Publishing, 2001.
16. C. F. Juang. A tsk-type recurrent fuzzy network for dynamic systems processing by neural network and genetic algorithms. *IEEE Transactions on Fuzzy Systems*, 10(2):155–170, Apr 2002.
17. N. Kasabov. *Evolving connectionist systems – methods and applications in bioinformatics, brain study and intelligent machines*. Springer Verlag, London-New York, 2002.
18. N. Kasabov. Evolving fuzzy neural networks for on-line supervised/unsupervised, knowledge-based learning. *IEEE Trans. SMC – part B, Cybernetics*, 31(6):902–918, December 2001.
19. N. Kasabov and D. Dimitrov. A method for gene regulatory network modelling with the use of evolving connectionist systems. In *ICONIP' 2002 – International Conference on Neuro-Information Processing*, Singapore, 2002. IEEE Press.
20. N. Kasabov and Q. Song. Ga-parameter optimisation of evolving connectionist systems for classification and a case study from bioinformatics. In *ICONIP'2002 – International Conference on Neuro-Information Processing*, Singapore. IEEE Press, 2002.

21. N. Kasabov and Q. Song. Denfis: Dynamic, evolving neural-fuzzy inference systems and its application for time-series prediction. *IEEE Trans. on Fuzzy Systems*, 10(2):144–154, April 2002.
22. N. Kasabov, Q. Song, and I. Nishikawa. Evolutionary computation for parameter optimisation of on-line evolving connectionist systems for prediction of time series with changing dynamics. In *Int. Joint Conf. on Neural Networks IJCNN'2003*, USA, 2003.
23. J. R. Koza. *Genetic Programming*. MIT Press, 1992.
24. D. A. Linkens and H. O. Nyongesa. Genetic algorithms for fuzzy control.2. online system development and application. *IEE Proceedings of Control Theory and Applications*, 142(3):177–185, May 1995.
25. H. Liu and R. Setiono. Incremental feature selection. *Applied Intelligence*, 9(3):217–230, 1998.
26. J. David C. MacKay. Bayesian methods for neural networks: Theory and applications. Technical Report Neural Networks Summer School lecture notes, Cavendish Laboratory, Cambridge, 2001.
27. S. Mishra, P. K. Dash, P. K. Hota, and M. Tripathy. Genetically optimized neuro-fuzzy ipfc for damping modal oscillations of power system. *IEEE transactions on Power Systems*, 17(4):1140–1147, Nov 2002.
28. Radford M. Neal. *Bayesian Learning for Neural Networks*. Lecture Notes in Statistics. Springer-Verlag New York, Inc., 1996.
29. I. K. Nikolos, K. P. Valavanis, N. C. Tsourveloudis, and A. N. Kostaras. Evolutionary algorithm based offline/online path planner for uav navigation. *Transactions on Systems, Man, and Cybernetics – Part B: Cybernetics*, page 1, 2003.
30. R. Palaniappan, P Raveendran, and S. Omatu. Vep optimal channel selection using genetic algoithm for neural network classification of alcoholics. *IEEE Transactions on Neural Networks*, 13(2):486–491, March 2002.
31. S. Perkins, K. Lacker, and J. Theiler. Grafting: Fast, incremental feature selection by gradient descent in function space. *Journal of Machine Learning Research*, 3:1333–1356, 2003.
32. D. Quagliarella. *Genetic algorithms and evolution strategy in engineering and computer science : recent advances and industrial applications*. John Wiley and Sons, New York, 1998.
33. A. Rajapakse, K. Furuta, and S. Kondo. Evolutionary learning of fuzzy logic controllers and their adaptation through perpetual evolution. *IEEE Transactions on Fuzzy Systems*, 10(3):309–321, Jun 2002.
34. Silicon Recognition. Zisc78 neural network silicon chip hardware reference manual. Technical report, May 20th 2003.
35. G. Rudolph. *Convergence Properties of Evolutionary Algorithms*. Kovac, Hambury, 1997.
36. H. P. Schwefel. *Numerische Optimierung von Computer-Modellen mittels der Evolutionsstrategie*, volume 26 of *Interdisciplinary Systems Research*. Birkhauser, Basel, Germany, 1977.
37. Q. Song and N. Kasabov. Weighted data normalization and feature selection for evolving connectionist sstems. In *International Conference on Neuro Information Processing (submitted)*, 2003.

38. J. Tippayachai, W. Ongsakul, and I. Ngamroo. Parallel micro genetic algorithm for constrained economic dispathc. *IEEE Transactions on Power Systems*, 17(3):790–797, Aug 2002.
39. Xin Yao and Yong Liu. A new evolutionary system for evolving artificial neural networks. *IEEE Transactions on Neural Networks*, 8(3):694–713, May 1997.

Part II

From Applications To Methods

10

Monitoring

J. Strackeljan

TU Clausthal, Institute of Applied Mechanics
38678 Clausthal-Zellerfeld, Graupenstr.3 Germany,
jens.strackeljan@tu-clausthal.de

Part 2 of the book is entitled "From applications to methods" as a global heading. The purpose is to describe applications from various fields of industry, rather than special methods, as Part 1. A common feature of all these applications is the requirement for selecting the best possible solution on the basis of an existing problem situation and the resulting problem definition. The examples described here should provide concrete indications or criteria for the selection of an appropriate method in compliance with a "best practice" guideline.

The present chapter is dedicated to the problem of monitoring plants, systems, and machines, as well as quality control methods. This combination is expedient, since the techniques applied to problems in these two categories are very often identical. In both cases, deviations with respect to a condition designated as "good" must be continuously recognised and indicated. The question of whether the problem involves a malfunction in a machine or a defect in a manufactured product is only of subordinate importance for the present purpose. In a manufacturing plant, a machine which does not operate properly is frequently the cause of poor product quality. For instance, wear and tear on a tool, such as a milling cutter on a milling machine, can directly impair the surface quality of a machined component; consequently, the condition of the tool must be monitored. The extent to which the rotor of an industrial centrifuge is contaminated must also be determined, in order to achieve the best possible separation and thus high quality of the final product. Thus, the two problem fields, "monitoring" and "quality control" are closely related. The range of possible applications for technical monitoring functions is practically unlimited and extends to all fields of industry. Hence, it is clear from the beginning that the examples presented here constitute only a very small selection. Methods of this kind are also suited for medical applications, for instance, monitoring of patients in intensive care wards.

10.1 Introduction

The use of advanced pattern recognition systems in assuming an objective perspective in statements concerning the state of technical systems has gained increasing importance. Thus the economy of highly automated and cost intensive machines can be guaranteed only upon the high availability of these machines. The use of advanced, high-performance monitoring and diagnosis systems can make a significant contribution to this. Certain processes can be carried out safely for man and the environment only by means of reliably operating machines, particularly in fields where safety and environmental aspects play an important role.

In the automatic control of technical systems, supervisory functions serve to indicate undesired or non-permitted machine or process states and to take appropriate actions in order to maintain operation and to avoid damage or accident. The following basic functions can be distinguished:

1. Monitoring
   Measurable variables are checked with regard to tolerances, and alarms are generated for the operator.
2. Automatic protection
   In the case of a dangerous process state, the monitoring system automatically initiates an appropriate counteraction.
3. Monitoring with fault diagnosis
   Based on measured variables, features are determined and a fault diagnosis is performed; in advanced systems, decisions are made for counteractions.

The advantage of the classical level-based monitoring (1 and 2) is simplicity, but these are only capable of reacting to a relatively large-scale change in a given feature. If the early detection of small faults and a fault diagnosis are desired, advanced methods based on Fuzzy Technology, Neural Networks or combinations could be used.

A general problem which is very often cited as an argument against the application of adaptive methods in monitoring is the risk that the system may gradually and unnoticeably adapt to undetected changes. As a result, the system may fail to detect real faults, if these faults likewise develop slowly. Problems could occur if the transition from state A to C proceed over a long time (Fig. 10.1), because the classifier adaptation could follow the transition in small steps. Consequently, adaptive monitoring systems have hardly become established in fields which are especially critical, such as safety-relevant applications.

The second limitation arises from problems to generate a learning data set for the classifier design. In most cases only few or sometimes even no real measurement data are available to represent information about all possible machine or process states. Consequently the simulation of fault classes becomes more important. This chapter will emphasise main aspects regarding

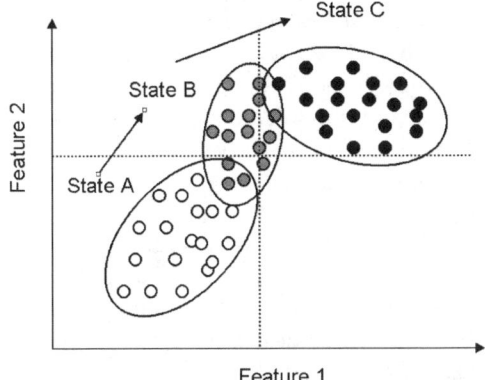

Fig. 10.1. Transition of machinery condition from a healthy state A to a fault state C

the integration of simulation results to improve adaptive behavoir of monitoring systems.

10.2 Components of a Monitoring and Quality Control System

For allowing an unambiguous definition of terms, the essential components of a technical monitoring system are briefly described. The condition of a plant (chlorine plant of a chemical company), of a machine (chlorine pump), or of a component (roller bearing of the pump) is monitored by means of appropriate sensor technology.

1. Sensor technology
   For this purpose, the measurement of machine vibrations, for instance, has proved to be a diagnostic approach very well suited for determining variations in machine behaviour. However, the temperature, pressure, or other process parameters, as well as oil analysis with respect to the number of suspended particles, etc., can also be employed for monitoring.
2. Analogue data pre-processing
   The signals from the sensors must frequently be conditioned by offset elimination, filtering, etc.
3. A/D conversion
   As a rule, a sensor generates an analogue voltage or current signal, which is then converted to a digital signal for further use.
4. Feature calculation and/or feature extraction
   In the next step, status parameters (features) must be determined from the digital signal or from the available individual values, which are unambiguously correlated with the monitoring task as far as possible. In the

simplest case, an integral effective value of the vibratory acceleration can be calculated from the time signal of a machine vibration. The transformation to the frequency range is more elaborate, but also more expedient, since a multitude of spectral components are then available as features. Other advanced transformations, such as wavelets, have also been successfully applied in machine monitoring.

5. Feature selection
   As a rule, the features which are best suited for accomplishing the task concerned must be selected from those available from the calculations. If limitations on sensor technology prevent the generation of more than a very small number of features (less than 5), the selection of features is not necessary [21, 22].

6. State Identification/Classification
   The function of the classifier is to perform an automatic appraisal of the status on the basis of the selected features. In the simplest case, a current characteristic value is compared with a predefined limiting value. On the other hand, the classification can also be performed on the basis of learning systems (neural networks).

7. Adaptation to new or altered situations
   An advanced system should be capable of recognising new, unknown situations for which no trained classifier is available, and of independently deriving appropriate reactions.

As seen before the computer based monitoring consists of different steps to come from the raw process data to the final inference about the state of the machine or process. Figure 10.2 illustrate the essential steps. While a human expert or operator is only able to monitor a limited number of measured variables a computer could solve this task easily but even an advanced software cannot make reliable inference from a large number of measurements without a vast amount of history data to base the inference on. This growing need of data due to the growing number of measured data is the well known phenomenon called curse of dimensionality. It is the major reason for reducing the number of input variable by integration feature extraction and feature selection methods in the monitoring process [14].

10.3 Smart Adaptive Monitoring Systems

The concept of "smart, adaptive monitoring system" must first be defined. Furthermore, if a system is to be considered "smart" and "adaptive", the requirements imposed on such a system must be specified.

10.3.1 Generell Properties of Smart Monitoring Systems

In the present part, the following four properties are employed as necessary criteria for a smart system [3]:

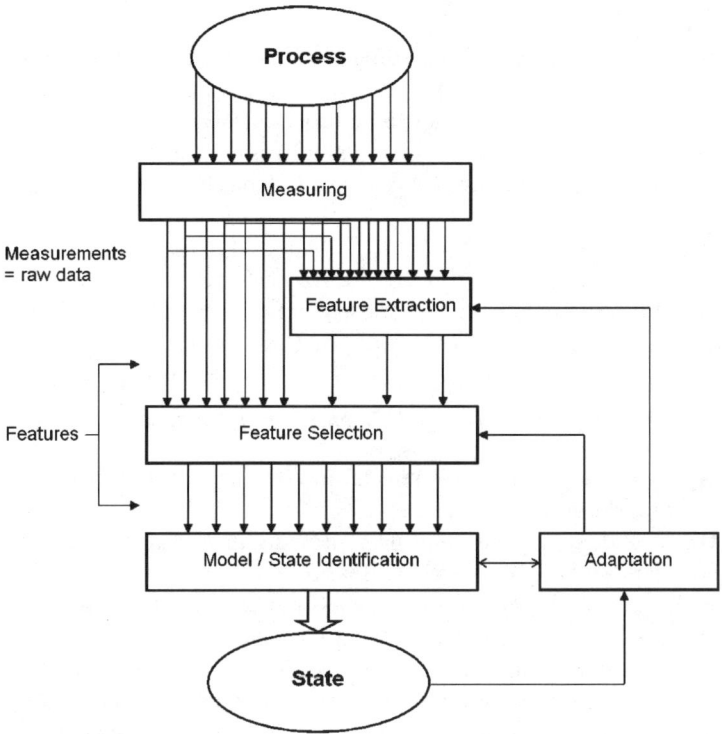

Fig. 10.2. Process of monitoring and state identification

1. Adapting
   Ability to modify the system behaviour to fit environment, new locations and process changes. The aspect is identified as the most important feature for a smart adaptive monitoring systems.
2. Sensing
   Ability to acquire informations from the world around and response to it with consistent behavior. Chemcial and nuclear power plant monotoring are large scale sensing systems. They accept input from hundreds of sensors to regulate temperature, presssure and power throughout the plant. In spite of the fact that different sensors exist more and better sensors remain a continous need. Laser vibrometer which allows a precise measurement of displacement and velocity, fiber optic sensors and micro-electronic sensors should be mentoined as examples of promising new sensore technologies. Along with a proliferation of sensors comes a greatly increased need to combine, or fuse data from multiple sensors for more effective monitoring. Need range from processing of multiple data streams from a simple array of identical senosrs to data from sensors based entirely different physical phenomena operating asynchonously at vastly different rates.

3. Inferring

Ability to solve problems using embedded knowledge and draw conclusions. These expert knowledge is in general a combination of theoretical understanding and a collection of heuristic problem solving rules that experience has shown to be effective. A smart monitoring system should be able to detect sensor faults to prevent the use of nonsensical input values for a classification.

4. Learning

Ability to learn from experience to improve the system performance. Learning and adaptive behaviour are closely combined and without the implementation of learning strategies adapting will work.

10.3.2 Adapting, What does It Mean for a Monitoring System?

The property of "adapting" is of special interest here and is therefore considered in greater detail.

At this point, it may be helpful to provide an example of monitoring in which no adaptive capability exists. The roller bearings of a rotating industrial centrifuge must be monitored for possible damage. For this purpose, the vibration signals are frequently recorded by means of an acceleration sensor. If the value exceeds a limit which has been manually preset on the basis of experience or as specified in standards and guidelines, an alarm is triggered. The alarm threshold can be adjusted, of course, but its value then remains constant for the monitoring phase. A system of this kind, which is still employed for most monitoring applications in industry today, is not adaptive and certainly not smart. Which prerequisite for this designation is lacking in such a system? A change in vibratory behaviour can be due to a wide variety of factors which do not result from a fault in the machine itself. The sensor output may drift in the course of operation. Emissions from other machines, such as electromagnetic fields or superimposed mechanical vibrations, may cause interference in the system. Minor conversion work may have been performed on the machine, for instance, in the course of maintenance and servicing; such measures can cause a change in behaviour without resulting in a malfunction. Furthermore, the product being processed in the centrifuge can also affect the behaviour as a result of a variation in composition or in operation, for instance, an increase in throughput rate, without causing a defect in the machine. The preset limiting value will perhaps yield acceptable results for clearly defined process conditions and a sufficiently short monitoring period. However, it certainly provides no guarantee for operation without false alarms.

10 Monitoring

The following levels of adaptability are conceivable in conjunction with monitoring systems.

1. Level 1
   An adaptive monitoring system is capable of recognising variations in the surrounding and process conditions. Modifications, such as the adaptation of limiting values, can be performed automatically by the system itself. The previously mentioned monitoring system for roller bearings remains unaltered on a machine.
2. Level 2
   An adaptive monitoring system can be transferred from one machine to another without the need of readjustment by an expert. Any necessary adjustment work should be reducible to an absolute minimum. However, the monitoring task itself should not be altered in this connection. That is, monitoring of a roller bearing is still the specified task, although the type of roller bearing and the parameters are different.
3. Level 3
   An adaptive monitoring system can be employed for other monitoring tasks without the need of altering the basic structure. Necessary limiting values or control parameters of the classification algorithm are, to a large extent, specified independently. At this third level, the monitored object itself can also be varied. The system which had previously been employed for detecting damage to roller bearings should now be employed for recognising unbalance in the rotor or for monitoring the process. For allowing the system to function at this level, learning strategies are implemented, rather than pre-programmed algorithms for calculating problem-specific features, such as the effective value of the vibratory acceleration.

In its present status, the technology usually does not even attain the first of the levels just defined. This situation may at first be surprising, and perhaps also somewhat disappointing, but is easy to understand from an engineering standpoint. The decisive external parameters which cause variations in the monitoring parameters are highly diversified, and the mutual interactions among them are often unknown; consequently, a consideration of these parameters in a diagnostic model is difficult or impossible. Once trained, a system is capable of performing a monitoring task as long as the prerequisites for the training status are satisfied. If these conditions change, however, problems will occur in the monitoring system, and the system must be retrained. For example, a defective roller bearing generates a typical fault signal. The results of experimental investigations indicate that the level of vibration generated by a bearing with identical damage can vary by a factor as high as 100 with different machines. The main reason for this variation is the difference in conditions of installation and paths for structure-borne sound with various machines. Consequently, all approaches which depend on limiting values must be excluded, since these limits likewise fluctuate by a factor as high as 100. Thus, they cannot be applied for attaining level 2, that is, transferability to other machines. However, methods of this kind are typical of industrial monitoring to the present day and are applied in a wide variety of fields. So it is known that certain electric motors will display higher vibration when operating under low loads, than when they are operating under high load. Yet in the traditional methods of vibration analysis these variations are not effectively taken into account, except perhaps in a qualitative manner. Slight variations in vibration results from one survey to the next are assumed to be due to "process conditions", and regarded as not significant. If we were able to collect relevant quantitative data regarding the "process conditions" existing at the time that the vibration data was collected, and correct the vibration data for these conditions, then our diagnostic capability would become far more accurate and adaptive [9]. The aforementioned level 3 currently exists only as a vision in the minds of researchers and is further considered only in perspective here.

10.3.3 Needs and Acceptance of Adaptive Monitoring Systems

If monitoring problems are already being solved sufficiently well today without adaptive behaviour, a logical question is why there is a demand for smart systems at all and whether this demand can be justified?

For this purpose, the type of monitoring which is currently applied in industry must first be considered. From a methodical standpoint, this kind of monitoring no longer satisfies all of the requirements which must be imposed on a modern monitoring system. The concept of preventive maintenance implies the application of techniques for the early detection of faults and thus the implementation of appropriate maintenance measures in due time. Maintenance of a machine should not depend merely on a specified operating time or distance, or on a specified time interval. This revolution in maintenance strategy has already occurred in the automotive field during the past five years. The notion that an automobile must be inspected for the first time after 15,000 km or one year, whichever comes first, after the original purchase, is obsolete. As far as monitoring of machines in industry is concerned, however, a change of this kind has hardly taken place at all in practice. The potential for increasing the operating time of machines by decreasing the frequency of failures and for entirely avoiding unplanned down time is evident here. However, the acceptance of such monitoring methods is possible only if the probability of false alarms can be kept extremely low. Changes in the process or in the machine which do not result from a fault must be distinguished from those which do result from a fault. Precisely this adaptivity at level 1 still presents serious problems for many systems, however. If a given system functions correctly, weeks or even months are often necessary for transferring this property to similar machines. Support for the expert in charge by the system itself is severely limited or completely absent. The requirement for human experts and the associated labour costs still severely restrict the acceptance of monitoring systems. On the other hand, precisely this situation offers a special opportunity for those who develop and supply monitoring systems, since an adaptive system becomes independent of particular applications and can thus provide specialised solutions at acceptable prices. For controlling the quality of products during flexible manufacturing, the systems for monitoring the machines and products must also be flexible. In this case, flexibility can be equated to adaptivity.

In this connection, a further reason for the hesitation to apply advanced monitoring systems should be mentioned. In this case, however, the problem is not so much of a technical nature. During the past 15 years, the allegedly "ultimate" system has repeatedly been offered to industry as a solution to monitoring problems. Many managers invested money and were then disappointed because the programs did not perform as promised. At present, genuine innovations would be feasible with the application of computational intelligence methods, but the willingness to invest again has decreased markedly.

10.4 Learning Procedures for Smart Adaptive Monitoring Systems

Besides adaptivity, learning ability is a decisive factor for a smart technical system. To a certain extent, these two properties are inseparably related, since the previously mentioned concepts of leaning anew or relearning for achieving adaptive behaviour are not possible without the implementation of learning algorithms. As described in Chap. 6, the two fundamentally distinct approaches of "supervised" and "unsupervised" learning also apply to monitoring systems.

The following three learning objectives are essential for a monitoring system:

1. Number of fault classes and quality states
   Especially in the case of machine monitoring, faults are often the result of a gradual transition from a good to a poor condition, rather than instantenous occurrence. Consequently, the number of status classes to be selected is not clearly evident. For classical monitoring with limiting values, only two classes are employed. The transition then occurs very quickly, and the operator's indicating lamp suddenly changes from green to red. However, one would expect an entirely different kind of information behaviour from a smart system. A possible solution is the introduction of additional, intermediate states. Thus, an alternation class can improve the transition between "good" and "poor". Another possibility is the use of fuzzy (diffuse) classifiers, which allow the appraisal of a condition not only for a status class, but also for gradual differentiations (fuzzy classification). In this case too, however, the number of fuzzy classes to be defined must first be specified. If one prefers to apply an algorithm for controlling the performance of this task, rather than doing it oneself, the use of "Kohonen

feature maps" is advisable. The appropriate number of class parameters can be determined automatically with the use of these maps. A change in the number of classes is also conceivable, if new error classes occur. An adaptive system must allow such an extension.

2. Limiting values and classification parameters

For dealing with these parameters, training is certainly the most important task for monitoring systems. Manual entry and management of several hundred limiting values is not practicable; however, the need for such a large number of values is quite normal with the use of many features. Of course, a system which is capable of learning and which has recognised that new classes are necessary for the monitoring task can also be trained to employ the associated classifiers in a second step.

3. Features

An adaptive system should recognise the condition that the previously applied features no longer suffice for solving a monitoring problem with sufficient reliability. Hence, a further learning objective is the calculation – or at least a new selection – of new features which are better suited for the purpose. For achieving this objective, the application of automatic selection methods is necessary. The selection have to consider constraints of the classification algorithm. In this sense only so called wrapper approaches for the feature selection process will find suitable feature combinations.

One of the major problems is the manual setting of alarm levels. As two identical machines will not run or wear out in the same time the levels need to be adjusted during the machine lifetime. There are serval approaches to calculate the levels automatically based on historical data concerning outlier elimination 9. Even the use of a moving average with adjustable window length and different methods of exponential weighting of data in the window will provide very sufficient results. Comparing the two bold printed solid lines in Fig. 10.3 it can be seen that the levels set by the algorithm are approximately the same as those set manually. Events where the vibration level was higher than the current alarm level are identified in both cases. The markers indicate measurement points where the vibration is below or above the level.

The performance of a monitoring system can be improved if a broader frequency range of the input signal is used (Fig. 10.4). By using a reference mask an alert is activated if an amplitude of the power spectra is rising over the predefines reference line. For many rotating machinery the consideration of amplitude and phase is necessary. The modification of centre and radius of alert circles allows adaptation to changes in process and machinery state. The ability of setting limits and defining alert circles automatically could be seen as a minimum requirement to name a monitoring system adaptive.

For these methods the existence of real measurements as a learning data set is necessary. Independently from these data we could use the expert knowledge which consists of expert information on the problem area in the form of facts and rules. A monitoring system should simulate the conclusions made by a

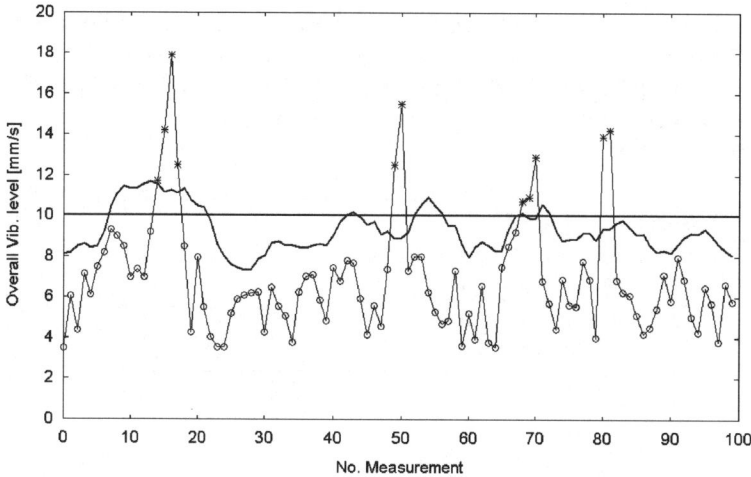

Fig. 10.3. Trendline showing manually set alarm levels and adaptive levels (*bold lines*) and the real overall vibration measurement data of a rotating machinery

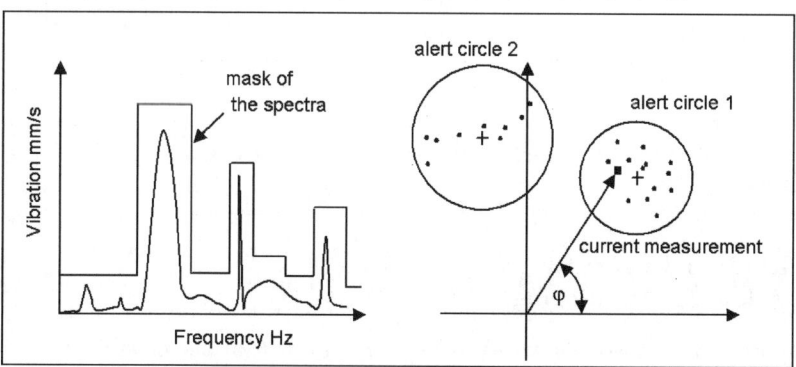

Fig. 10.4. Masc of a power spectra and alert circles if more than one feature is required

human expert in the case of relationships difficult to define using algorithms. For this purpose, it is necessary to record as much information as possible as rules in the form:

if premise is fulfilled then conclusion is valid.

Figure 10.5 shows, with reference to the detection of cracks in a rotating shaft, a typical rule which can be regarded as state of the art. Curves represent the relative values obtained by means of first (f), second (2 × f) and third (3 × f) running speed components and the RMS value when the measurements are performed using shaft displacement. The relationship between the structure and the amplitudes of a vibration signal and the cause of the defect in the technical object can scarcely be described completely and sufficiently in the form of rules. Classification procedures that arrive at a decision on the basis of pre-assigned, thus learned examples are frequently superior here. In practice, knowledge involves uncertainty factors that only allow for approximating inferences.

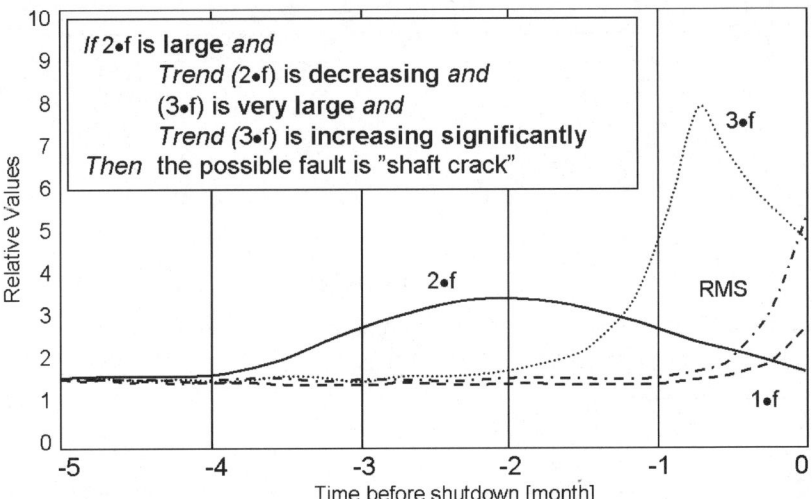

Fig. 10.5. Trends in some characteristic values for the detection of cracks in rotating shafts

10.4.1 Signal Processing Techniques for Adaptive Monitoring Sytems

Without going in details it has to be pointed out that the classical technique to represent time signals in the frequency domain by calculating a normal Fourier transformation (FFT) is an inadequate technique if adaptive behavoir as to be considered. The main disadvantage of the FFT is the lack to distinguish time variant effects. Transient malfunctions or the analysis of non-stationary effects requires other transformations than the ordinary FFT. One approach to improve the resolution in time is the Short Time Fourier Transformation (STFT). But with shorting the time window for each FFT the time resolution could be improved but that implies a lower accuracy in the frequency domain (Fig. 10.6). While the FFT has no temporal resolution in STFT the resolution is fixed for the whole time-frequency plane. The Wavelet analysis (WA) is now able to adapt both the time and frequency resolution. WA can provide a high-frequency resolution at low-frequencies and maintain a good time localization at high-frequency end.

This, coupled with the incapacity of FFT to detect non-stationary signals, makes WA analysis an alternative for machine fault diagnosis. WA provides multi-resolution in time-frequency distribution for easier detection of

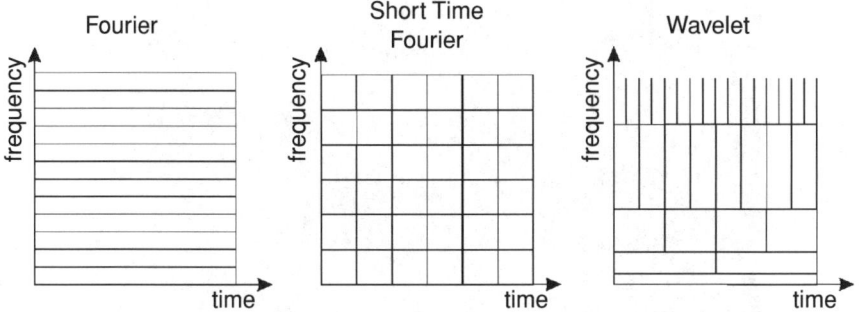

Fig. 10.6. Comparison of FFFT, STFT and WA in time-frequency plane [14]

abnormal vibration signals. Recently extensive research has applied WA to analysis of mechanical vibration signals. In [25] you could find a good summery about the effectiveness of WA for rolling element bearing diagnosis and the authors in [8] have used WA in processing non-stationary signals for fault diagnosis in industrial machines.

10.5 Applications

You will find a lot of research papers describing new developments in the field of condition monitoring using advanced techniques. This applications were selected because the author was involved in the design of the monitoring concept and its realization.

10.5.1 Roller Bearing Diagnostic

As a practical application of the diagnostic system, we describe here the early detection of defects on roller bearings. The initial stage of bearing damage is characterized by dislodging of very small material particles from the running surface, by the appearance of fatigue cracks near the surface, or by rupture or bursting of bearing components (Fig. 10.7). The geometrical alteration of the race is revealed by the generation of a short shock when the roller passages over the site of damage. This shock propagates in the form of an approximately spherical wave front from the site of origin and excites nearly all bearing components capable of vibration to characteristic vibrations in the corresponding frequency range.

For the detection of bearing faults the system calculates 24 features from the vibration signal. The implemented classifier calculates the membership values and provides a complete appraisal of the current state of the bearing at intervals of 1 min. For the convenience of the operating personnel, the membership values are determined only for the four global classes, "shutdown", "intact state", "change in condition", and "fault". The shut-down

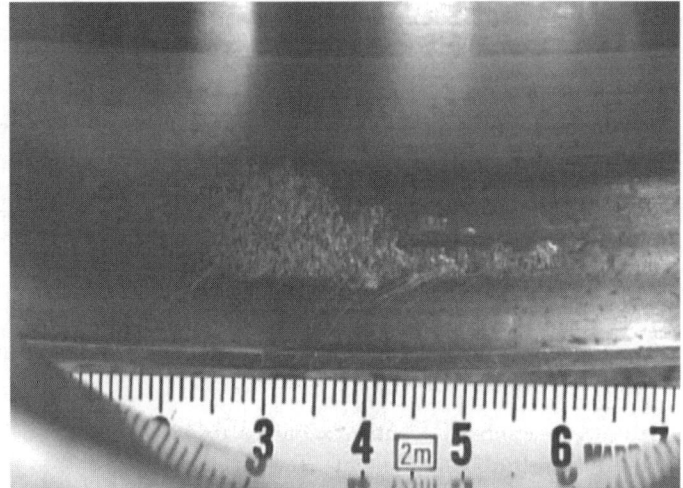

Fig. 10.7. Typical bearing fault in the outer race of a roller bearing

class describes a situation status where the machine does not operate. In Fig. 10.8 the evolution of the membership values is plotted over a period of 19 hours. The membership m_1 for class Ω_1 is plotted in the upper section of Fig. 10.8. This membership is determined in a fuzzy manner, but only the values 0 and 1 are employed for indicating membership in the "'shut-down"' class. The lower section of Fig. 10.8 shows the class membership values for the classes Ω_1 to Ω_4.

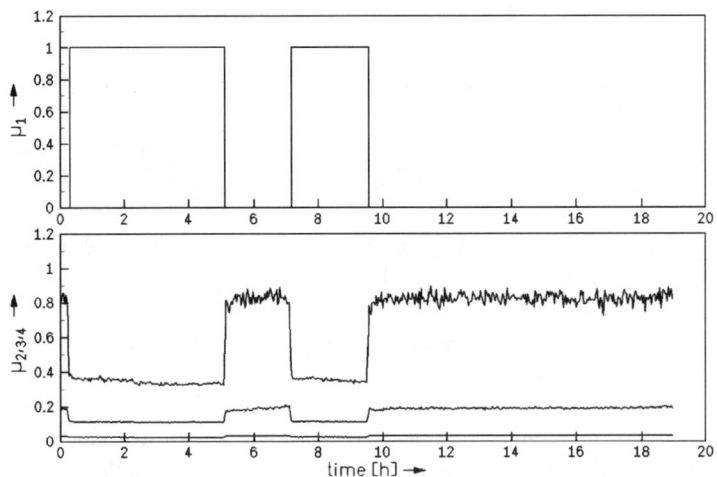

Fig. 10.8. Membership values for the classes Ω_1 to Ω_4

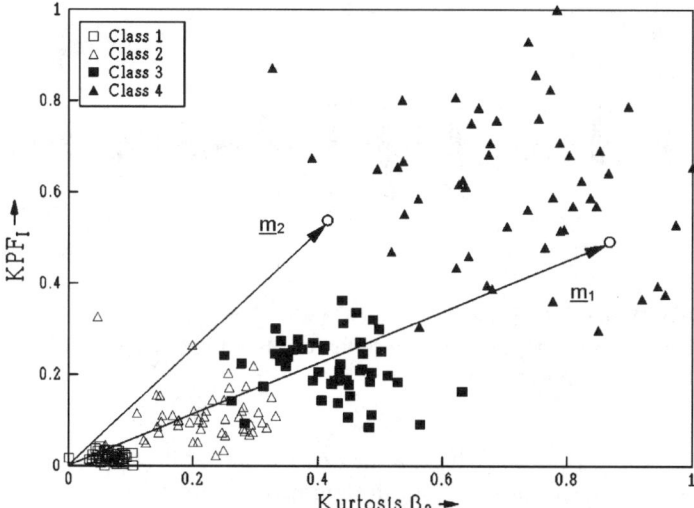

Fig. 10.9. Learning samples for the diagnosis of roller bearings

The advantage of projecting 24 characteristic values onto a single value, which can be followed without difficulty on a trend plot, is evident. If the feature vector is considered more exactly, of course, the system can provide information which extends beyond the pure representation of class membership. Figure 10.9 illustrates the learning set for the four defined classes for the sole consideration of the Kurtosis features and the KPF-value.

Kurtosis, refers to the fourth order moment around the average normalized by s4 where s is the standard deviation, measured the peakiness of a signal. KPF is a characteristic damage frequency which can be calculated from the bearing geometry and shaft rotational frequency. A bearing fault can be detected from frequency spectrum at the KPF i.e. by using the KPF-value. The index I in Fig. 10.9 denotes that the corresponding frequency is calculated for the inner ring. Only very slight scatter of the feature vectors occurs for the exactly defined shut-down class, whereas the scatter increases considerably for the "change" and "defect" classes. An example of the importance of recording more than one process parameter during diagnosis is presented in Fig. 10.10. The left part of Fig. 10.10 shows a typical time signal for a roller bearing with a outer race damage. On the right side is the signal for the same roller bearing and defect with the exception that its nature is different from the result of cavitation. In this case the kurtosis value decreases to the value for a nearly intact bearing because the caviation lowers the peakiness and now it would not permit any conclusion concerning the machine condition. In combination with the KPF-values, however, the diagnostic system provides a correct appraisal which is largely independent of external effects.

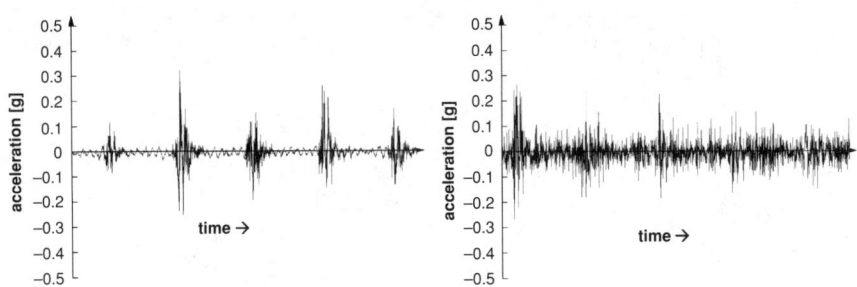

Fig. 10.10. Time record showing typical fault signature (*left*) and a signal of a defect roller bearing with cavitation interference Learning samples for the diagnosis of roller bearings (*right*)

The example presented here indicates the suitability of a fuzzy pattern recognition approach, precisely for designing a monitoring system for rotating components for which an analytical description of the relation between cause and effect cannot be provided. At a plant of the primary chemical industry, the described diagnostic system has been in operation for more than four years to monitor several continuously operating centrifuges, and the effectivity has been proved repeatedly by early detection of 18 roller bearing faults.

The complete system is implemented in LabVIEW [23] which is a powerful program development environment using a graphical programming language to create programs in block diagram form. LabVIEW also includes extensive library functions fpr signal processing. However for the design of an automatically working classification system you have to develop additional functionality, combined in SubVI-libraries. LabVIEW programs are called virtual instruments (VIs) because their appearance and operation can imitate actual instruments. A VI consists of an interactive user interface and a data flow diagram that serves as the source code. For the process of the feature generation for example by the calculation of a power spectra standard VI's from the LabView Toolbox could be used. A new VI-library has been developed for all elements of the automatic classifications process including the:

- Automatic feature selection using a wrapper approach,
- Design of the fuzzy classification algorithm,
- Determination of membership functions and calculation of class membership values for current signals. The algorithm based on a fuzzy weighted distance classifier,
- Automatically adaptation if the current signals are not covered by existing membership function (including the triggering of a new feature selection process).

10.5.2 The Human Interface Supervision System (HISS) System

The Human Interface Supervision System (HISS) [16] which monitors different plants in all fields of industry where a great hazard prevails, where unacceptable working conditions exist, or where long distances make it difficult for humans to check plants or processes at regular intervals. HISS is an automatic self-learning supervision system with integrated video and audio supervision and an electronic nose. The HISS project was born in order to find a solution to this problem by imitating human sense organs – eyes, ears, nose – with the help of sensors. All sensor signals are integrated by an educable logic in such a way that events in technical-process facilities which cannot be identified by conventional process-control systems can be logically evaluated and the resulting information passed on (Fig. 10.11).

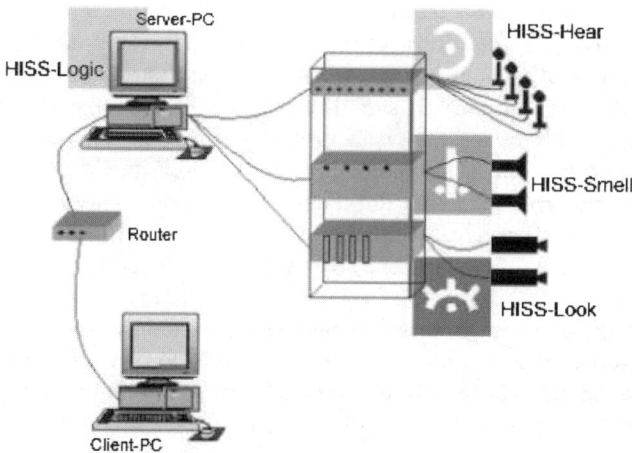

Fig. 10.11. The Human Interface Supervision System is an automatic self-learning supervision system with integrated video and audio supervision and an electronic nose [16]

1. HISS Seeing component
   consists of one or more video cameras linked to an image-processing system.
2. The HISS Smelling component
   provides a measuring system for the continuous monitoring of chemical plants on the basis of an electronic nose. Released gases and gas mixtures can be detected with a high degree of sensitivity and classified by pattern evaluation. These characteristics enable a rapid response to possible process-gas leaks combined with the low probability of false alarms even when there is interference by other gas mixtures.
3. The HISS Audio component
   is an independent acoustic supervision module in the total HISS system working with different types of microphones and acceleration sensors. The

technology of HISS Audio allows it to be used not only in natural-gas exploration but also in any other area of industrial plant technology in which acoustic signals can provide information about the state of a production process. As a result of the independent acoustic pattern recognition integrated in HISS Audio it is possible to recognize and distinguish different plant situations. For example, very small gas leaks or the early signs of motor or pump defects can be detected by acoustic means. The analysis of flow noises permits further insight into the process situation. Here a special focus is given to the audio monitoring of a natural gas extraction plant.

In an alarm situation HISS Audio automatically reports irregular conditions in the facility to a control station. For this purpose the most varied acoustic measuring devices monitor the field of air and structure noise and so allow a highly accurate analysis of the total acoustic picture of an industrial facility. In alarm or suspected-alarm situations acoustic facility scenarios are stored in a multi-channel library. HISS Audio makes it possible to listen to important parts of a facility from a distant control station. Both stored signals from the past and current signals can be called up remotely in real time. HISS Audio is able to learn new facility scenarios and to make assessments. In order to be capable of distinguishing undesirable facility situations from normal ones the system uses a database of stored acoustic facility scenarios. This database is interactively expandable and grows with the active life of the system. All unknown facility scenarios are assessed by control station staff. In this way, the system is in a position to carry out very exact signal analysis and scenario assessment in an extremely short period of time.

The systems shows a high robustness in relation to ambient noises but nevertheless detects the smallest changes. Because of elaborate multi-channel digital signal pre-processing and a two-stage pattern-recognition process the accuracy of the analytical process is not influenced by sporadically occurring ambient signals. Ambient noises which persist for a long period are incorporated into the normal state. Creeping changes within the normal range of a facility are taken into account and do not lead to alarm messages.

In existing installations gas leaks (escaping natural-gas) down to a leak volume of 5–40 m^3/h are detected by microphones (Fig. 10.12). The accuracy of the assessment achieved dependents on the background noise of the facility and the location of the listening post. The influence of ambient noises on the classification of facility noises is minimized by digital signal pre-processing and spectral distribution of the signals. From the aspect of the pattern recognition task a crucial point is the normalisation of the sensor signal. The loudness of the signal depends on the leak volume and the unknown distance between the location of the damage and the position of the microphone. The concept fulfil the minimal requirements which could be defined for an level 1 adaptive monitoring system because the leak detection system operates successfully more or less independent from the plant specific background noise. The time

Fig. 10.12. Typical sensor constellation for the monitoring of a sweet gas plant. Microphones and acceleration sensors are used for the detection of leaks and machine faults

for adapting and adjusting the system to a changing environment has been reduced to an acceptable level.

The HISS Logic unit comprises a facility model in which the tasks and characteristics of the process facility (e.g. whether it is a sweet or acid gas facility) are stored 13. In an event-related database new situations are stored and are available for a comparison with current measurements. Impotant is the consideratation of process variables. An increase in the speed of production leads to a change in the background noise of the facility. If the process variables are included in the Logic unit this change does not lead to an alarm signal. Information is pre-processed in the individual components and transferred via interfaces to the central computer unit. In the central computer unit the situation is registered, evaluated and a decision about further processing is made (Fig. 10.13).

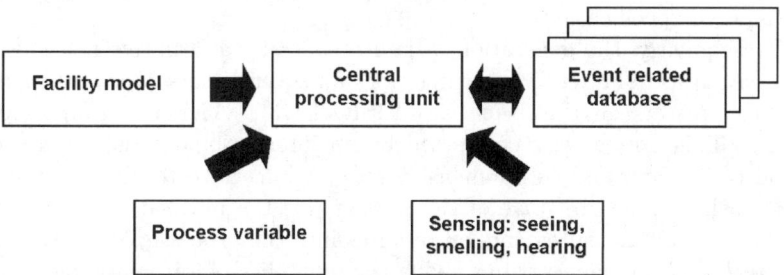

Fig. 10.13. Current measurement situation is compared to stored scenarios. The systems is learning from these experiences

HISS Logic receives information from each of the three measurement systems if one of these systems have detected a fault. The Logic unit compresses the information with respect to time and location parameters and gives an appropriate message to the operator. Two cases must be distinguished: Either the system recognizes the situation which has led to the fault, in which case it can issue a message which points directly to this situation or it does not recognize the situation, in which case the message is unknown scenario. In this case the operator has to give a feedback, stating what the current situation is and the system uses this feedback in order to learn. If this or a similar situation will occur again in the future the system recognizes the situation and can also give an expert message. By and by HISS Logic thus acquires an increasingly comprehensive case database and consequently an ever greater knowledge about the facility and its experience increases over time.

10.6 Introduction of Simulation Results into Monitoring Concepts

10.6.1 Status of Simulation Technology with the Application of FEM

A general problem in designing classifiers on the basis of learning data is the need to provide a sufficiently large learning set. In the ideal case, the learning set should include all states which will occur during later operation, as far as possible. This requirement also includes all fault situations. For machine monitoring and process control tasks, this demand can hardly be fulfilled in practice. During the normal operation of real plants, fault states very seldom occur, and if they do occur, the undivided interest of the plant operator is to correct them as quickly as possible. A prolongation of this undesirable situation for the purpose of obtaining learning patterns is unrealistic, since hazards to humans, to the environment, and to the machines concerned may be associated with a process situation of this kind. Furthermore, no one would allow the manufacture and the loss of products which do not conform with the imposed specifications.

Consequently, the generation of fault classes by alternative methods has been the subject of intensive research work for several years. One of the largest research projects on this topic was VISION [13], which was supported by the EU. This project was conducted by an international consortium over a period of 6 years, and comprehensive publications have resulted from these investigations. The objective of the project was the integration of computer simulations and measurements on test rigs into the process of accepting data from real measurements for improving the reliability of automatic monitoring systems (Fig. 10.14). Hence, the simulation had to perform two functions:

1. The machine behaviour must be simulated, and the results must be compared with real data for adaptation of the simulation models applied.

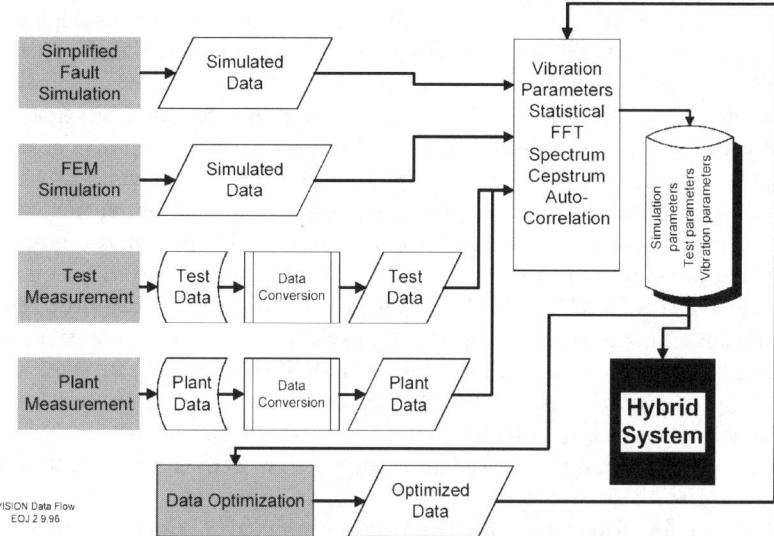

Fig. 10.14. Diagnosis integration overview [3]

Essentially, finite element programs (FEM) were employed for the purpose. Since boundary conditions, such the variation of rotational speed and dimensions of machine components, can also be taken into account on the computer, the adaptive behaviour can be simulated.

2. On the basis of this result, machine states are generated with artificially introduced faults, such as roller bearing damage, alignment faults, gear defects in transmissions.

The data from the fault simulation were compared with data from very simple test rigs, for which the specific implementation of defective machine components is feasible, in contrast to real industrial plants. The basic idea of this approach is not new in itself, but the objective of the project was to achieve a new dimension in the quality of the simulation. From a conceptual standpoint, promising trial solutions have certainly resulted from this project, but these are concerned essentially with integration strategies for the different data sets to yield a standardised learning set. However, the overall result can never be better than the simulation which is taken as basis. Precisely this weakness limits the applicability of FEM in the field of machine monitoring at present. Modelling of the extremely complex, frequently non-linear processes of contact between machine components in motion is feasible only in simple cases. As an example consider the contact of a roller bearing with pitting in the track. In principle, an expert can also determine the structure of the signal thus generated without the application of FEM programs. A periodic excitation of natural vibrations in the system always occurs when a rolling element is in contact with the uneven surface of the track. In practical tests, amplitude information immediately at the track can also be correlated with the magnitude of the damage present. However, if the constant presence of lubricant grease or oil is considered, and the fictive concept of ideal rolling is abandoned, the problem becomes much more difficult. Consideration of the usually unknown transmission distance within the bearing housing and the following sites of contact with the adjoining machine components is not feasible in practice. Precisely this approach is necessary for the simulation of really complete machines as the sensors are usually not installed at the immediate site of damage. How can they be, if the site where the damage will occur is not even known? Problems in FEM application include, for instance, the realistic description of connecting elements. Rigid restraints in the computer model, by means of which individual nodes of the FEM network can theoretically be rigidly fixed with respect to all degrees of freedom, do not exist in real systems with connections which can be loosened. Consequently, the systems represented are usually too stiff, and the natural frequencies usually do not correspond to the real conditions.

Hence, simulations by FEM have not yet been developed to the extent necessary for direct application with neural networks as a part of learning sets for individual faults. Are there any fields of application in which this approach can still be useful?

10.6.2 Hybrid Systems for Monitoring

An approach of this kind can be useful especially in applications where hybrid monitoring systems utilise various trial solutions for intelligent data processing in a parallel structure. The following components have already been realised for improving the adaptivity of monitoring systems:

1. Neural networks (NN) Neural networks are certainly suited for applications in machine monitoring. Numerous reports on successful applications have been published. Because of their capability of operating with noisy signals and of representing the non-linear relationships between sources of error and the resulting signal, neural networks are predestined for this field of application. A further advantage is the high potential for generalisation, that is, the capability of reaching a sensible decision even with input data which have not been learned. In contrast to closed-loop control applications, it is not necessary to cover the entire input data space for classification tasks with neural networks. However, the representation of a classification decision is not transparent. As the level of network complexity increases, the interpretability decreases drastically. Because of this disadvantage, it has proven expedient to avoid working with large networks which can represent a large number of faults; instead, smaller networks should be taught to deal with a single fault or fault group. There are many different kinds of Neural networks available. Specially learning vector quantization (LVQ) which based on prototypes is particularly suited for applications where

prior knowledge is available. You could find very interesting approaches where authors have modified the LVQ for condition monitoring tasks. In [5] a modified Generalized relevance learning vector quantization algorithm is used for monitoring a piston compressor allowing a stable prototype identification which is invariant with respect to small time shifts.
2. Rule-based fuzzy systems
   Despite the general reservations with respect to expert systems, the application of rule-based systems for machine monitoring still makes sense. Especially the possibility of expanding an initially rudimentary system speaks in favour of this approach. During the initial stage of installation, such a system is very well suited for representing expert knowledge by means of simple rules in the form of "if-then" relationships. The rule base is frequently very broad and is applicable to a wide variety of machines and plants. The degree of adaptivity is therefore relatively high. A general disadvantage of such a system is the fact that it cannot operate more efficiently than the rules which are introduced by the human expert. The loss of information which results from the conversion of knowledge for deriving the rule base must be taken into consideration here. Thus, it can be concluded that the diagnostic reliability of such control systems is no better than that of a human expert with average experience. However, advantages result from the fact that the expert can define or establish rules even for classes and states for which no data are available. In contrast to neural networks, moreover, the form of knowledge presentation is usually interpretable and allows the validation of a classification decision. Rules are brittle when presented with noisy, incomplete or non-linear data and they do not scale up very well. Inconsistencies and errors start to creep in as the knowledge base grows [11].
   The previously described situation concerning rule-based systems has been improved by the introduction of diffuseness with fuzzy logic. However, rule-based diagnostic systems have also received major impetus with the introduction of automatic rule generation by coupling of neural networks and fuzzy systems (neuro-fuzzy systems).
3. Neuro-fuzzy diagnostic systems
   Neuro-fuzzy systems are the key to the implementation of learning capability in intrinsically static rule bases. Many possible methods are available for the automatic generation of rules. A few of these approaches are described more closely in the first part of this book. For instance, the NEFCLASS [15] software is capable of performing this function. The rule bases thus derived are often considerably better than those which result from a purely linguistic description of expert knowledge. Furthermore, the dependence of process parameters can be modelled in this form even if it is not possible for the experts to express it in such a form. Vibratory signals can frequently be analysed only as functions of process situations. These relationships can be recognised and represented by a neuro-fuzzy system if the process parameters are recorded together with the vibratory signals as a learning set.

With the application of a neuro-fuzzy system, a high priority also results for the simulation, since the generally valid fuzzy rules thus derived are of special interest here, rather than the exact numerical values, for instance, of an individual natural frequency of a machine. For this purpose, an error of a few per cent in the result of a numerical simulation is not important. The decisive features are the variation of the value in modelling of a fault and the conclusions which can be reached from this result.

The data mining technique described by McGarry in the paper [10] reinvolves the process of training several RBF networks on vibration data and then extracting symbolic rules from these networks as an aid to understanding the RBF network internal operation and the relationships of the parameters associated with each fault class.

A criticim of neural network architecture is their susceptibility to the so called catastrophic interference which should be understood as the abilty to forget previous learned data when presented with new patterns. A problem which should never occur when a neural network is used as a classification algorithm in condition monitoring applications. To avoid this some authors have described neural network architectures with a kind of memory. In generell two different concepts are promising: Either the network possess a context unit which can store pattern for a later recall, or the network combining high-levels of recurrence coupled with some form of back-propagation [2].

10.7 Best Practice, Tips and Hints

Best practices. These two words represent benchmarking standards- nothing is better or exceeds a best practice [17]. The area of monitoring is much too broad to give a general best practice guideline. But the questions in the following catalogue should be answered before the selection of a method or combination of various methods, in order to ensure that the methods considered are really suited for the problem involved. That will not assure the best practice but the use of suitable methods which represent the state-of-the-art.

1. Are enough data available for all (as many) conditions or qualities (as possible)?
   This question must be answered in order to decide, for instance, whether automatic learning of the classifiers can be accomplished with neural networks. For a neural network which is employed as a monitoring system, the situation is more favourable than for one employed as a process controller. With a monitoring system, an unlearned condition results either in no indication at all or, in the worst case, in a false indication, but the process itself is not affected. If only data for a "good" condition are available, these data can still be employed as starting material for a monitoring operation.

All significant deviations from this condition are then detected. The adaptive system then evolves in the course of the operation with the continuing occurrence of new events.
2. How many features are available?
   Features should be closely related to the monitored object. How many features of this kind can be provided ? The determination of the number of the features from that on a feature selection is necessary, can be decided only under the considering of the current monitoring task. As a provisional value you can take a number of 10 features. If more features are available a selection should be carried out. Notice the existence of the close relationship between the number of feature and the number of necessary random samples for a lot of classifiers. If a reduction of dimensionality is necessary prefer feature selection if possible. The interpretability of new features for instance calculated by a principal component analysis is difficult for both an expert in the field and more than ever for an operator.
3. Can every set of process or machine data be unambiguously correlated with a class attribute (quality or damage)?
   If this is possible, methods of "supervised learning" should be applied. If this is not possible, or if class information is available only for portions of the data sets, cluster methods, such as Kohonen networks, should be employed. The problem is especially difficult if the task involves a very large number of available features, and the associated data bases cannot be classified by an expert. In such cases, two questions must be considered in parallel: the question of feature selection and that of clustering. Nearly all methods for the selection of features utilise the classification efficiency of the learning or test data set as selection criterion. However, if no class designation is available, the classification efficiency is not calculable, and so-called filter methods must then be employed. As a rule, these methods yield features which are decidedly less efficient.
4. Do I have enough theoretical background, system information and a suitable software to carry out a precise simulation which should include fault simulation?
   The development of a model based prognostic and diagnostic model requires a proven methodology to create and validate physical models that capture the dynamic response of the system under normal and faulted conditions [4]. There are a couple of model update algorithms which could improve the accuracy of a FEM-Model but you should never underestimate the work load and expert knowledge which is necessary for realization of the corresponding experiments and program usage.
5. What is of better quality or more reliable? The system model or the response of your sensors indicating changes in the machine or process?
   Having both would be the best situation because you are free in the decision to use a forward model which calculate a prognosis of the system response and compare the estimation with the actual sensor signal (Fig. 10.15). This procedure allows an estimation of the time to failure of the system.

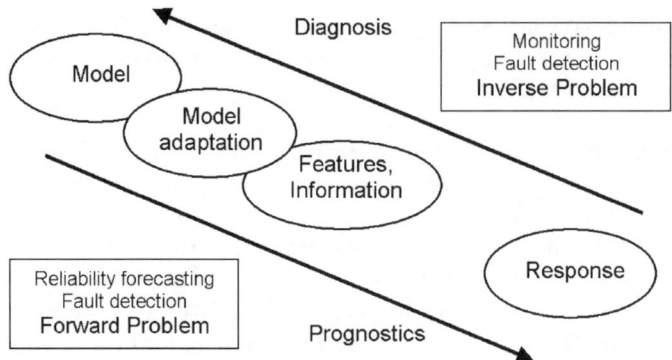

Fig. 10.15. Prognostic and diagnostic approach for a montoring task [1]

Additionally the difference between sensor and model will give you useful information about possible faults. For the inverse problem the quality of the model is less important because the describing features and all condition information are extracted from the sensor signal independently from the model. For the diagnostic of roller bearings we have a rough model about the signal structure but in most cases not enough information for the prediction.

6. Do I have a safety relevant application?

    Until now there are a couple of problems concerning self adaptive monitoring system in safety relevant applications. These limitations are a result from problems which could occur during the evaluation and testing phase of such systems in all conceivable situations. The behaviour today and in the future after a system adjustment could differ significantly. This will hit legal aspects because national and international testing institutions might have problems to certificate a system. Today in nuclear power plants advanced self-learning and adaptive systems are additional information sources but in general not the trigger for an automatic shout-down.

7. The cost factor. Do I have sensor technique and communication infrastructure for an on-line monitoring?

    Current commonly used sensor technology only permits the most rudimentary form of signal processing and analysis within the sensor. The use of advanced data analysis techniques will need more signal data. This means that large quantities of data must be transmitted from the sensor to a separate data collector for subsequent processing and analysis. If permanent on-line vibration monitoring is required – and for adaptive techniques it is absolute necessary – then at present, for anything other than an overall vibration alarm, the cost of providing the required communications infrastructure and the data collection and analysis equipment far outweighs the benefits to be obtained [9].

10.8 Conclusions

Analyzing the current status a gap between new research developments in monitoring ideas, methods and algorithms and real world applications in industry could not be neglected. New hybrid diagnostic systems which use combinations of expert knowledge, data driven modelling and advanced soft computing algorithms for classification and prediction have proven their performance in the past. The combination of different techniques is the key to implement learning ability and in consequence adaptive behaviour. No expert in this field will argue the predominance of these techniques in comparison to standard methods which will use a simple threshold monitoring or a fixed classification algorithm. But there are still a lot of problem to solve. The automatically determination, extraction and selection of features which are best suited for a given process is a challenge field of research in general and also in condition monitoring. Even future research activities will not lead to an exclusive monitoring strategy because the field of application for the different monitoring applications is to broad. But the permanent improvement in the simulation techniques will offer new possibilities by replacing the experimental learning part through the computer based fault simulation. Than we will find adaptive techniques also in areas where safety relevant aspects nowadays prevent a real fault simulation and therefore broader possible applications of adaptive monitoring systems.

References

1. Adams D.E. (2001) Smart Diagnostics,prognostrics, and self healing in next generation structures, Motion Systems Magazine, April 2001.
2. Addison J.F.D., Wermter St., Kenneth J McGarry and John MacIntyre (2002) Methods for Integrating Memory into Neural Networks Applied to Condition Monitoring, 6th IASTED International Conference Artificial Intelligence and Soft Computing (ASC 2002), Banff, Alberta, Canada July 17-19, 2002.
3. Adgar A., Emmanouilidis C., MacIntyre J. et al. (1998) The application of adaptive systems in condition monitoring, International Journal of Condition Monitoring and Engineering Management, Vol. 1, pp. 13-17.
4. Begg C.D., Byington, C.S., Maynard K.P. (2000) Dynamic Simulation of mechanical fault transition, Proc. of the 54th Meeeting of the Society for Machinery Failure Prevention, pp. 203-212.
5. Bojer T., Hammer B. and Koers C. (2003) Monitoring technical systems with prototype based clustering. In: M.Verleysen (ed.), European Symposium on Artificial Neural Networks'2003, pp. 433-439, 2003
6. CSC (2001) Get smart. How Intelligent Technology will enhance our world. CSC consulting. www.csc.com
7. Debabrata P. (1994) Detection of Change in Process Using Wavelets, Proc. of the IEEE-SP Int. Symposium on Time-Frequency and Time-Scale Analysis, pp. 174-177.

8. Dowling H.M. (1993) Application of Non-stationary Analysis to Machinery Monitoring, Proceedings of IEEE Int Conf. On Acoustics, Speech and Signal Processing, pp. 59-62.
9. Dunn S. Condition Monitoring in the 21st Century, http://www.plant-maintenance.com/articles/ConMon21stCentury.shtml.
10. McGarry K. and MacIntyre J. (2001) Data mining in a vibration analysis domain by extracting symbolic rules from RBF neural networks, Proceedings of 14th International Congress on Condition Monitoring and Engineering Management, pp. 553-560, Manchester, UK.
11. McGarry K. and MacIntyre J. (1999) Hybrid diagnostic system based upon simulation and artificial intelligence, Proceedings of the International Conference on Condition Monitoring, 593-600, April 12th-16th, University of Wales, Swansea.
12. MacIntry J. and Jenning I. (1997) Condition Monitoring Software that Thinks for Itself. Proceedings of the MAINTEC 97 Conference, Birmingham, March 1997.
13. MacIntyre J., Adgar A., Emmanouilidis C. and McGarry K. (1999) The VISION project: a European collaboration on vibration analysis, Proceedings of 12th International Congress on Condition Monitoring and Engineering Management, 253-258, Sunderland, UK.
14. Rinta-Runsala E. (2000) Drive System monitoring: Requirements and Suggestions. VTT Case Study Report, Finland.
15. Nauck D. (2001) Intelligent Data Analysis with NEFCLASS. Proc. European Symposium on Intelligent Technologies (eunite 2001), Tenerife. Verlag Mainz, Aachen (on CD-ROM).
16. Siemens (2000) The HISS System, www.hiss-system.de.
17. Smith R. (2002) Best Maintenance Practices, http://www.mt-online.com.
18. Steppe J.M. (1994) Feature and Model Selection in Feedforward Neural Networks. PhD Thesis, Air Force Institute of Technology.
19. Strackeljan J. and Weber R. (1998) Quality Control and Maintenance. In: Fuzzy Handbook Prade u. Dubois (Eds.), Vol. 7, Practical Applications of Fuzzy Technologies, Kluwer Academic Publisher.
20. Strackeljan J. and Lahdelma S. (2000) Condition Monitoring of rotating machines based on Fuzzy Logic, In: Proceedings of the Int. Seminar Maintenance, Condition Monitoring and Diagnostics, Pohto Publications, pp. 53-68.
21. Strackeljan J. (2001) Feature selection methods – an application oriented overview. In TOOLMET'01 Symposium, Oulu, Finland, pp. 29-49.
22. Strackeljan J. and Schubert A. (2002): Evolutionary strategy to Select Input Features for a Neural Network Classifier. In: Advances in Computational Intelligence and Learning: Methods and Applications (eds.: Zimmerman H-J., Tselentis G.)
23. Strackeljan J. and Dobras J. (2003): Adaptive Process and Quality Monitoring using a new LabView Toolbox, Eunite Symposium 2003, Oulu.
24. Tse P. (1998) Neural Networks Based Robust Machine Fault Diagnostic and Life Span Predicting System, Ph.D. Thesis, The University of Sussex, United Kingdom.
25. Tse P., Peng Y.H. and Yam R. (2001) Wavelet Analysis and Envelope Detection for rolling element bearing fault diagnosis – their effectiveness and flexibilities, Journal of Vibration and Acoustic, Vol. 123, pp. 303-310.

11

Examples of Smart Adaptive Systems in Model-Based Diagnosis

K. Leiviskä

University of Oulu, Department of Process and Environmental Engineering, P.O. Box 4300, 90014 Oulun yliopisto, Finland
Kauko.Leiviska@oulu.fi

This chapter looks at model-based diagnosis from the problem-oriented point of view: how to go from actual diagnosis problems to intelligent methods applicable for solving them. Because this area is very large, it is approached from four cases studied during few latest years in Control Engineering Laboratory, University of Oulu.

Two cases are from process industry and two from manufacturing industries. The technical content in applications is described in a quite general way. More details are found in the referenced articles. Emphasis is on the problems and how to solve them using intelligent methods. Adaptivity is conveyed as the second theme throughout this chapter. However, the number of Smart Adaptive Systems in diagnosis is low, even though there is a clear need for them.

Some warnings are due to the terminology in this chapter. Four special processes are considered and it is rather impossible to write about them without using also the special terminology. Otherwise, the text would look very primitive to real process specialists. On the other hand, it would require a couple of textbooks to explain the terminology, so it has not been even attempted here.

The chapter is organised as follows: Section 11.1 is an introduction to fault diagnosis as a research area referring to different problems inside it and the most usual ways, how they are solved. Section 11.2 introduces the problems: paper machine web break detection, nozzle clogging problem in continuous casting of steel, diagnosing faults in screw insertion in electronics production and the localisation of faults on printed circuit boards in electronics production. Section 11.3 shows, how these problems have been solved using intelligent methods concentrating on giving a wider view to the topic. Section 11.4 gives some general hints and tips for solving the common problems in model-based diagnosis and Sect. 11.5 is a (very) short summary.

11.1 Introduction

Fault diagnosis is already a classical area for intelligent methods, especially for fuzzy logic applications [1, 2]. Compared with algorithmic-based fault diagnosis the biggest advantage is that fuzzy logic gives possibilities to follow human's way of fault diagnosing and to handle different information and knowledge in a more efficient way. Applications vary from troubleshooting of hydraulic and electronic systems using vibration analysis to revealing the slowly proceeding changes in process operation. The essential feature is the combination of fuzzy logic with model-based diagnosis and other conventional tools. Three of four applications concerned in this chapter belong to this category.

Fault diagnosis can be divided into three separate stages or tasks:

1. Fault detection that generates symptoms based on the monitoring of the object system,
2. Fault localisation that isolates fault causes based on the generated symptoms, and
3. Fault recovery that decides upon the actions required for bringing the operation of the object system back to the satisfactory level.

The fault detection problem is, in principle, a classification problem that classifies the state of the process to normal or faulty. This is based on thresholds, feature selection, classification and trend analysis. Trend analysis is a good area for intelligent methods, because there the threshold between normal and faulty situations is difficult to tell and the target is to reveal the slowly developing fault situations. Cases 1 and 2 below include only the fault detection stage.

The frequency range of phenomena studied in fault detection varies from milliseconds in vibration analysis to hours and days in analysing the slowly developing process faults. This sets different requirements for the system performance and computational efficiency in real-time situations. Also different methods are needed in data pre-processing. Spectral and correlation analysis are used in vibration analysis to recognise the "fingerprints" of faults whereas regression analysis, moving averages, temporal reasoning or various trend indicators are used in revealing slowly developing faults.

Fault localisation starts from the symptoms generated in the previous stage and it either reports fault causes to the operator or starts automatically the corrective actions. In the first case, the man/machine interface is important. Actually this stage includes the defining of size, type, location and detection time of the fault in question. If no information is available on the fault-symptom causality, classification methods, e.g. fuzzy clustering, is used. If the fault-symptom causality can be given in the form of if...then rules, the reasoning methods are used. Cases 3 and 4 below include also the fault location stage.

Fault recovery is a case for (fuzzy) expert systems. Based on fault location and cause, corrective actions are recommended to the user. This stage is not included in any of the cases below. In most cases it is left for the operator to choose or it is obvious from the problem statement.

Environmental changes, together with the changes in quality and production requirements lead to the need for adaptation. For instance, in nozzle clogging problem in Sect. 11.2.2, change in the steel grade changes the patterns of single faults and it can also introduce totally new faults. Reasons for faults can be traced back to earlier stages of production, to dosing of different chemicals, bad control of process operations, and also to different routing of customer orders through the production line. In this case, the new steel grade change would require new models to be taken into the use.

Fault diagnosis is in many cases essential to quality control and optimising the performance of the machinery. For example the overall performance of paper machines can be rated to the operating efficiency, which varies depending on breaks and the duration of the repair operations caused by the web break.

Leiviskä [1] analysed several cases for using intelligent methods in the applications developed in Control Engineering Laboratory, University of Oulu. Two applications in model based diagnosis from process industries were considered: nozzle clogging in continuous casting of steel [4] and web break indicator for the paper machine [5]. These cases are revisited here from another perspective together with two further examples from electronics production; screw insertion machine [6] and PCB production [7].

11.2 Example Problems

11.2.1 Case 1: Paper Machine

The runnability of the paper machine depends on the amount of web breaks in comparison with the paper machine speed [5]. The paper web breaks when the strain on it exceeds the strength of paper. When the runnability is good, the machine can be run at the desired speed with the least possible amount of breaks. The web break sensitivity stands for the probability of web to break, predicting the amount of breaks during one day. Paper web breaks commonly account for 2–7 percent of the total production loss, depending on the paper machine type and its operation. According to statistics only 10–15 percent of web breaks have a distinct reason. The most of the indistinct breaks are due to dynamical changes in the chemical process conditions already before the actual paper machine. In this area the paper making process is typically non-linear with many long delays that change in time and with process conditions, there are process feedbacks at several levels, there are closed control loops, there exist factors that cannot be measured and there are strong interactions between physical and chemical factors [8].

The aim of this case study is to combine on-line measurements and expert knowledge and develop the web break sensitivity indicator. The indicator would give the process operator continuous information on the web break sensitivity in an easily understandable way. Being able to follow the development of the break risk would give the operator a possibility to react on its changes in advance and therefore avoid breaks. In this way, the efficiency of the paper machine could be increased and also the process stability would improve with fewer disturbances. In this case a lot of mill data from several years' operation was available.

Paper machines produce a great amount of different paper grades with varying raw materials and process conditions. This leads to the fact that reasons for web breaks can vary from one grade to another. This calls for adaptive systems. Leaning capability would also shorten the start-up period for the new machines.

11.2.2 Case 2: Continuous Casting

Submerged entry nozzle connects the tundish and the mold in the continuous casting of steel (Fig. 11.1) [4]. It is a tube with a diameter of about 15 cm and it divides the molten steel into two directions on the both sides of the nozzle. The nozzle transfers molten steel from the tundish to the mold and separates steel from the atmosphere.

Steel is cast in batches (heats). Usually one cast series includes 3–6 successive heats. Casting of a heat takes about 35–50 minutes. Each series is designed to consist of the same steel quality. The nozzle is changed after each series and the new series is started with a new nozzle. About 360–720 tons of steel goes through a nozzle during its lifetime. Nozzle clogging stops the

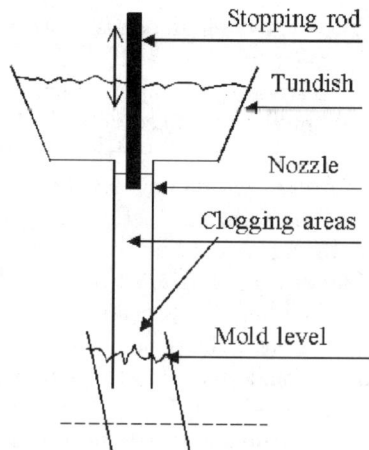

Fig. 11.1. Submerged entry nozzle in the continuous casting process [4]

cast and causes production losses and extra work. The risk of nozzle clogging increases all the time so that when about 360 tons of steel has been cast, the nozzle condition must be checked carefully. The operators estimate, based on vision and the casting plan, if the clogging should be removed or the nozzle changed in the middle of the cast series.

This case study aims to develop tools for predicting the occurrence of nozzle clogging phenomenon. It is known that better management of casting brings along considerable potential for production increase and quality improvement. Also in this case a lot of mill data was available.

Also in this case, the production of a big variety of grades calls for adaptive systems. In its simplest form, it would mean building of separate models for few grade categories.

11.2.3 Screw Insertion Process

Screw insertion is a typical automated assembly task in modern end-of-line electronics manufacturing and its purpose is to join two parts together [6]. In case of automated screw insertion units intended for flexible and fast mass production, already small processing deviations may result in a high number of defective products. Deviations from normal process conditions may also affect productivity and increase the downtime. Therefore screw insertion aims to zero defects. Typical reasons leading to a defective product are cross-threading, no screw advancement and hole misalignment.

Existing systems focus on classification between successful and failed insertions. This case study aims to develop a system that could also detect the type of the defect, because this is necessary for early fault recovery operations. In this case no actual mill data was available.

Changing of raw materials and components may introduce new type of faults and the features of existing faults may also change. Adaptation requires that the systems are able to learn or can be trained to new situations.

11.2.4 PCB Production

In electronic manufacturing, malfunctions at a component level on a Printed Circuit Board (PCB) will have an effect on the state of the final product. A defect free product depends on all the components, whereas a faulty product can be the result of one single component. This means that even the smallest defect on one single component can have a significant impact on the overall performance of a PCB [7].

This case study focuses on defect detection during the final testing of the PCB boards. The developed system is able to detect and trace a defect into a small area of the PCB. This information is then sent back to the operators who are in charge of repairing the boards. In this case, the production is done in short series and therefore the biggest limitation is a small amount of data available.

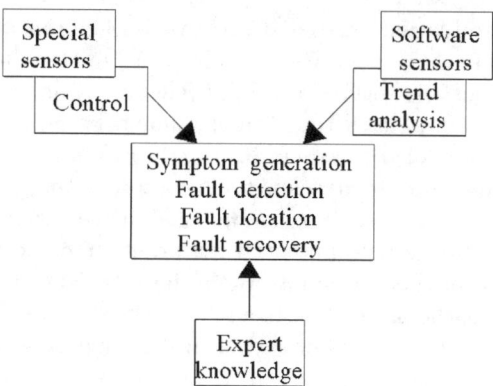

Fig. 11.2. Origin of data in model based-diagnosis

Once again, change in components or even in the whole design of PCB leads to the need for adaptation. This is also a very strict requirement because of short series production that gives only a short time for learning.

11.3 Solutions

Model-based diagnosis is a flexible tool for making the decisions early enough to guarantee smooth operation. Intelligent methods provide additional tools for generating, analysing, modelling and visualising in different stages in diagnosis. Expert knowledge is necessary in applications where only limited amount of data is available. Neural networks are useful in extracting features in data-intensive applications and in tuning of the systems. As mentioned before, fuzzy systems are conventional explanatory tools for diagnosis. Fig 11.2 shows the origin of data in model-based diagnosis.

11.3.1 Case 1: Paper Machine

The web break indicator problem is highly complex, and various techniques are available for handling different aspects of the system [8, 9, 10, 11]:

1. Principal component analysis (PCA) has been used in data reduction and interpretation, i.e. to find small amount of principal components, which can reproduce most of the information. This technique can reveal interesting relationships, but it has difficulties with non-linearity and variable delays.
2. Fuzzy causal modelling provides methods for extracting expert knowledge and also rules can be generated. This is a suitable method for restricted special cases, but the influence diagrams become very complex resulting in huge rule bases.

3. Statistical process control (SPC) provides information about the stability of variables. On-line monitoring of the current situation is important for the paper mills. However, selecting the critical variables and handling false alarms require other techniques.
4. Neural networks work fairly well for limited special cases where the quality of data can be kept on sufficient level. The variety of operating conditions, and possible causes of web breaks, are far too high to be included directly into a neural network.
5. Decision tree techniques require large data sets for statistical tests. Although the overall amount of data is huge, insufficient data becomes a problem in handling of various operation situations.

In this case, the web break sensitivity indicator was developed as a Case-Based Reasoning type application [12] with linguistic equations approach [13] and fuzzy logic (Fig. 11.3). The case base contains modelled example cases classified according to the amount of the related breaks. For example, breaks can be classified to five categories, depending on how many breaks occur during a period of one day: No breaks (0), a few breaks (1–2), normal (3–4), many breaks (5–10) and a lot of breaks (>10). For example, if there are five cases in each category the case base includes 25 cases as a whole. Each case includes one or more models written in the form of linguistic equations [13] and describing the interaction between process variables. This hierarchy is described in Fig. 11.4.

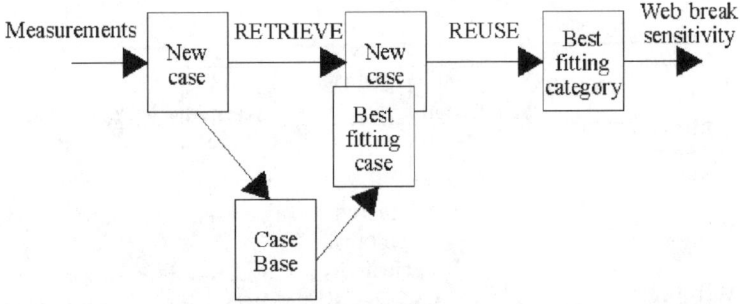

Fig. 11.3. Structure of the system [5]

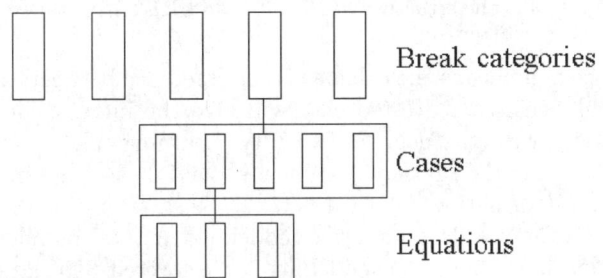

Fig. 11.4. Structure of the Case Base

The new case is presented to the system as a set of measurements. In *retrieve stage* these measurements are substituted to every model in the case base and for each case, all the equations are considered. If an equation is true, the degree of membership for that equation is one. All deviations reduce this degree. The degree of membership for the each case is evaluated by taking the weighted average from the degrees of membership of the individual equations. The degree of membership of each web break category is generated from degrees of membership calculated for all the cases included in the category. Once again, the weighted average is used. The indicator gives the predicted amount of web breaks in a day as output. The best result has been reached by selecting the break level of the best fitting break category as the output of the system in *reuse stage*. This numerical value is calculated in periods of one minute, but the reported value is filtered over two hours to avoid too rapid variations.

The on-line version of web break sensitivity indicator operates primarily as a case *retrieval* and reuse application. The evaluation of the predicted break sensitivity is based on the real break sensitivity calculated with the information of break occurrence in paper machine. This *revise* and *retain* stages (adaptation and learning) are performed off-line with simulator using real process measurements. The structure of this analysis stage is presented in Fig. 11.5.

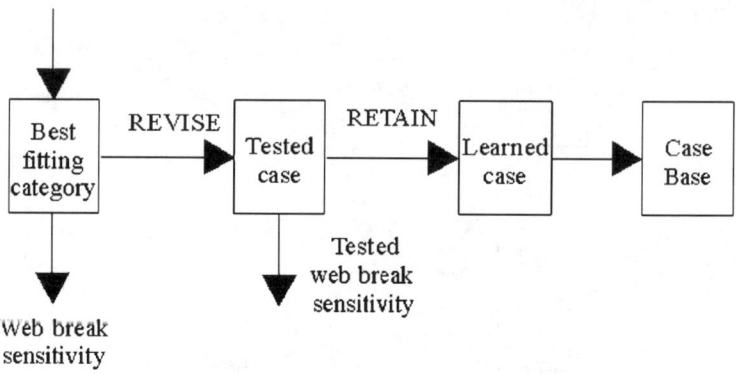

Fig. 11.5. The structure of *REVISE* and *RETAIN* stages [5]

The revise stage means the calculation of difference between the predicted and real break sensitivity. Tested case will give the information of the web break sensitivity with the degree of quality. Learning takes place when the quality degree is poor. In *retain* stage new potential cases are modelled with the break class information. Learned new case is saved in the case base. This case is an example of a data-intensive modelling. A total number of 73 variables were considered and the modelling data covered 300 days. Data was analysed in the periods of 24 hours and in the real application the number of break categories was 10. Case selection was done interactively. The main analysis tool was FuzzEqu Toolbox in Matlab-environment [5].

Hints and Tips

- In this case, data pre-processing is of primary importance. The first important point is the elimination of delays that are caused by long process chains and variable-sized storages between process stages. Cross correlation was used as a major tool. The second point is the case selection, which is done interactively with the automatic selection tool where user's role is to define the selection criteria and fill in the missing values.
- In order to get acceptance from the end user, models must be clear and transparent. The models were checked together with the mill staff and explanations for possible inconsistencies were found.
- System testing on-line might also be problematic because of long time required. The simulation tools would enhance the checking of system consistency.

11.3.2 Case 2: Continuous Casting

There are several factors that contribute to nozzle clogging [4]:

1. metallurgical factors: steel cleanness and aluminium content,
2. hydrodynamic factors: steel flow rate and nozzle geometry,
3. thermodynamic factors: cold steel and heat transfer inside the nozzle,
4. nozzle material, and
5. unpredictable disturbances and operational faults.

Two process variables tell about the increased risk for nozzle clogging: cast-ing speed and stopping rod position. Experience has shown that these two variables can give the first indication of nozzle clogging. These variables cannot, however, answer the question, how long time the casting can continue and when the nozzle should be changed. This is done using neural network models. Figure 11.6 shows a block diagram for the system that aims to estimate also the time available for undisturbed casting.

Nozzle clogging was modelled using neural networks [4]. Data was collected from two casters, 5800 heats from each caster and 67 variables per a heat. All the heats were not used in modelling, because clogging is a problem only in the case of certain grades. Four steel grade groups were concerned (grades 1, 2, 3 and 4). Training utilised conventional back propagation procedure. Most of the inputs were scaled to the average value of zero and the standard deviation 1. The serial number of the heat was scaled inside the interval 0.16–1.16. The output variable of the network was the amount of cast tons with a certain nozzle. Its values were scaled inside the interval 0.16–1.16 by dividing the actual measurement value by 720 that is the average total amount of six heats in tons. Modelling used feedforward networks with only one hidden layer. These networks model steady-state dependencies between input and output variables.

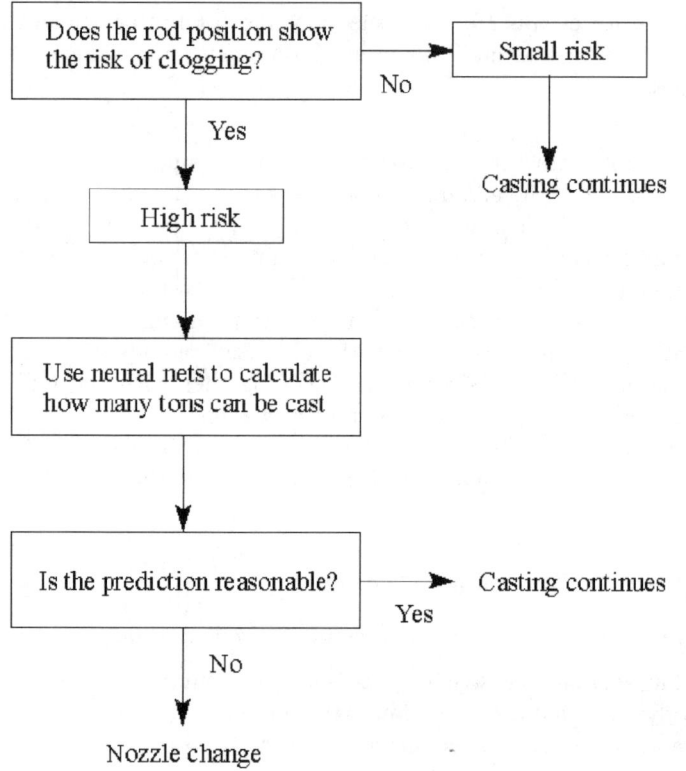

Fig. 11.6. Nozzle clogging detection based on the stopping rod position and neural network model [4]

Modelled variables were chosen with correlation analysis [4]. The amount of data limits the size of the network. The number of training parameters should not exceed the number of training points. In practice, network modelling is difficult if the number of data point is less than 60, because training requires 30 points at minimum and almost the same number is needed for testing. With five inputs and 30 training points, a conventional network can include only 5 neurons. In addition to normal testing procedures, cross-testing considered using the model developed for one caster to the data from another [4]. Only grade 1 showed reasonable results. This confirms that different variables effect on the nozzle clogging in different casters and adaptation is needed in moving the model from one caster to another.

Modelling used only data from cases where clogging occurred. Models were also tested using data from corresponding successful heats [4]. The result was as expected. It is impossible to tell how many tons could have been cast in successful cases, if casting had continued with the same nozzle. Therefore, the cast tons given by the models remain in these cases lower than actual. The result is important from the safety point of view: mod-els never pre-dict too high cast tons.

Hints and Tips

- Pre-processing was essential also in this case. The variable selection used correlation analysis, but the final selection of variables was done after testing several possible combinations. In many cases, expert opinion played a crucial role.
- Updating of neural network models was seen as problematic in practice, and some kind of adaptive mechanism is required. Models must also be re-trained when moving from one caster to another or from a grade group to another.

11.3.3 Case 3: Screw Insertion Process

There have recently been a few attempts to investigate the use of the soft computing techniques for the screw insertion observation [14, 15, 16]. However, the research has mainly focused on classification between successful and failed insertions. In this case, the identification of the fault type was also required.

This case study consists of quality monitoring and fault identification tasks. According to Fig. 11.7, residuals of the data-based normal model are utilised for quality monitoring using fuzzy statistical inference. A learning algorithm is applied to the parameter estimation of the normal model. Specified fault models use fuzzy rules in fault detection.

The system was developed both for electric and pneumatic screwdrivers. In electric case, a simulated torque signal was used and in pneumatic case the measured pneumatic signal from the actual insertion machine was used. In the latter case, a full two-level factorial design program was carried out together with three center-point runs [6]. The system was programmed and tested on-line in Matlab-environment.

Fuzzy rule bases were constructed to achieve a generic fault detection and isolation database for both signal types. In case of the torque signal, the fuzzy rule base was developed on the basis of expert knowledge and theory. In the other case, real experiments were conducted to define a fault model – a rule base for the pneumatic screw insertions. The fault models seem to detect known faults. It was also found out during the tests that the normal model detects unknown faults – i.e. faults not represented in the fault models. The operation of the normal model was also robust against unknown situations and smaller changes in process parameters. This is very important, because process conditions usually change when products change. Also when starting new production lines, this approach seems to work after short ramp-up times.

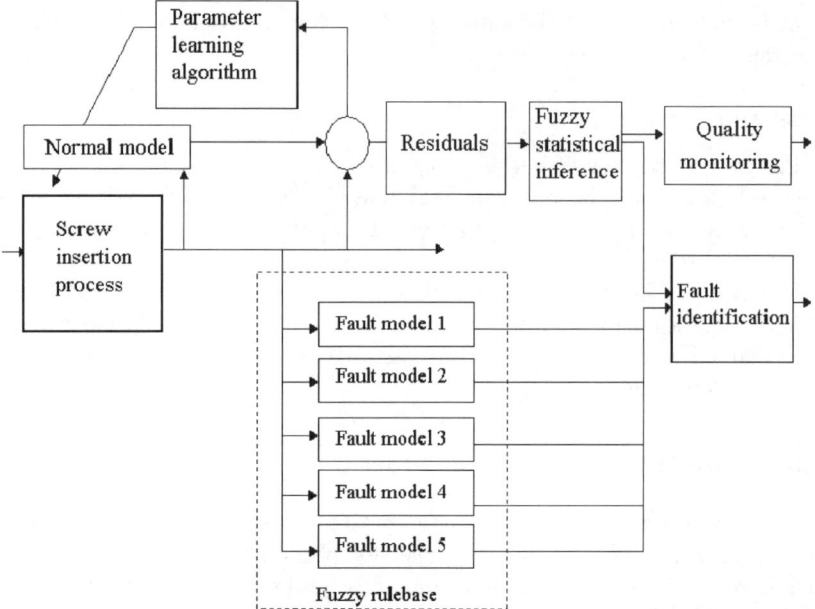

Fig. 11.7. The structure of quality monitoring and fault detection scheme for screw insertion process [6]

Hints and Tips

- Systematic experiment design helped in the training of the system. In practice, this may require some work and care, but the results are good. The procedure might be repeated in new installations and when changing to totally new products.
- This is a "few data" case and expert knowledge is required in constructing the rule bases. This is, however, the case in all short series production and the combination of experimental work, simulation and expert knowledge seems to lead to good results.
- The system could also be run as a simpler fault detection case using only the normal model. It behaves in a robust manner and it tolerates some changes in the production environment.

11.3.4 Case 4: PCB Production

Two kinds of tests can be made to detect faults in electronics production: in-circuit tests and functional tests [17]. In-circuit tests are more and more left aside because of reliability problems and a growing difficulty to physically reach a component. A general approach to functional testing is to develop fault models according to the kind of measurements that are possible to collect and then to define test strategies using for example fault trees [18, 19]. Expert systems often work well only with a very simple PCB architecture and capturing of human expertise is a difficult task in changing environments. The system presented in this case relies only partly on expert knowledge. In order to be flexible the system must be able to correct "inaccurate expert knowledge" and learn from experience. This is achieved by using Linguistic Equations approach as in Sect. 11.3.1 [7].

For the analysis, the PCB board is divided into distinct areas and measurements describing the condition of the area (OK or faulty) are identified. Then linguistic equations tying together the measurements are formed. This includes also the tuning of membership definitions [7, 13]. All these stages can be based on data or on expert knowledge. By a reasonable combination of both approaches, it was found that even data from 15 prototype PCBs was enough for tuning the system.

The use of the systems resembles very much the case in Sect. 11.3.1. The actual measurements from the functional testing are introduced to the system. The system calculates the equations one by one. If the equations are "close" to zero, the area is in normal condition. If not, there is some fault in this area, which is reported to the user. The "closeness" depends on the standard deviations of the linguistic variables [7].

Hints and Tips

- Combining data-driven and knowledge-driven approaches is problematic in many cases. Linguistic equations give a good environment for doing it. For continuous input of user knowledge, a special knowledge acquisition tool was developed [7]. This is useful especially during ramp-up of new products or whole production lines when the knowledge cumulated only gradually during a longer time. It is also useful when starting the fault localisation system.
- The optimal way to decrease ramp-up times is to start knowledge acquisition already in the product design stage and combine the product specifications and data-based models in the same case base. Also here Linguistic Equations provide the user with a systematic environment for model building.

11.4 Hints and Tips

11.4.1 Variable Selection

Correlation analysis combined with expert knowledge is a good tool in variable selection. Calculated variables must be used with care and they should preferably have some physical meaning – it makes the application more transparent to the user. If discontinuous information (rankings, on-off data, ordinal numbers) is used, it must be checked in advance that the modelling method can use them.

Mathematical models and simulators can give hints what variables to choose. Once again, the physical meaning of the variables should be confirmed, together with their measuring possibilities. Experiments with the process should be used, if possible. The systematic design of experiments should be used both in actual tests and in screening designs. These strategies usually lead also to the minimum number of experiments necessary in practical cases.

11.4.2 Data Preparation

Data intensive cases require a lot of data pre-processing. Two cases can be separated: preparation of the modelling data set and the assuring similar data quality for the on-line use of the system. Both are difficult in complicated cases.

Preparation of the modelling data set requires versatile tools. Data consistency must be guaranteed and in many cases more than removing of outliers and replacing missing values is needed. Simulation and balance calculations can be used in assuring the correctness of data. Combining information from several databases requires also good time-keeping in information systems. Time-labelled values (e.g. from the mill laboratory) require extra care. In practice, a lot of pre-processing has already been done in automation and monitoring systems. Sometimes it is difficult to tell what kind of filtering has already been done when data has passed during several systems before introducing it to the fault diagnosis system. Anyhow, each filtering may have an effect to the value used in models. Also different systems have different means to handle e.g. missing values, outliers, and cumulated values. Experience has, however, shown that the original measured value is in many cases difficult, or even impossible, to access.

Scaling is needed almost always. It is elementary with neural networks and the starting point in applying Linguistic Equations approach. The same scaling must be installed in the on-line systems.

11.4.3 Model Selection

Fuzzy models seem to be superior in many cases; be them rule-based or any other variant. They seem to behave best especially when combining data-driven approaches with expert knowledge. The experiences from these four applications show that neural networks lead to systems that are difficult to tune and update and they also remain the black-box systems to the user. Concerning smart adaptive diagnosis systems, Case-Based Reasoning paradigm seems to offer a good backbone to include also the learning stage into the system structure. This is, however, possible with conventional fuzzy systems as shown in the screw insertion example.

11.4.4 Training

It was very confusing to find in Case 2 that after data preparation the data intensive problem was converted almost to "too few data" problem. This must be checked in all cases and the number of parameters in models should be carefully chosen. Automatic tools for training exist in abundance, they all have their special features and must be studied before applying them.

When combining data intensive systems with expert knowledge the tool requirements are increased. A tool is needed that easily combines the user-given rules with automatically generated ones, that gives possibilities to simulate and evaluate the different rules and rule bases and finally optimally combines them. This process is complicated by the fact that in fault diagnosis the size of the rule base easily increases to tens and hundreds, even thousands, of rules and the conventional way of presenting the rules looses its visibility.

11.4.5 Testing

It is clear that testing must be done with independent data, preferably from a totally different data set compared with the training data. Before going on-line, the system must be tested in test bench using simulators.

11.4.6 Evaluation

The best technical evaluation strategy is to run the system in on-line environment using real mill systems in data acquisition and user interface for the longer time. This is not always possible and from the three cases above only Cases 1 and 4 were tested on-line at the mill site. Case 2 was tested at the development site, but with the data collected from the mill and Case 3 with simulated and experimentally acquired data. In off-site testing, it is difficult to evaluate all data preparation and user interface tools in the correct way.

In order to estimate the economic effects of the diagnosis system, the baseline must be defined. This might be difficult in cases, where products or production is changing or a totally new production line is started. Then the earlier experiences or available theoretical values for i.e. the defect tendency must be used.

11.4.7 Adaptation

All cases above show the need for adaptation, but only Cases 1 and 3 show how this adaptation can be realised – so they are real Smart Adaptive Systems. Anyway, intelligent model-based systems offer a good chance for adaptation. The question is, when it is necessary and how to do it in practice. Fuzzy logic gives the most versatile possibilities and adaptation can consider rule bases, membership functions or separate tuning factors. Adaptation itself can also utilise rule-based paradigm (See chapter on Adaptive Fuzzy Controllers).

Combining fuzzy logic and Case-Based Reasoning is also a relevant approach for Smart Adaptive Systems.

11.5 Summary

This chapter described four different cases where model-based diagnosis has been applied in industrial systems. Two cases came from process industry and two from manufacturing industries. In spite of totally different application areas, synergies are easy to see and similar approaches can be used in both areas. It is also clear that both areas can learn from each other.

Fuzzy logic seems to be the tool that most easily applies to diagnosis task. It also offers good possibilities for adaptation, especially when applied together with Case-Based Reasoning.

References

1. Ayoubi M., Isermann R., Neuro-fuzzy systems for diagnosis. Fuzzy Sets and Systems 89(1997)289–307.
2. Isermann R., Supervision, fault-detection and fault-diagnosis methods. XIV Imeko World Congress, 1–6 June 1997, Tampere, Finland.
3. Leiviskä K. Adaptation in intelligent systems – case studies from process industry. Euro CI Symposium, 2002, Kosice, Slovakia.
4. Ikäheimonen, J., Leiviskä, K., Matkala, J. Nozzle clogging prediction in continuous casting of steel. 15th IFAC World Conference, Barcelona, Spain, 2002.
5. Ahola, T., Kumpula, H., Juuso, E. Case Based Prediction of Paper Web Break Sensitivity. 3rd EUNITE Annual Symposium, Oulu, Finland, July 10–11, 2003.
6. Ruusunen M., Paavola M. Monitoring of automated screw insertion processes – a soft computing approach. 7th IFAC Workshop on Intelligent Manufacturing Systems, Budapest, Hungary, 6.–8. 4. 2003.
7. Gebus S., Juuso E.K.K. Expert System for Defect Localization on Printed Circuit Boards. Automation Days, Sept. 2003, Helsinki, Finland.
8. Oinonen K. Methodology development with a web break sensitivity indicator as a case. Proceedings of TOOLMET'96 – Tool Environments and Development Methods for Intelligent Systems. Report A No 4, May 1996, Oulu, Finland, 1996, pp. 132–142.
9. Miyahishi, T., Shimada, H. (1997): Neural networks in web-break diagnosis of a newsprint paper machine. In: 1997 Engineering & Papermakers Conference, Tappi Press, Atlanta, pp. 605–612.
10. Oyama, Y., Matumoto, T., Ogino, K. (1997): Pattern-based fault classification system for the web-inspection process using a neural network. In: 1997 Japan TAPPI Annual Meeting Proceedings, Japan TAPPI, pp. 575–580. (Abstract in ABIPST 68(10), 1120).
11. Fadum, O. (1996): Expert systems in action. In: Fadum, O. (Ed.), Process, Quality and Information Control in the Worldwide Pulp and Paper Industry, Miller Freeman Inc., New York, pp. 23–34.

12. Aamodt A., Plaza E. Case-Based Reasoning: Foundational Issues, Methodological Variations and System Approaches. AICom – Artifical Intelligence Communications, IOS Press, Vol. 7: 1, 1994, pp. 39–59.
13. Juuso E.K. Integration of intelligent systems in developing of smart adaptive systems. International Journal of Approximate Reasoning, in press.
14. Lara B., Seneviratne L.D., Althoefer K. Radial basis artificial neural networks for screw insertions classification. In: Proceedings of the 2000 IEEE International Conference on Robotics & Automation, 2000, pp. 1912–1917.
15. Seneviratne L.D., Visuwan P. Weightless neural network based monitoring of screw fastenings in automated assembly. In: Proceedings of ICONIP '99. 6th International Conference, 1999, pp. 353–358.
16. Mrad F., Gao Z., Dhayagude N. (1995). Fuzzy logic control of automated screw fastening. Conference Record of the 1995 IEEE, 1995, 2, pp. 1673–1680.
17. McKeon A., Wakeling A. Fault Diagnosis in analogue circuits using AI techniques, International Test Conference, 1989, Paper 5.2, pp. 118–123.
18. Stout R.D.L., Van de Goor A.J., Wolff R.E. Automatic fault localization at chip level, Microprocessors and Microsystems 22, 1998, pp. 13–22.
19. Tong D.W., Walther E., Zalondek K.C. Diagnosing an analog feedback system using model-based reasoning, Proceedings of the Annual AI Systems in Government Conference, 1989, pp. 290–295.

12

Design of Adaptive Fuzzy Controllers

K. Leiviskä and L. Yliniemi

University of Oulu, Department of Process and Environmental Engineering,
P.O. Box 4300, 90014 Oulun yliopisto, Finland
Kauko.Leiviska@oulu.fi

This chapter shows how to design adaptive fuzzy controllers for process control. First, the general design methods with the necessary steps are referred to and then a simpler case of adaptive tuning of the scaling factors in the fuzzy logic controller is presented. A case study concerning the adaptive fuzzy control of the rotary dryer visualises the approach.

Fuzzy logic is particularly suitable for process control if no model exists for the process or it is too complicated to handle or highly non-linear and sensitive in the operation region. As conventional control methods are in most cases inadequate for complex industrial processes, fuzzy logic control (FLC) is one of the most promising control approaches in these cases.

Designing of the fuzzy controller follows the general principles shown in Chap. 5, but the on-line application and requirements for the dynamical system response set some additional requirements that are commented upon in this chapter. A need for adaptive tuning is one of these features.

The terminology used in this chapter deserves some comments, especially for non-control specialists. Control means the mechanism needed for two different tasks: to keep the system in its target state or to move it from one target state to another. In control textbooks, this target state is usually called the set point and we speak about stabilising control and trajectory control, depending on the task in question.

In technical systems we can separate two control principles. In feedback control, the controlled variable is measured, its value is compared with its set point and the control action (the change in the manipulated variable) is calculated based on this deviation, the control error. In feedforward control, the measurement is exercised to the main disturbance variable and the manipulated variable is set according to the changes in it. The common problem of the room temperature control gives good examples for both principles. If the heat input to the room is controlled based on the room temperature itself, we speak about feedback control, and if it is set according to the outside temperature, we have the case of feedforward control. (Note that the authors are coming from the North).

K. Leiviskä and L. Yliniemi: *Design of Adaptive Fuzzy Controllers*, StudFuzz **173**, 251–265 (2005)
www.springerlink.com © Springer-Verlag Berlin Heidelberg 2005

Feedback is a very strong principle, and examples are found in nature (predator-pray systems), in biology (hormonal control mechanisms), in sociology, and in economy.

Controller tuning is important in process control. In feedforward control the accuracy totally depends on how good model we have between the disturbance variable and the control variable. In feedback control, the tuning effects on the control speed, accuracy and oscillation tendency. In conventional feedback control the controller takes into account the actual control error, its integral and its derivative, and therefore we speak about three-term controllers or PID-controllers (PID coming from proportional, integral and derivative). This means that in tuning three parameters must be set, one for each term. Actually, the digital control brings one more term along, the control interval, meaning how often the control action is exercised. There are several other combinations possible – P-, PI-, and PD-controllers together with some special algorithms that take the controlled system properties into account [1, 2].

12.1 Introduction

There are several possibilities to implement fuzzy logic in control. The controller itself can be replaced by fuzzy model, the fuzzy logic controller. Fuzzy model can also be used to replace the feedforward controller or it can be run parallel to or in series with the conventional controller, thus helping to take some special situations into account. Fuzzy model can also be used in on-line tuning of the conventional controller or the fuzzy controller, as shown later in this chapter.

Figure 12.1 shows the block diagram of the two-term fuzzy logic controller used as an example later in the text; two-term meaning that only control error and the change in error are taken into account. Following notations are used: $sp(k)$ is the set point, $e(k)$ is the error, $ce(k)$ is the change in error, Ge, Gce and Gcu are the tunable scaling factors, $se(k)$ is the scaled error, $sce(k)$ is

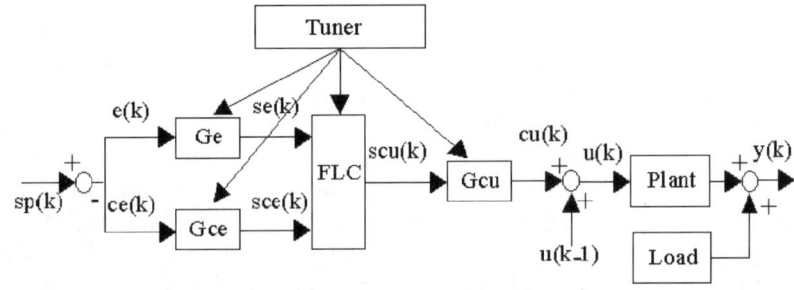

Fig. 12.1. Basic form of the two-term fuzzy logic controller

the scaled change in error, $scu(k)$ is the crisp increment in the manipulated variable, $cu(k)$ is the de-scaled increment in the manipulated variable, $u(k)$ is the actual manipulated variable, $u(k-1)$ is the previous manipulated variable and $y(k)$ is the controlled variable. The variable k refers to the discrete control interval. The fuzzy logic controller (FLC) in Fig. 12.1 consists of blocks for fuzzification, fuzzy reasoning and defuzzification as shown in Fig. 12.1 in Chap. 5. The on-line tuning problem can be defined as follows: *"Select the scaling factors and/or the parameters of FLC so that the requirements set for the controlled variable $y(k)$ are fulfilled in the changing process conditions"*. The solution of this tuning problem leads to adaptive fuzzy control. Following relationships are valid between the variables shown in Fig. 12.1 when the denotations are as before:

The error is the difference between the set point and the controlled variable

$$e(k) = sp(k) - y(k) \tag{12.1}$$

The change in the error is the deviation between two successive errors

$$ce(k) = e(k) - e(k-1) \tag{12.2}$$

If the set point does not change, combining (12.1) and (12.2) gives for the change in the error

$$ce(k) = y(k-1) - y(k) \tag{12.3}$$

For the scaling, the following equations can be written

$$se(k) = [sp(k) - y(k)] \times Ge(k) = e(k) \times Ge(k) \tag{12.4}$$
$$sce(k) = [s(k) - e(k-1)] \times Ge(k) = ce(k) \times Ge(k) \tag{12.5}$$
$$cu(k) = scu(k) \times Gcu(k) \tag{12.6}$$

The rules in the FLC are of the form

$$R_{FLC} : \text{If } e \text{ And } \triangle e \text{ Then } \triangle u \tag{12.7}$$

The case study at the end of this chapter concerns with the adaptive fuzzy control of the rotary dryer and it is based on [3]. Reference [4] deals with the same tuning method, but instead of PI-controller PID-controller is used and also an improved supervisory control strategy is tested.

12.2 Design Approaches

The design of fuzzy logic controllers returns often to the design of the fuzzy model. The fuzzy model includes many tuning factors as membership functions, scaling factors, the type of the fuzzifier, rule base and type of the defuzzifier. These factors have already been considered in Chap. 5 for fuzzy expert systems and the results shown there are not repeated here. This section introduces some general design approaches and more detailed discussion and practical guidelines for the design of adaptive fuzzy controller are given in Sect.12.3. Ali and Zhang [5] have written down the following list of questions for constructing a rule-based fuzzy model for a given system:

- How to define membership functions? How to describe a given variable by linguistic terms? How to define each linguistic term within its universe of discourse and membership function, and how to determine the best shape for each of these functions?
- How to obtain the fuzzy rule base? In modelling of many engineering problems, usually, nobody has sufficient experience to provide a comprehensive knowledge base for a complex system that cannot be modelled physically, and where experimental observations are insufficient for statistical modelling. Moreover, human experience is debatable and almost impossible to be verified absolutely.
- What are the best expressions for performing union and intersection operations? In other words, which particular function of the t-norms and s-norms should be used for a particular inference.
- What is the best defuzzification technique for a given problem?
- How to reduce the computational effort in operating with fuzzy sets, which are normally much slower than operating with crisp numbers?
- How to improve computational accuracy of the model? Being fuzzy does not mean inaccurate. At least, accuracy should be acceptable by the nature of the engineering problem.

Isomursu [6] and later Frantti [7] have also approached the design of fuzzy controllers using basically the general fuzzy modelling approach. They have modified the systematics for fuzzy model design originally given in [8] for on-line adaptation purposes. The approach given in [7] divides the design of fuzzy controllers into following parts:

- Data pre-processing and analysis. Assuring that data used in the design fulfils the requirements set by the design methods. This means mostly eliminating or replacing erroneous measurement values and compensating for process time delays.
- Automatic generation of membership functions. Making the fast adaptation and tuning of fuzzy model (controller in this case) possible. On-line use requires fast and robust algorithms.
- Normalization (Scaling). The input values of the physical domain are normalized to some appropriate interval.
- Fuzzification. Representation of input values in the fuzzy domain.
- Reasoning. Interpretation of the output from read rules.
- Defuzzification. Crisp representation of the fuzzy output.
- Denormalization (De-scaling). Normalized output value is scaled back to the physical domain.

12 Design of Adaptive Fuzzy Controllers

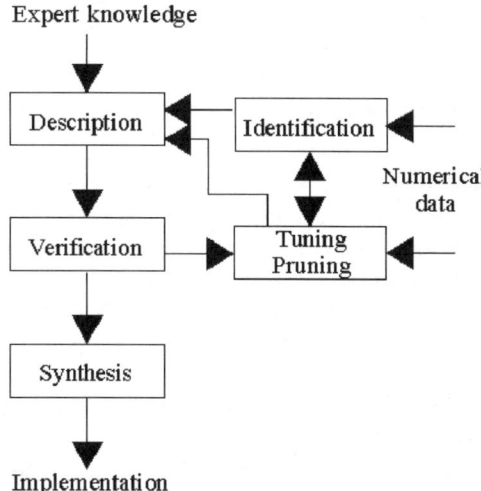

Fig. 12.2. Flow chart of the design methodology [9]

- Representation. Representation of the model result(s).

Several parameters are involved in the design and it is mentioned in [8], that the tuning of the knowledge base (rules) and database (membership functions) is often a non-trivial and time-consuming part of fuzzy modelling. Mostly data-intensive methods must be used and the exploitation of expert knowledge in the traditional way is very difficult when fast adaptation and tuning of the fuzzy model is needed. The third approach has been given in [9] and shown in Fig. 12.2. The following stages are included:

- Description describes the whole fuzzy system according to the expressive, extensive, and modular model. The structure of each module is determined from the expert linguistic knowledge or identified automatically from numerical data.
- Identification provides description from data.
- Verification. Possible errors and deviations have to be detected in order to correct them.
- Tuning techniques are used in correcting deviations from a desired behaviour. These techniques can be exploited in advance at the identification stage so as to already obtain the description of a module with small deviations.
- After tuning and also identification stages, the resulting module descriptions might be simplified by applying pruning techniques.
- The last step is synthesis and it represents the system in an executable format suitable for the application.

12.2.1 Hints and Tips

From the viewpoint of designing on-line adaptive fuzzy controllers, following practical comments can be made.

- Data pre-processing
  On-line applications must rely on on-line data coming from the process automation systems. This means that data has been filtered, its validity has been checked and also some corrections might have been done. Before applying it, it must be clear to the user, how it has been processed. Some systems mark unreliable or erroneous data with some specific symbols or values (e.g. -1). These values must be replaced, for instance, by interpolation. Some extra checking is still needed, e.g. to guarantee that the measured value is inside the normal range of the variable in question.
- Generation of membership functions
  There is a big difference, if the membership functions are adapted during the use or not. If membership functions remain constant, the choice is in their number, location, spread and shape.
  The number of membership functions depends on the application, but some general comments can be made. Use as few membership functions as possible and try to assure that they have a clear physical meaning. This decreases the computational load, and also helps the user to understand the application. Sometimes compromises must be made; the good enough performance cannot be gained with a low number of membership functions. Increasing the number of membership functions of input variables means tighter and more accurate control, but with the expense of the rule base and computing load.
  In many algorithms the locations of membership functions are evenly distributed over the range of the variable. This seems to be good enough to provide reasonable results. In some application tools it is possible to tune the location and spread of membership functions manually during the start-up. Most often this is done by trial and error and the system performance guides, how this should be done. Locating more membership functions closer to the set point means tighter control around the set point and a lazier one when the controlled variable if far from it. If, however, we want more peaceful control around the set point, we can replace the triangular membership function describing the zero error by the trapezoidal one. This introduces the dead band where small errors do not lead to any reaction in the manipulated variable. In practical applications, triangular or trapezoidal membership functions are used.
  If membership functions are subjected to adaptation, or if they are automatically generated, several additional things must be taken into account. The period length and spread of source data must be chosen so that it represents the whole range of variables, but it is collected in the similar process conditions. The adaptation mechanism must guarantee that the adaptation results in the complete set of membership functions and the

system is realisable. The size of the rule base depends on the choice of the number of membership functions for each input variable. Complete rule bases are used and their tuning is mostly done experimentally. Some data based methods exist, but the problem may rise concerning the consistency and completeness of the rule base. Graphical tools are of great help in designing both the membership functions and rule bases.

Definition of fuzzy relations, compositional operators and inference mechanism together with the defuzzification strategy follow the guidelines given in Chap. 5.

12.3 Adaptive Fuzzy Control

Adaptive control originates already from 50's and it is widely used and established area in Control Engineering. Both structural and parameter adaptation have been researched and applied. Concerning parameter adaptation, three main approaches have been developed:

1. Gain scheduling, that adapts to known (or measurable) changes in the environment
2. Model Reference Adaptive Control that adapts the system parameters so that it follows the model behaviour
3. Self-Tuning Control that adapts the controller so that the system response remains optimal in the sense of an objective function.

12.3.1 Literature Review

Despite the field of adaptive fuzzy logic control (FLC) is new, a number of methods for designing adaptive FLCs has been developed. Daugherity et al. [10] described a self-tuning fuzzy controller (STFLC) where the scaling factors of the inputs are changed in the tuning procedure. The process in which the tuning method was applied was a simple gas-fired water heater and the aim was to replace an existing PID-controller with a fuzzy controller.

Zheng et al. [11] presented an adaptive FLC consisting of two rule sets: The lower rule set, called the control rule set, forms a general fuzzy controller. The upper rule sets, called tuning rule sets, adjusted the control rule set and membership functions of input and output variables. Lui et al. [12] introduced a novel self-tuning adaptive resolution (STAR) fuzzy control algorithm. One of the unique features was that the fuzzy linguistic concepts change constantly in response to the states of input signals. This was achieved by modifying the corresponding membership functions. This adaptive resolution capability was used to realise a control strategy that attempts to minimise both the rise time and the overshoot. This approach was applied for a simple two input-one output fuzzy controller.

Jung et al. [13] presented a self-tuning fuzzy water level controller based on the real-time tuning of the scaling factors for the steam generator of a nuclear power plant. This method used an error ratio as a tuning index and a variable reference tuning index according to system response characteristics for advanced tuning performances. Chiricozzi et al. [14] proposed a new gain self-tuning method for PI-controllers based on the fuzzy inference mechanism. The purpose was to design a FLC to adapt on-line the PI-controller parameters. Miyata et al. [15] proposed a generation of piecewise membership functions in the fuzzy control by the steepest descent method. This algorithm was tested on a travelling control of a parallel parking of an autonomous mobile robot. Chen and Lin [16] presented a methodology to tune the initial membership functions of a fuzzy logic controller for controlled systems. These membership functions of the controller output were adjusted according to the performance index of sliding mode control.

Mudi and Pal [17] presented a simple but robust model for self-tuning fuzzy logic controllers (FLCs). The self-tuning of a FLC was based on adjusting the output scaling factor of a FLC on-line by fuzzy rules according to the current trend of the controlled process. Chung et al. [18] propose a self-tuning PI-type fuzzy controller with a smart and easy structure. The tuning scheme allows tune the scaling factors by only seven rules. The adaptive fuzzy method developed by Ramkumar and Chidambaram [19] has the basic idea to parameterise Ziegler-Nichols like tuning formula by two parameters and then to use an on-line fuzzy inference mechanism to tune the parameters of a PI-controller. This method includes the fuzzy tuner with the error in the process output as input and the tuning parameters as outputs. The idea to use the Ziegler-Nichols formula is originally presented in [20].

12.3.2 Design of On-line Adaptive Controllers

The adaptive control problem becomes somewhat easier, especially in on-line tuning, if tuning includes only the scaling factors of Fig. 12.1. Yliniemi [3] and Pirrello et al. [4] have compared different self-tuning fuzzy controllers, which utilize the tuning of the scaling factors.

The procedure followed in all cases is as follows:

- Definition of performance requirements for the controller. These can be defined e.g. as overshoot, settling time, rise time, integral absolute error (IAE) and integral-of-time absolute error (ITAE) in case of both responses due to stepwise set point changes and load disturbances.
- Definition of the fuzzy controller. In the case of PI-controller, the input variables are usually the control error and the change of the error. The output variable from the controller is usually defined as the incremental change in the manipulated variable. The selection of the number and shape of the membership functions follows the comments before.
- Also the formation of the control rules the comments in Sect.12.2.
- Definition of the tuning variables. This selection depends on which variables effect on the dynamic behaviour of the control system. If different

control actions are required at different error levels, the input variable to the tuning algorithm is the error. If the controller performance depends on some other variable, e.g. the production rate, this variable is chosen as the input variable. In the latter case we usually speak about gain scheduling. Tuning output variables are, in this case, the incremental changes of the scaling factors.
- Definition of the tuning rule base. This is once again case-dependent and can be based on experience or theoretical considerations.
- Selection of the defuzzification method. Usually, the centre of gravity/area method is used.
- Testing with simulations.

In order to achieve the best possible control there are a number of constants, which vary from one method to another. The selection of them calls for knowledge and experience. Their effects can also be studied using simulations.

12.3.3 Hints and Tips

- The selection of tuning inputs and their membership functions calls for process knowledge. The control error is the most common one because in non-linear cases bigger errors require stronger control actions to get the controlled variable inside its favourable region. If some other variable is chosen, it must be reliably measured and its effect to the control performance must be known.
- Also here as few membership functions as possible must be chosen. This keeps the rule bases reasonable small. Because the actual processes are always non-linear, rule bases are usually asymmetrical (see Sect.12.4). Expert knowledge helps, but usually the design is by trial and error.
- Graphical design environments help. Drawing control surfaces help in finding possible discontinuities in rule bases or in definitions of membership functions and also in evaluation of local control efforts.

12.4 Example: Self-tuning Fuzzy Control of the Rotary Dryer

12.4.1 Control of the Rotary Dryer

The experimental work described in this section was done with a direct air-heated, concurrent pilot plant rotary dryer shown in Fig. 12.3 [4]. The screw conveyor feeds the solid, calcite (more than 98% $CaCO_3$), from the silo into a drum of length 3 m and diameter 0.5 m. The drum is slightly inclined horizontally and insulated to eliminate heat losses, and contains 20 spiral flights for solids transport. Two belt conveyors transfer the dried solids back into the silo for wetting. Propane gas is used as the fuel. The fan takes the flue gases

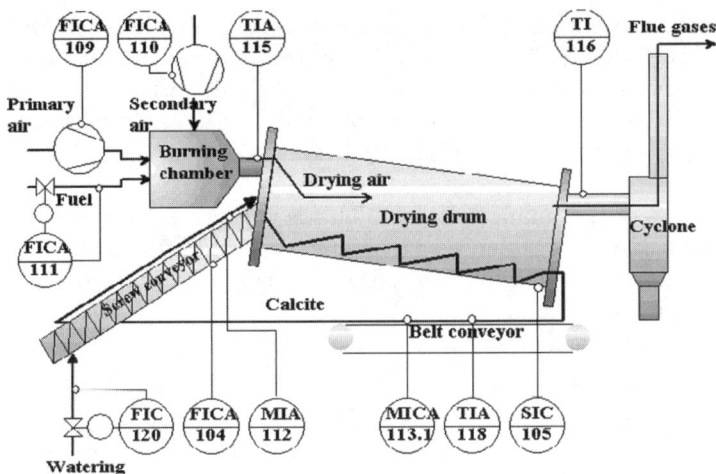

Fig. 12.3. The dryer and its instrumentation

to the cyclone, where dust is recovered. The dryer can operate in a concurrent or counter-current manner.

The dryer is connected to an instrumentation system for control experiments. In addition to measurements of temperature and flow of the solids and drying air, the input and output moisture of the solids is measured continuously by the IR-analysers. The flow of fuel and secondary air are controlled for keeping the input temperature of drying air in the desired value. The velocity of the solids is controlled by the rotational speed of the screw conveyor. It is also possible to control the delay time of the dryer by the rotational speed of the drum.

It is difficult to control a rotary dryer due to the long time delays involved. Accidental variations in the input variables as in the moisture content, temperature or flow of the solids will disturb the process for long periods of time, until they are observed in the output variables, especially in the output moisture content. Therefore, pure feedback control is inadequate for keeping the most important variable to be controlled, the output moisture content of the solids, at its set point with acceptable variations. Increasing demands for uniform product quality and for economic and environmental aspects have necessitated improvements in dryer control. Interest has been shown in recent years in intelligent control systems based on expert systems, fuzzy logic or neural nets for eliminating process disturbances at an early stage.

The basic objectives for the development of dryer control are [3]:

- To maintain the desired product moisture content in spite of disturbances in drying operation,
- To maximize production with optimal energy use and at minimal costs so that the costs of investment in automation are reasonable compared with other equipment costs,

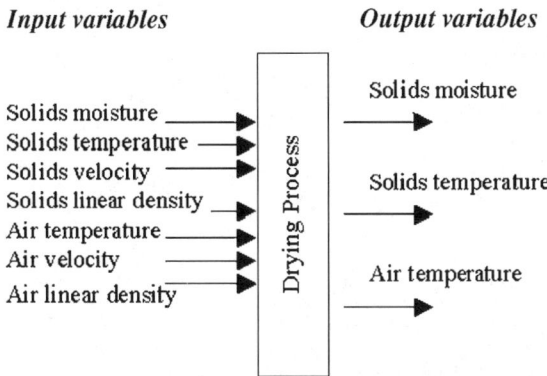

Fig. 12.4. Different input and output variables of the rotary dryer [4]

- To avoid overdrying, which increases energy costs and can cause thermal damage to heat-sensitive solids, and
- To stabilize the process.

Figure 12.4 shows the main input and output variables of the dryer.

12.4.2 Design Procedure for the Pilot Plant Dryer

In this case the method described in [17] is used. This method is a simple but robust model for self-tuning fuzzy logic controllers. The self-tuning of a FLC is based on adjusting the scaling factor for the manipulated variable (Gcu in Fig. 12.1) on-line using the fuzzy rule base according to the current state of the controlled process. The rule-base for tuning the scaling factor was defined based on the error (e) and the change of error (Δe) using the most common and unbiased membership function. The design procedure follows the steps shown in the previous section.

In this case the design procedure goes as follows:

1. Definition of the performance requirements of the controller. The aim of the STFLC is to improve the performance of the existing fuzzy controllers by utilizing the operator's expert knowledge.
2. Definition of the fuzzy controller. The main controlled variable is the output moisture of solids and the main manipulated variable is the input temperature of drying air, which correlates with the fuel flow. The velocity of solids, which correlates with the rotational speed of the screw conveyor can be used as an auxiliary manipulated variable, but it is not included here. The main disturbances are the variations in the input moisture of solids and the feed flow. The existing fuzzy controllers [21] were used.
3. Definition of the tuning variables. In the method of Mudi and Pal [17] the only tuned variable is the scaling factor for the manipulated variable, which is called the gain updating factor, Gcu. The input variables to the tuner

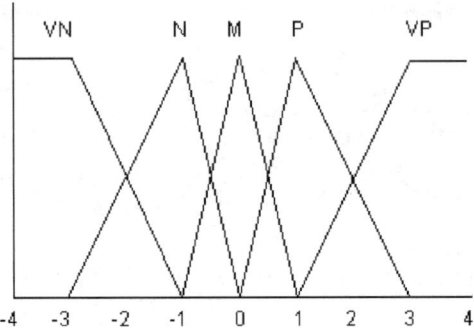

Fig. 12.5. Membership functions for the control error [3]

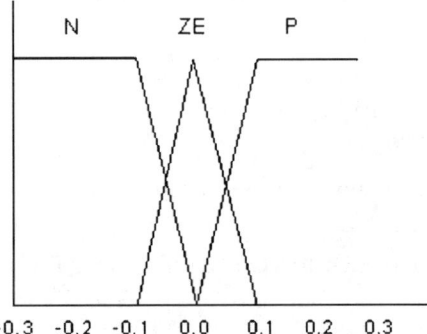

Fig. 12.6. Membership functions for the change of error [3]

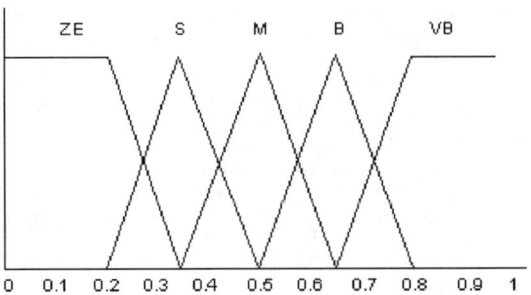

Fig. 12.7. Membership functions for the gain updating factor [3]

are the control error (e) and the change in error (Δe). The number and location of membership functions were chosen experimentally. According to Figs. 12.5 and 12.6, five membership for the control error and three for the change of error were found best in this case. Triangular membership functions were chosen. The output from the tuner is the gain updating factor with five membership functions (Fig. 12.7).

Table 12.1. The rule base for gain updating factor [3]

$e\backslash \Delta e$	N	ZE	P
VN	VB	M	S
N	VB	B	S
M	B	ZE	B
P	S	M	B
VP	S	M	VB

4. Definition of the tuning rule base. The rules based on the analysis of the controlled response and on the experience are shown in Table 12.1.
1. Selection a method of defuzzification. The centre of area method (COA) is used to transform the output of the fired rules into the crisp value.
2. Simulation tests. The controller has been implemented in Matlab®, using Simulink and Fuzzy Logic Toolbox. Simulations have been carried out using the hybrid PI-FLC and the direct FLC as actual fuzzy logic controller [21] and making step changes in the main disturbance variable i.e. the input moisture of solids. Two examples are shown in Figs. 12.8 and 12.9 [3].

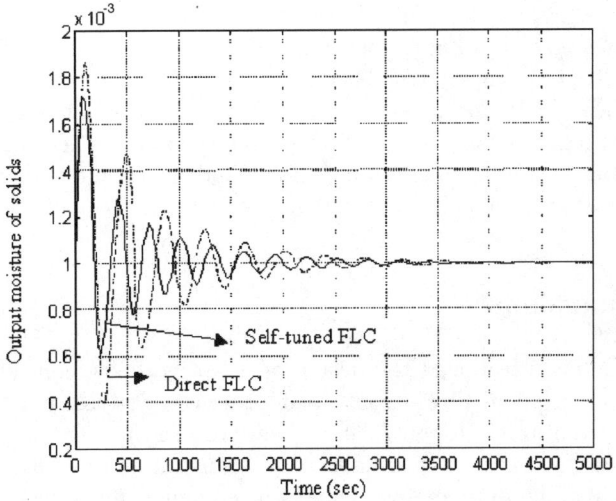

Fig. 12.8. The behaviour of the output moisture of solids; the self-tuned direct FLC and the direct FLC for a step change in the input moisture of solids from 2.4 m-% to 3.4 m-%

12.4.3 Hints and Tips

Some general conclusions can be drawn from Table 12.1. If both e and Δe are big and have the same sign (the situation is bad and getting worse), the

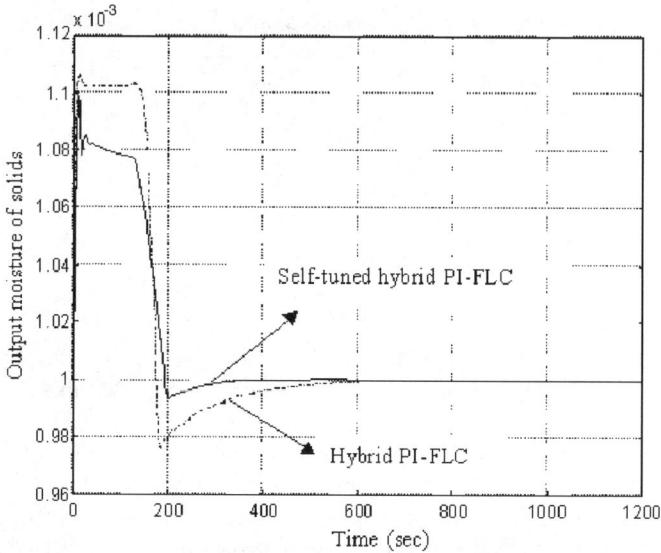

Fig. 12.9. The behaviour of the output moisture of solids; the self-tuned hybrid PI-FLC and the hybrid PI-FLC for a step change in the input moisture of solids from 2.4 m-% to 3.4 m-% [3]

gain updating vectors is big (upper left and lower right corners of the table). This means that strong control actions are carried out to get the controlled variable inside the recommended region. If, on the other hand, both input variables are big, but have opposite signs (the situation is getting better, the gain updating vector is small (upper right and lower left corners). If both variables are small, the gain updating vector is zero.

12.5 Conclusions

This chapter has introduced the design of adaptive fuzzy controllers. These controllers are used in direct control of process variables and in this way replacing conventional PI-controllers. Good results have been gained with rather simple structures that are also easy to install and maintain. However, their design includes quite many manual stages that require expert knowledge.

References

1. Leiviskä K. Control Methods. In "Process Control" edited by K. Leiviskä. Book 14 in Series on Papermaking Science and Technology, Fapet Oy, Finland, 1999.
2. Corripio A.B. Digital Control Techniques. AIChE, Washington, Series A, vol. 3, 1982.

3. Yliniemi L. Adaptive Fuzzy Control of a Rotary Dryer. In Leiviskä K., (editor): Industrial Applications of Soft Computing. Paper, Mineral and Metal Processing Industries. Physica-Verlag, Heidelberg, New York, 2001, 109–132, 2001.
4. Pirrello L., Yliniemi L., Leiviskä K., Galluzzo M. Self-tuning fuzzy control of a rotary dryer. – 15th Triennal World Congress of the International Federation of Automatic Control, Barcelona, 21–26 July 2002. CD IFAC 2002. s. 1–6, 2002.
5. Ali Y. M., Zhang L. A methodology for fuzzy modeling of engineering systems. Fuzzy Sets and Systems, 118, (2001), 181–197, 2001.
6. Isomursu P. A software engineering approach to the development of fuzzy control systems. Doctor Thesis, University of Oulu, Department of Electrical Engineering. (Published as Research Report/95, Technical Research Center of Finland), Oulu, 1995.
7. Frantti T. Timing of fuzzy membership functions from data. Doctoral Thesis. Acta Universitatis Ouluensis, Technica C159. University of Oulu, 88 p., 2001.
8. Driankov D., Hellendoorn H., Reinfrank M. An Introduction to Fuzzy Control. 2nd Edition, New York, Pringer-Verlag, 1994.
9. Moreno-Velo F.J., Baturone I., Barrio-Lorite F.J., Sánchez-Solano S., Barriga A. A Design Methodology for Complex Fuzzy Systems. Eunite 2003, July 10–11. 2003, Oulu, Finland, 2003.
10. Daugherity W.C., Rathakrishnan B., Yen Y. Performance evaluation of a self-tuning fuzzy controller. In: Proc. IEEE International Conference on Fuzzy Systems, 389–397, 1992.
11. Zheng J., Guo P., Wang J.D. STFC – self-tuning fuzzy controller. In: Proc. IEEE International Conference on Systems, Man and Cybernetics 2, 1603–1608, 1992.
12. Lui H.C., Gun M.K., Goh T.H., Wang P.Z. A self-tuning adaptive resolution (STAR) fuzzy control algorithms. In: Proc. 3rd IEEE World Congress on Computational Intelligence. 3, 1508–1513, 1994.
13. Jung C.H., Ham C.S., Lee K.I. A real-time self-tuning fuzzy controller through scaling factor adjustment for the steam generator of NNP. Fuzzy Sets and Systems 7: 53–60, 1995.
14. Chiricozzi E., Parasiliti F., Ptursini M., Zhang D.O. Fuzzy self-tuning PI control of PM synchronous motor drives. In: Proc. International Conference on Power Electronics and Drive Systems 2, 749–754, 1995.
15. Miyata H., Ohki M., Ohikita M. Self-tuning of fuzzy reasoning by the steepest descent method and its application to a parallel parking. IEICE Transactions on Information and Systems e79-d (5), 561–569, 1996.
16. Chen J.Y., Lin Y.H. A self-tuning fuzzy controller design. In: Proc. IEEE International Conference on Neural Networks 3, 1358–1362, 1997.
17. Mudi R.K. Pal N.R. A robust self-tuning scheme of PI-type and PD-type fuzzy controllers. IEEE transactions on Fuzzy Systems 7(1), 2–16, 1998.
18. Chung H.Y., Chen B.C., Lin C.C. A PI-type fuzzy controller with self-tuning scaling factors. Fuzzy Sets and Systems 93, 23–28, 1998.
19. Ramkumar K.B., Chidambaram M. Fuzzy self-tuning PI controller for bioreactors. Bioprocess Engineering 12(5), 263–267, 1995.
20. He S.Z., Tan S., Xu F.L. Fuzzy self-tuning of PID controllers. Fuzzy Sets and Systems 56, 37–46, 1993.
21. Yliniemi L. Advanced control of a rotary dryer. Doctoral Thesis, University of Oulu, Department of Process Engineering, 1999.

13

Optimal Design Synthesis of Component-Based Systems Using Intelligent Techniques

P.P. Angelov[1], Y. Zhang[2], and J.A. Wright[2]

[1] Dept of Communication Systems, University of Lancaster, Lancaster, Lancashire, LA1 4YR, UK
P.Angelov@Lancaster.ac.UK
[2] Dept of Civil and Building Engineering, Loughborough University, Loughborough, Leicestershire, LE11 3TU, UK
<Y.Zhang,J.A.Wright>@Lboro.ac.uk

In this chapter we have considered a special case of application of intelligent techniques, namely evolutionary algorithms and fuzzy optimization for automatic synthesis of the design of a component-based system. We have also considered the problem of adaptation of the solutions found and the ways it can be done "smartly", maximizing the reuse of the previous knowledge and search history. Some useful hints and tips for practical use of these techniques in their application to different problems has been made.

The approach we have presented for automatic synthesis of optimal design of component-based systems is novel and original. It can be applied to similar problems in automatic design of complex engineering systems, like petrochemical pipelines, electronic circuits design, etc. The complex problem of design synthesis, which involves human decisions at its early stages, is treated as a soft constraints satisfaction problem. EA and problem-specific operators are used to numerically solve this fuzzy optimization problem. Their use is motivated by the complex nature of the problem, which results in the use of different types of variables (real and integer) that represent both the physical and the topological properties of the system. The ultimate objective of this fuzzy optimization problem is to design a feasible and efficient system.

The important problem of the design adaptation to the changed specifications (constraints) has also been considered in detail. This adds to the flexibility of the approach and its portability. The approach which is presented in this chapter has been tested with real systems and is realized in Java. It fully automates the design process, although interactive supervision by a human-designer is possible using a specialized GUI.

P.P. Angelov, Y. Zhang, and J.A. Wright: *Optimal Design Synthesis of Component-based Systems using Intelligent Techniques*, StudFuzz **173**, 267–283 (2005)
www.springerlink.com © Springer-Verlag Berlin Heidelberg 2005

13.1 Introduction

Many complex engineering systems, such as HVAC, electronic circuits, petrochemical installations, pipelines etc. can be considered as a set of a number of interconnected components of several different types.

During the late 70s and 80s the component-based approach has been widely used for computer-based simulation, in particular for HVAC systems [4, 5, 6, 7, 8]. This made possible the investigation of changes in HVAC system operating variables and parameters. Studies considering a fixed system configuration are now widely available [9]. Existing approaches allow the designer to chose a particular configuration from a list of prescribed systems [6] or to develop a user-defined system [4, 8]. These options, however, are not flexible, since they offer a limited range of possible solutions [6] and use time-consuming and error-prone procedures [4, 8].

The problem of the design synthesis traditionally involves heavy human participation. Recently, some specialized computer software is used mostly to visualize or optimize the strategic design decisions, which are already taken by the human designer based on his experience and knowledge. The difficulty in the automation of the design process is in the creative and subjective nature of this process, and especially its early stage (so called "conceptual design phase").

An alternative approach has been recently developed [1, 2], which offers a new degree of freedom in the design process. This approach combines the use of intelligent techniques (fuzzy sets for some design constraint definition, evolutionary algorithm for design synthesis and optimization problem solving) and treats the strategic design decisions concerning the system type and configuration as a part of the design optimization process, instead of being predefined [1, 2]. This approach allows new alternative system configurations to be explored automatically and hence, the optimal design to be determined. The important problem of the design adaptation to the changes in the requirements is investigated in this chapter.

Since for a given design problem, a number of feasible systems exist, the proposed approach combines the automatic generation of alternative system configurations with the optimization. The system configuration including the set of typical components and their topological links is encoded into a set of chromosomes. This has been combined with the physical variables representing the mass and energy transfer through the system. A "genome" data structure (combining several types of chromosomes) has been considered which is convenient from both algorithmic and software point of view. This multi-modal problem is solved by EA, because they are particularly appropriate for tackling complex optimization problems [1, 2].

The objective has been to design a feasible and energy efficient system, which satisfies to the greatest possible extent the fuzzy (soft) constraints of the design specification and which is able to adapt to the change in the design specification. The system input is a "design specification" also called "design brief", a standardized description of the design requirements and constraints. The output from the system is a ranked list of suggested design schemes ordered by their energy cost and sizes. The process of design is fully automated. A computational module has been produced in Java, given a performance specification and a library of system components, can synthesize functionally viable systems designs. In addition, the specialized GUI allows interactive supervision of the design process. An example of automatic design of HVAC systems is given as an illustration of the technique.

13.2 Intelligent Synthesis of the Optimal Design of a Component-based System

The component-based systems are composed of interlinked and interacting components. They represent the transfer and elementary operations over the energy, material flow, control signals etc. The functionality of a component is predefined. The design of component-based system includes three main steps:

1. Selection of the collection of components to be present in the system
2. Connection of the components by linking and matching the inputs and outputs
3. Optimize the functionality and performance of all linked components

These three steps have to be performed in this specified order, because steps 2 and 3 depend on the successful outcome of the step 1; step 3 depends on the successful outcome of the steps 1 and 2. For example, if a component is neither selected for the system, neither connecting it, nor setting it up, is possible. Most of the present approaches to system design automation try to decompose the problem and to address the problems on the lower level first, then to aggregate the results. In this chapter, an integrated approach is studied, to simultaneously solve these as a combinatorial structure and parameter optimization problem . More importantly, the sub-tasks 1 and 2 are usually left for the human designer [4, 5, 6, 7], while they have been included in the optimization problem for the first time in [2]. In the present chapter some of the constraints of the design brief are defined as a fuzzy (soft) constraints and the design adaptation aspect is studied.

13.2.1 HVAC System Design Problem

An HVAC system is a typical component-based system. Each component in an HVAC system has certain functionality. For example, heating coils, cooling coils, and steam injectors are air-handling components for heating, cooling, and humidifying air to preferred conditions. Diverging and mixing tees are air distributions components that transport air through the system. And ambient and zones are the source and consumers of conditioned air. Each type of components has a unique air-conditioning process associated to it. The air-conditioning processes are modelled as nodes in the network representation of HVAC system. The inputs and outputs of the components are streams of airflow, which goes in and come out from the components. The connections among the components in the HVAC system are very important. First of all, the components have to be linked together by their input/output streams to provide a closed circuit in which air can circulate. The arrangement of the psychrometric processes is also critical. For example, if air in the system needs to be dehumidified, then the output from a cooling coil needs to be connected to the input of the heating coil. This ensures that the cooling process first chills the air and removes the extra moisture and then the heating process restores the temperature of the air. If the cooling coil and the heating coil are arranged otherwise, the system does not work at all.

When the components are connected, their operational parameters have to be optimized to ensure that the system meets the design requirements. For the HVAC systems, the requirements for design are normally to maintain the thermal comfort, and to provide enough ventilation to the occupied zones. To satisfy the design requirements, the parameters of the components have to be carefully determined. Such parameters include cooling/heating capacity of coils, the amount of steam, which a humidifier provides, and stream distribution factors of airflow diverting tees.

Formulation of the HVAC system design as an optimization problem results in the search for optimal combination of component collection, their connections, and their parameter values. The optimal design is defined as the system that consumes minimal resources, while at the same time, meets all design requirements provided by the user. In this chapter, the objective of minimal energy consumption for the optimal system design is considered. The design requirements are formulated as constraints, which are further discussed in Sect. 13.3.

The optimization problem for HVAC system design is a highly constrained mixed integer-real multi-modal optimization problem. The viable system configurations are distributed in a discontinuous manner throughout the search space; the number of performance optimization variables strongly depends on the system configuration, and therefore, the location of the solution in the search space.

The EA has been selected as a computational approach suitable to address such problems. Because of the specifics of this particular optimization problem, modifications have been made, which concern data structure (see Sect. 13.2.2), problem-specific operators (more details are given in [1]), and constraints handling (see Sect. 13.3). A flowchart of the overall approach considered in this chapter is given in Fig. 13.1.

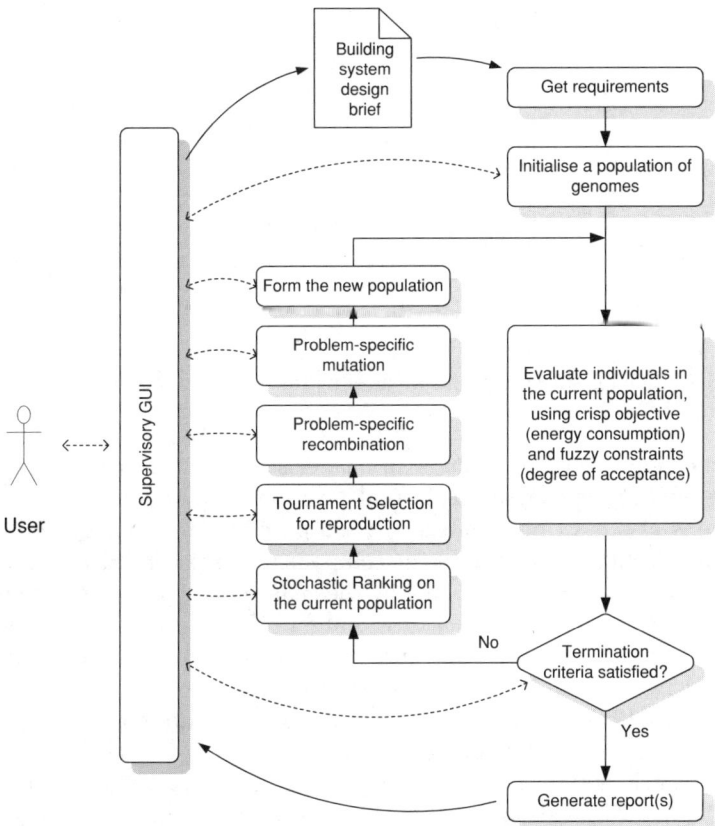

Fig. 13.1. A general flowchart of the approach

13.2.2 Data Structure

The integration of the design synthesis (structure optimization) and design optimization (parameter optimization) into a soft constraint satisfaction problem requires the formulation of a respective integrated data structure (the genotype). We use the genome (a collection of a specifically structured chromosomes), which is represented in Fig. 13.5 for the specific configuration represented in Fig. 13.4. As required for the HVAC system design problem (steps 1–3), the genome consists of three member chromosomes, encoding the collection of components (CompSet), the network (Topology), and the parameters (DutyFlow) respectively. There are normally a number of instances of the DutyFlow chromosome in a design Genome. Each of them represents a set of parameters under certain design condition. In Fig. 13.5 for the particular configuration represented in Fig. 13.4 there are three different design conditions: winter, summer, and swing seasons.

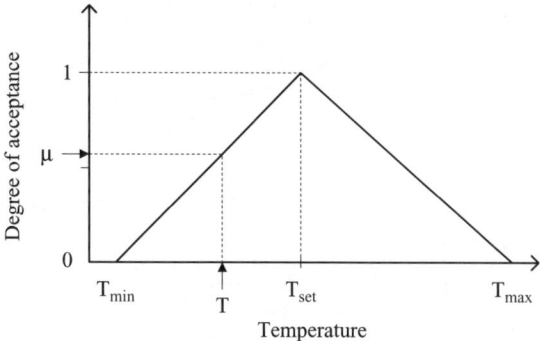

Fig. 13.2. Membership function for the zone temperature requirement formulation

The CompSet chromosome is an integer string of which each gene identifies the type of the component in that position. The Topology member chromosome is also a string of integers. Each gene of the Topology chromosome identifies an outgoing link from the component in the current position, to the component at a position indicated by the gene value. For the components like airflow diverting tees, which has two outgoing links, the secondary links are appended to the list of primary links of all components. This encoding has a variable length and provides easy access to the graph.

The DutyFlow chromosomes are real-encoded chromosomes. Each gene encodes a parameter of a component. These parameters include airflow rates in the network, and duties on the components. Each DutyFlow chromosome is actually a set of operational setting points on which the system performs under certain operating conditions. Sufficient information is provided by each of the DutyFlow chromosomes for system performance simulation in the operating condition.

As whole, the genome structure provides information about the set of components present in the HVAC system configuration, how they are connected, and how the system operates in various running conditions. The quality of the configuration design can then be evaluated based on the performance of the system, and observation of a number of engineering requirements.

13.3 Design Synthesis Using Fuzzy Optimization

The design synthesis problem has been formulated as a fuzzy (soft) constraint satisfaction problem, which in general can be described as:

$$J = f(X_{comp}, Y_{topo}, Z_{param})$$
$$= \sum_{conds} \left(h_{Cond.i} \cdot \left(\sum_{comps} Q_{comp.i} + W_{circ} \right) \right) \to \min \quad (13.1)$$

Subject to:
$$h_{configuration.i}(X_{comp}, Y_{topo}) = 0 \tag{13.2}$$
$$h_{energybalance.j}(X_{comp}, Y_{topo}, Z_{param}) \cong 0 \tag{13.3}$$
$$h_{operation.l}(X_{comp}, Y_{topo}, Z_{param}) \cong 0 \tag{13.4}$$

Where:

$X_{comp} = (x_1, \ldots, x_n) \in \mathfrak{I}^n$ denotes the vector of integer variables, which represent the set of components of a configuration

$Y_{topo} = (y_1, \ldots, y_{n+m}) \in \mathfrak{I}^{n+m}$ is the vector of integer variables representing the topology of a configuration

$Z_{param} = (z_1, \ldots, z_{m+1+k}) \in \mathfrak{R}^{m+1+k}$ denotes the vector of real variables, which represent the operational setting in the system under a design condition, including airflow rates and coil duties

h_{Cond} denotes the nominal occurrence (hours) of a design condition in a typical year

Q_{comp} denotes the duty (KW) on a component under the design condition

W_{circ} denotes the fan power (KW) for air circulation in the system under a design condition

The objective is to minimize the estimated annual energy consumption (kwhr) of an HVAC design. The estimated annual energy consumption is the integration of a selection of typical design conditions and their occurrence in a typical year.

The first set of constraints is crisp and binary. It concerns the viability, feasibility and physical realization of the configuration of the HVAC system. It includes availability of components, connectivity of the topology, and accessibility to particular components. Any violation of this group of constraint results in an infeasible and unsolvable configuration therefore receives a "zero" degree of acceptance.

The other set of constraints ensures that the energy balance will be achieved in an HVAC system under an operational condition. The energy balance in the system is checked as the satisfaction of operational set points for the air-conditioned zones in the system. These constraints are normally categorized as "equality constraints". Since this operational conditions allow slight violations (for example, the zone temperature can be in a range $[T_{min}, T_{max}]$) they are formulated by appropriate fuzzy sets. Triangular membership functions have been used for definition of the zone temperature requirements as more adequate representations of the human perception, which tolerates slight violations of the standard requirements.

$$\mu = \begin{cases} 0, & \text{if } T \leq T_{\min} \\ \frac{T_{\min}-T}{T_{\min}-T_{set}}, & \text{if } T_{\min} < T \leq T_{set} \\ \frac{T-T_{\max}}{T_{set}-T_{\max}}, & \text{if } T_{set} < T < T_{\max} \\ 0, & \text{if } T \geq T_{\max} \end{cases}$$

The last, third set of (inequality) constraints concern the working range of components of the HVAC system for a feasible operating condition. Depending on the type of the component, these constraints can include feasible ranges of airflow, temperature, and humidity. Taking the feasible range of air temperature of the airflow entering a cooling coil for example, the degree of acceptance can be formulated by a trapezoidal membership function as show in Fig. 13.3.

$$\mu = \begin{cases} 0, & \text{if } T \leq T_{\min} \\ \frac{T_{\min}-T}{T_{\min}-T_{lb}}, & \text{if } T_{\min} < T < T_{lb} \\ 1, & \text{if } T_{lb} \leq T \leq T_{ub} \\ \frac{T-T_{\max}}{T_{ub}-T_{\max}}, & \text{if } T_{ub} < T < T_{\max} \\ 0, & \text{if } T \geq T_{\max} \end{cases}$$

The aggregation of the fuzzy constraints (13.3), (13.4) is performed by min operator, which represents the logical "AND":

$$\mu_{ineq} = \min\{\mu_{bal.j}, \mu_{op.l}\} \tag{13.5}$$

The overall objective is to synthesize a feasible and optimal design. Therefore, the optimization problem (13.1–13.4)) includes both crisp (13.2) and

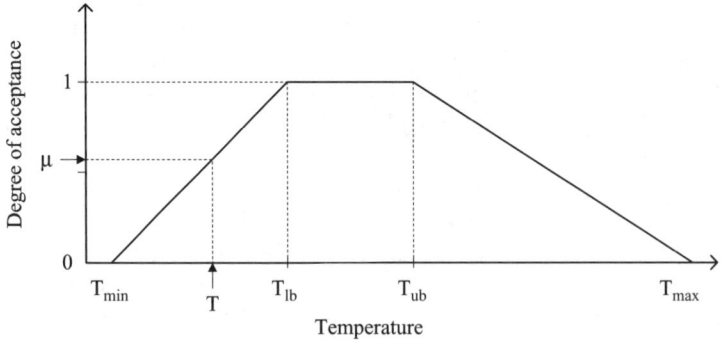

Fig. 13.3. Membership function for inequality constraints

fuzzy constraints and a crisp objective (13.1). The stochastic ranking method [14] is used for numerical solution of this fuzzy optimization problem. It selects probabilistically between the minimization of the energy consumption and the improvement of the degree of acceptance for an HVAC design.

It should be mentioned that the proposed approach is open and allows easily to add/modify any group of constraints.

13.4 Experimental Results

A set of experimental designs has been automatically generated in order to illustrate the ability to synthesize viable HVAC designs as well as to adapt to changing design requirements or weather conditions. Starting with specified design requirements, an (energetically) optimal HVAC system has been automatically synthesized. The set of the design requirements has been carefully selected to provide a good coverage of the operating condition of the HVAC system, therefore the optimal configuration is expected to perform well in real operational conditions. The next examples illustrate the ability of the approach to generate a design adapted to the changed design requirements and/or weather conditions and to make use of previously found "optimal" operation points to aid producing new control schedules (flow and duty). The significant improvement in terms of (energy) performance of optimization algorithm is demonstrated.

13.4.1 Design Synthesis

A simplified HVAC system design problem is set up for testing the capacity of this approach in synthesizing optimal designs. A virtual small office building that contains 2 air-conditioned zones is located in a continental climate where summer is hot and winter is cold. Three typical design conditions are selected to represent the operation in summer, winter, and swing season respectively. A conventional system design is selected from general design manual as the benchmark of performance. Figure 13.4 shows the configuration of the conventional design with its optimal operational setting for the summer condition. The genome encoding is shown in Fig. 13.5.

An optimal design for the given set of design conditions is then automatically designed using the software (ACG) written in Java. Figure 13.6 shows one of the generated designs, together with its operational setting in summer condition. The genome of the design is also shown in Fig. 13.7. It should be mentioned that the configuration of the synthesized design is significantly different compared to the conventional one: the two zones are arranged successively rather than in parallel. This allows better air distribution and energy reuse in the system.

The annual energy consumption associated with the conventional and generated configurations is compared in Table. 13.1. The operation throughout a

1.200
kg/s
25.8ßC
Heat (6)
Cool (9)
13.8ßC
1.227
kg/s
14.0ßC
Mix (3)
Mix (4)
Div (0)
0.842
kg/s
0.306
kg/s
38.9ßC
0kw
-14.68kw
14.0ßC
0.385
kg/s
Heat (8)
Heat (7)
Ambient
(13)
1.20kw
0.0kw
21.3ßC
21.3ßC
17.0ßC
14.0ßC
0.894
kg/s
0.027
kg/s
Zone (12)
[east]
Zone (11)
[west]
2.773kw
5.226kw
0.306
kg/s
24.0ßC
20.0ßC
21.3ßC
Div (2)
Div (1)
Mix (5)
1.200
kg/s
1.227
kg/s
0.842
kg/s

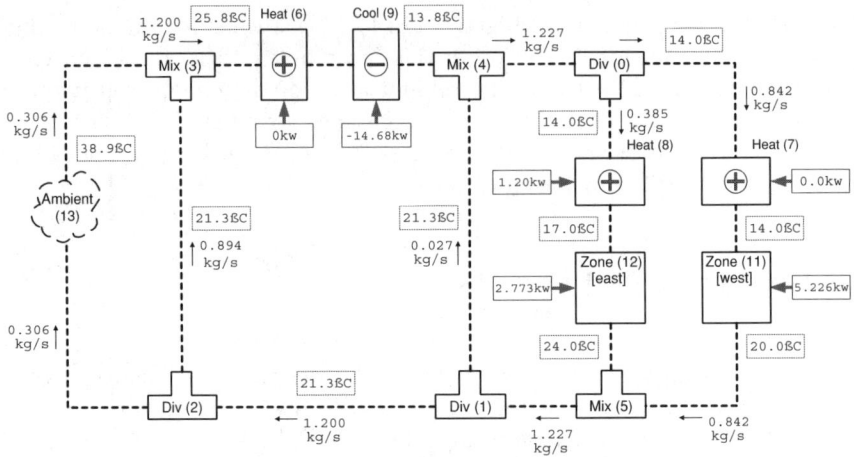

Fig. 13.4. A conventional HVAC system configuration (summer condition)

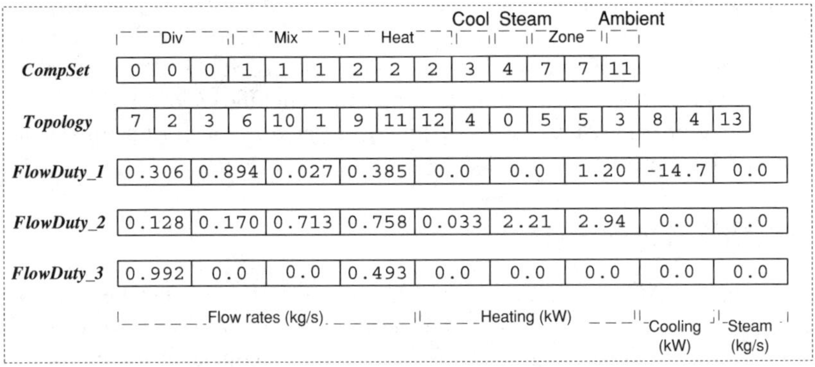

Cool Steam
Ambient
Div
Mix
Heat
Zone
CompSet
0 0 0 1 1 1 2 2 2 3 4 7 7 11
Topology
7 2 3 6 10 1 9 11 12 4 0 5 5 3 8 4 13
FlowDuty_1
0.306 0.894 0.027 0.385 0.0 0.0 1.20 -14.7 0.0
FlowDuty_2
0.128 0.170 0.713 0.758 0.033 2.21 2.94 0.0 0.0
FlowDuty_3
0.992 0.0 0.0 0.493 0.0 0.0 0.0 0.0 0.0
Flow rates (kg/s)
Heating (kW)
Cooling
(kW)
Steam
(kg/s)

Fig. 13.5. The genome for the reference configuration

year would result in significant energy saving (3575 kwhr, or 25%). It should be noted, however, that the actual saving highly depends on the particular design specification and therefore this figure should not be generalized.

13.4.2 Design Adaptation

Two sets of experiments are performed in order to evaluate the ability of the approach to adapt the synthesized design to changes in the design requirements and/or weather conditions. The first set involves a small change in the zone requirement. The temperature of zone (11) is reduced from 24°C to 22°C, which is very common in reality if an occupant tunes down the preferable room temperature. The new optimal operational setting for the generated design is

22.1ßC
1.166 kg/s
0.772 kg/s
22.1ßC
0.874 kg/s
Mix (4)
Div (0)
Mix (3)
21.0ßC
0.128 kg/s
0.394 kg/s
22.1ßC
Cool (8)
Heat (6)
13.2ßC
38.9ßC
Ambient (12)
20.0ßC
Div(2)
0.102 kg/s
Zone (11) [east]
24.0ßC
2.773kw
1.038 kg/s
0.292 kg/s
-0.93kw
0.0kw
24.0ßC
Cool (9)
Heat (7)
0.0kw
-9.32kw
0.128 kg/s
22.1ßC
13.6ßC
Zone (10) [west]
20.0ßC
Div (1)
Mix (5)
20.0ßC
1.166 kg/s
15.7ßC
0.874 kg/s
5.266kw

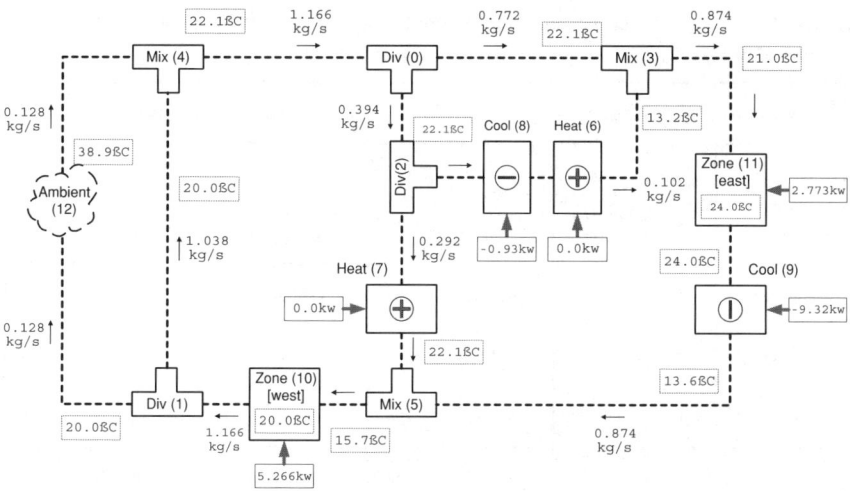

Fig. 13.6. Configuration synthesized automatically by the new approach (summer condition)

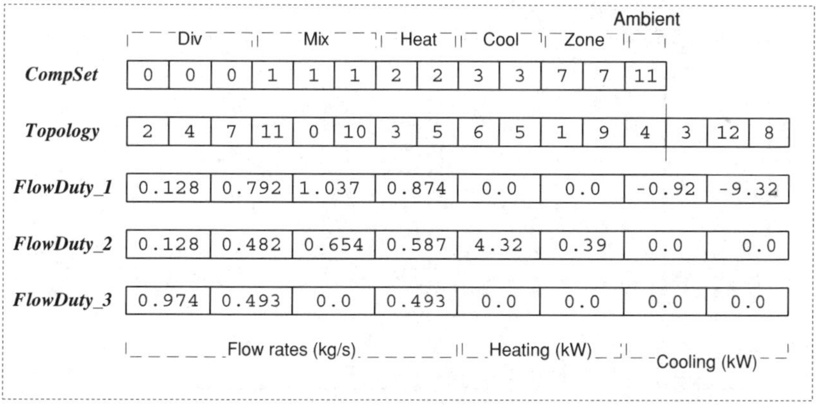

Div
Mix
Heat
Cool
Zone
Ambient
CompSet
0 0 0 1 1 1 2 2 3 3 7 7 11
Topology
2 4 7 11 0 10 3 5 6 5 1 9 4 3 12 8
FlowDuty_1
0.128 0.792 1.037 0.874 0.0 0.0 -0.92 -9.32
FlowDuty_2
0.128 0.482 0.654 0.587 4.32 0.39 0.0 0.0
FlowDuty_3
0.974 0.493 0.0 0.493 0.0 0.0 0.0 0.0
Flow rates (kg/s)
Heating (kW)
Cooling (kW)

Fig. 13.7. The genome for the automatically synthesized configuration

shown in Fig. 13.8, where the new zone temperature and the redistributed loads on the cooling coils are highlighted.

Three experiments have been performed to illustrate the ability of the approach to adapt the previously optimized operational conditions in the new optimization task:

1. EA starts with a randomly initialized population of solutions;
2. the random initial population is seeded with the best solution from the optimal design for the previous conditions;
3. EA starts using the whole population of solutions from the optimization for the previous design conditions.

Table 13.1. Energy consumption for conventional and automatically synthesized configurations

Design Condition		Conventional Configuration			Generated configuration			Total energy saving (kWh)
Condition	Number of Hours (h)	Total Heat/Cooling Consumption (kWh)	Total circulation consumption (kWh)	Total energy consumption (kWh)	Total Heat/Cooling Consumption (kWh)	Total circulation consumption (kWh)	Total energy consumption (kWh)	
Cond1 - Summer	650.0	10321.8	265.7	10587.5	6658.9	484.8	7143.7	3443.8
Cond2 - Winter	650.0	3368.7	105.8	3474.4	3061.4	147.7	3209.1	265.3
Cond3 - Swing	1300.0	0.0	296.2	296.2	0.0	430.3	430.3	-134.1
Annual Total				14358.1			10783.2	3574.9

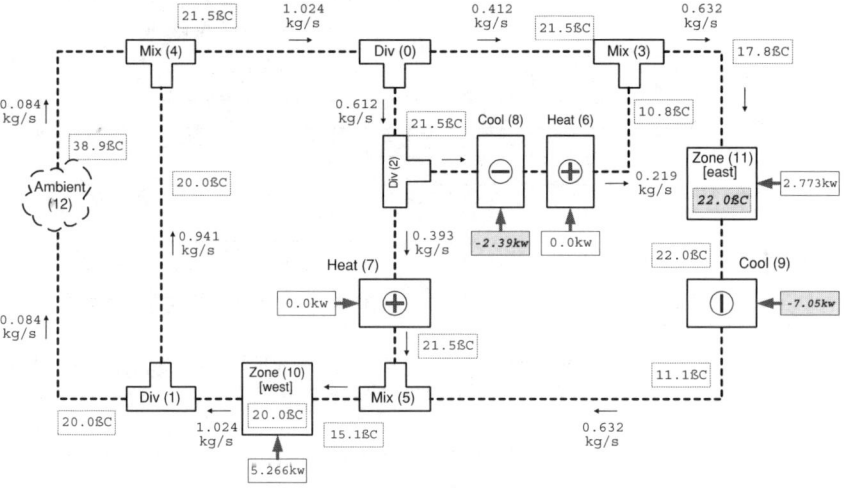

Fig. 13.8. Configuration synthesized automatically for different design requirements

The experiments are repeated 10 times each, and the averaged results are shown in Fig. 13.9. The results from the third experiment performs significantly better than the other two in terms of convergence of the objective (energy consumption), and the average number of generations required to find the first feasible (without constraint violation) solution. The second test result has an intermediate performance, though the difference in the average number of generations required to find the first feasible solution is insignificant compared to the first experiment.

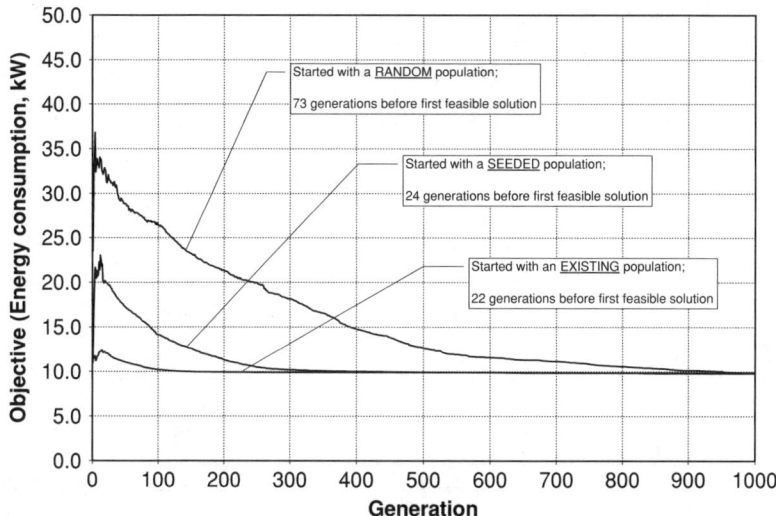

Fig. 13.9. Adaptation of the design synthesis to a small change in the design requirements

The second set of experiments are based on a significantly different design condition, in which the ambient temperature is dropped from 38.9°C to 28.9°C, and functionality of the 2 zones (supposedly) are exchanged. The heat gain and temperature settings of the 2 zones are swapped, compared to the original design condition. These changes and the load redistribution on the cooling coils in the system are highlighted in Fig. 13.10.

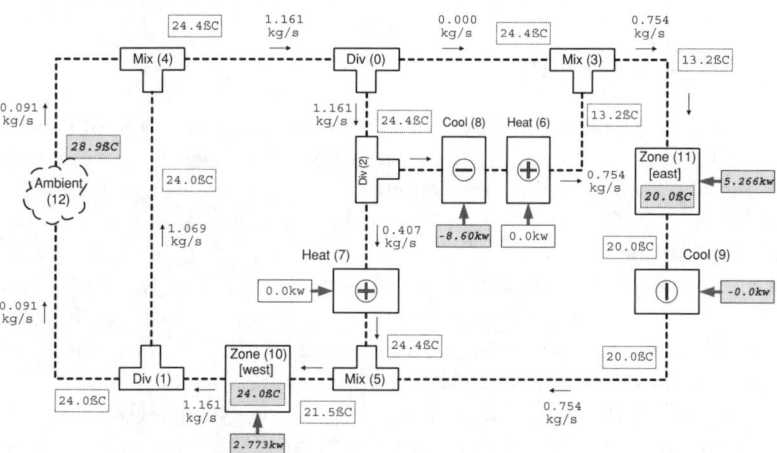

Fig. 13.10. Configuration synthesized automatically for different design requirements

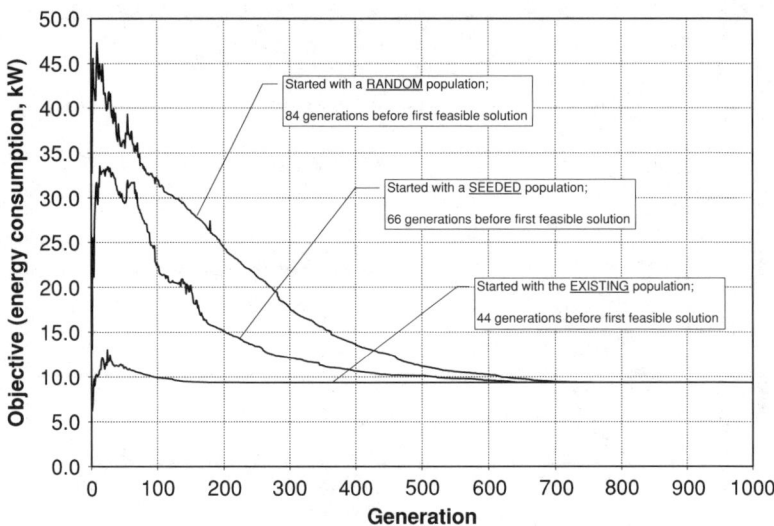

Fig. 13.11. Adaptation of the design synthesis to a larger change in the design requirements

All of the three experiments described above have been performed on the new design settings. The characteristics of the performance in all three experiments (Fig. 13.11) are similar to the characteristics when the change in the design requirements is small (Fig. 13.9). However, for this set of experiments, it takes significantly more generations for the EA to find the first feasible solution compared to the corresponding experiments in the first set.

13.5 Hints and Tips

In this chapter we have considered a special case of application of some intelligent techniques to the sophisticated and highly subjective process of design synthesis of a component-based system. This has been considered on the example of an HVAC system. It has been demonstrated that the evolutionary algorithms can be instrumental for the automation of the design process.

As a first step, the component-based system of interest should be encoded into a chromosome, or more generally, into a genome, which can consist of a number of chromosomes of different nature and length. All of the chromosomes in a genome, however, are linked by the physical representation of the system of interest. They describe different aspects of this system. For example, one chromosome may describe the set of components; the other one can describe the topology or links between the components; another one can describe the physical variables and parameters of the system, such as flows and duties in the case of HVAC. Some hints and tips:

- It should be noted that the evolutionary algorithms can be slow in time, especially when real values are considered;
- It may be advantageous to use mixing encoding (binary, integer and real-valued) or combination of two of these types; this may be closer to the nature of the variables (integers are more appropriate for components identification numbers and real values are needed for representation of flow rates, for example).

As a next step, the optimization of the encoded configuration with its parameters can be performed using a optimization search technique. Some hints and tips:

- Other search algorithms can be suitable for part of the problem (gradient-based search for the physical variables optimization; integer programming or combinatorial search for the configuration optimization), but there is no single algorithm or method that outperforms the others for the overall problem of design synthesis;
- It is very important to properly formulate the constraints in this optimization problem; it is highly constraint one and the result is only one single optimal configuration or a very small number of configurations with similar performance.

In this step, the constraints can be formulated "softly" considering some possible violations. Hints:

- This can be done using fuzzy sets theory and fuzzy optimization, respectively [15, 16];
- Be careful in formulation of the parameters of the membership functions of the fuzzy constraints. They can be logical physical values (zero flow rate, outside temperature etc.) or some tolerances (T_{min}, T_{max}), which are based on the experience, e.g. the comfort temperature is in the range [20°C, 22°C], it is not a fixed number;
- There is no significant difference in the results in respect to the type of membership functions (because the optimization leads to a single solution or a small number of optimal solutions); therefore, use Gaussian or triangular type, which are convenient for differentiation and encoding.

Finally, the portability or adaptability of the solution is practically limited because of the use of evolutionary algorithms, which are iterative, and because of the highly nonlinear nature of the optimization problem. The latter is also one of the reasons for the iterative nature of the solution. Nevertheless, using proper genome encoding and initialization, it is possible to reuse the solutions previously found, and start the search in a modified problem using the previously found solutions. This has been illustrated on the example of HVAC, but has more general nature. Some hints:

- It has been demonstrated that different configurations are needed for different seasons, but this can come as a result of switching off some of the capacities (coils) during of the seasons and other coils during the other season (compare Fig. 13.8 and Fig. 13.10, including the parameters values);
- It has been demonstrated that using the best solution from a similar problem increases significantly the search process (Fig. 13.9 and Fig. 13.11, the line in the middle), but it has also been demonstrated on a practical example, that if we use the whole population from a similar problem it speeds up the search even more (the same figures, the line in the bottom).

13.6 Conclusions

In this chapter we have considered a special case of application of intelligent techniques, namely evolutionary algorithms and fuzzy optimization for automatic synthesis of the design of a component-based system. We have also considered the problem of adaptation of the solutions found and the ways it can be done "smartly", maximizing the reuse of the previous knowledge and search history. Some useful hints and tips for practical use of these techniques in their application to different problems has been made.

The approach we have presented for automatic synthesis of optimal design of component-based systems is novel and original. It can be applied to similar problems in automatic design of complex engineering systems, like petrochemical pipelines, electronic circuits design, etc. The complex problem of design synthesis, which involves human decisions at its early stages, is treated as a soft constraints satisfaction problem. EA and problem-specific operators are used to numerically solve this fuzzy optimization problem. Their use is motivated by the complex nature of the problem, which results in the use of different types of variables (real and integer) that represent both the physical and the topological properties of the system. The ultimate objective of this fuzzy optimization problem is to design a feasible and efficient system.

The important problem of the design adaptation to the changed specifications (constraints) has also been considered in detail. This adds to the flexibility of the approach and its portability. The approach which is presented in this chapter has been tested with real systems and is realized in Java. It fully automates the design process, although interactive supervision by a human-designer is possible using a specialized GUI.

Acknowledgement

This particular problem of the optimal HVAC design synthesis has been supported by the American Society for Heating Refrigeration and Air-Conditioning Engineering (ASHRAE) through the RP-1049 research project.

References

1. Angelov P P, Zhang Y, Wright J A, Buswell R A, Hanby V I (2003) Automatic Design Synthesis and Optimization of Component-based Systems by Evolutionary Algorithms, In: Lecture Notes in Computer Science 2724 Genetic and Evolutionary Computation (E. Cantu-Paz et al. Eds.): Springer-Verlag, pp. 1938–1950.
2. Angelov P P, Hanby V I, Wright J A (1999) An Automatic Generator of HVAC Secondary Systems Configurations, Technical Report, Loughborough University, UK.
3. Michalewicz Z (1996) Genetic Algorithms + Data Structures = Evolution Programs. 3rd edn. Springer-Verlag, Berlin Heidelberg New York.
4. Blast (1992) Blast 3.0 User Manual, Urbana-Champaign, Illinois, Blast Support Office, Department of Machine and Industrial Engineering, University of Illinois.
5. York D A, Cappiello C C (1981) DOE-2 Engineers manual (Version 2.1.A), Lawrence Berkeley Laboratory and Los Alamos National Lab., LBL-11353 and LA-8520-M.
6. Park C, Clarke D R, Kelly G E (1985) An Overview of HVACSIM+, a Dynamic Building/HVAC/Control System Simulation Program, In: 1st Ann. Build Energy Sim Conf, Seattle, WA.
7. Klein S A, Beckman W A, Duffie J A (1976) TRNSYS – a Transient Simulation Program, ASHRAE Trans 82, Pt. 2.
8. Sahlin P (1991) IDA – a Modelling and Simulation Environment for Building Applications, Swedish Institute of Applied Mathematics, ITH Report No 1991:2.
9. Wright J A, Hanby V I (1987) The Formulation, Characteristics, and Solution of HVAC System Optimised Design Problems, ASHRAE Trans 93, pt. 2.
10. Hanby V I, Wright J A (1989) HVAC Optimisation Studies: Component Modelling Methodology, Building Services Engineering Research and Technology, 10 (1): 35–39.
11. Sowell E F, Taghavi K, Levy H, Low D W (1984) Generation of Building Energy Management System Models, ASHRAE Trans 90, pt. 1: 573–586.
12. Volkov A (2001) Automated Design of Branched Duct Networks of Ventilated and air-conditioning Systems, In: Proc. of the CLIMA 2000 Conference, Napoli, Italy, September 15–18, 2001.
13. Bentley P J (2000) Evolutionary Design by Computers, John Willey Ltd., London.
14. Runarsson T P, Yao X (2000) Stochastic Ranking for Constrained Evolutionary Optimisation, IEEE Transactions on Evolutionary Computation, 4(3): 284–294.
15. Angelov P (1994) A Generalized Approach to Fuzzy Optimization, International Journal of Intelligent Systems, 9(4): 261–268.
16. Filev D, Angelov P (1992) Fuzzy Optimal Control, Fuzzy Sets and Systems, v. 47 (2): 151–156.

14

Intelligent Methods in Finance Applications: From Questions to Solutions

M. Nelke

MIT – Management Intelligenter Technologien GmbH, Pascalstrasse 69, 52076 Aachen/Germany
martin.nelke@mitgmbh.de

In this chapter the application of intelligent methods in financial sector are discussed. The first part of the chapter is dedicated to the general description of the process of building smart adaptive system based on the data analysis process. The second part focuses on application of intelligent methods and building smart adaptive systems for two specific problems: factoring credit analysis and customer consultancy assignment. Finally the chapter concludes with a list of questions we have found very important time and time again when designing smart adaptive systems for our purposes. These questions can be regarded as basic guidelines and could be used together with information from other chapters in a general process of designing successful smart adaptive systems for other application domains.

14.1 Introduction

14.1.1 Intelligent Methods and Finance Applications

Smart adaptive systems are systems developed with the aid of Intelligent Technologies including neural networks, fuzzy systems, methods from machine learning, and evolutionary computing and aim to develop adaptive behavior that converge faster, more effectively and in a more appropriate way than standard adaptive systems. These improved characteristics are due to either learning and/or reasoning capabilities or to the intervention of and/or interaction with smart human controllers or decision makers. It is referred to adaptation in two ways: The first has to do with robustness i.e. systems that adapt in a changing environment. The second has to do with portability i.e. how can a system be transferred to a similar site with minimal changes. However, the need for adaptation in real application problems is growing. The end-user is not technology dependent: he just wants his problem to be solved [3].

The area of Business Intelligence is one of the most active ones in the late years. Business intelligence is a term describing new technologies and concepts to improve decision making in business by using data analysis and fact–based systems. Business Intelligence encompasses terminology including: Executive Information Systems, Decision Support Systems, Enterprise Information Systems, Management Support Systems and OLAP (On-Line Analytical Processing) as well as technologies such as Data Mining, Data Visualization and Geographic Information Systems. It also includes the enabling technologies of Data Warehousing and Middleware that are critical components of many Business Intelligence Systems. The need to develop customer relationship models, to mine large business data bases, to find effective interfaces of the hot topics in the area. Still, the business environment is characterized by complexity and is subject to external factors that make traditional models to operate only under ideal conditions. The problem of building systems that adapt to the users and in particular situations is present more than ever as competition is pressing in this area for effective and aggressive solutions.

Intelligent technologies, such as fuzzy logic and neural networks, have since long emerged from the ivory towers of science to the commercial benefit of many companies. In a time of ever more complex operations and tighter resources conventional methods are all too often not enough to maintain a competitive advantage. The technology race has accelerated its pace and impacts globally. This has led to increasing competitiveness, with many companies having seen themselves marooned. Intelligent technologies can be applied to the efficient solution of problems in many business and industrial areas. Both fuzzy technologies as well as neural networks are well proven and are already bringing rewards to those who have sought to adopt them.

Recent years have witnessed fundamental changes in the financial service sector with competition between banking and credit institutions intensifying as a result of the changing markets. The products and services offered by the various financial institutions have become interchangeable, as a result of which their perceived images have lost their sharp focus. Today, there are hardly any remaining banking institutions which specialize within only one specific customer group or with only one traditional banking product.

Against this background, the need to cater for the specific requirements of target customer groups has become the guiding principle behind the business strategies adopted by banking institutions. The importance of a trusting relationship between a bank and its customers, and the fostering of this relationship, have also become to be recognized as decisive factors in financial transactions. Accordingly, the building and nurturing of this trust between bank and customer is now at the center of a bank's strategic planning.

It is not simply the provision of services by the bank to the customer, but moreover the fulfillment of a customer's needs with the right banking products and services which has become to be seen as important. These customer needs, and subsequently the provision of the appropriate products and services to the customer, are dependent on the personal attributes of the individual, e.g. his age, income, assets and credit worthiness, see Sect.14.3.2.

Due to the steady growth of the number of business transactions on one side and of insolvencies on the other side, credit insurance plays an increasing role in the economy. Since the number of proposals which have to be decided by credit insurance companies is also steadily growing, automated systems for credit analysis are needed. Already running systems cannot model expert behavior well enough. So for the optimization of the credit analysis and for the making of objective decisions via an automated software system an approach by Business Intelligence technology is used, see Sect.14.3.1.

14.1.2 Requirements for Solutions Based on Typical Questions

The requirements for computer systems used for advanced analyses and applications in finance applications are derived by the necessity of adaptation to the dynamic change of the customer's behavior and situation and the target of increasing automation and better performance saving costs and fulfilling external directives (e.g. by laws and regulations for credits).

To develop a successful business solution based on a smart adaptive system, first of all it is necessary to focus on questions. Questions are asked by the end-user who wants to benefit from the system during his (daily) business. His questions like

- What will be my benefits in terms of time/cost savings or increased margins?
- Do the existing interfaces support an easy integration into our existing IT-systems?
- Can I compare different parameter settings in the way of simulation of business scenarios?
- Will I understand the behavior of the smart adaptive system?
- How much consultancy do I need when changes on the system are required caused by new boundary conditions?

Lead to basic requirements for the system development:

- Technology, e.g.
  - using of well-known established methods and technologies
  - supporting all necessary steps of the system development (interfaces, model development, model analysis, verification)
- Understanding, e.g.
  - rule-based models are easier to understand than non-linear formulas
  - reports showing the business results instead of data tables
  - user interfaces for end-users based on their business processes

- Adaptation, e.g.
    - data pre-processing finding appropriate parameter pre-settings
    - flexibility to adapt and compare different models to find the best one
    - running the same models over time for dynamic processes

In the next section the data analysis process as a path to develop a smart adaptive system solution is described.

14.2 Building a Smart Adaptive System Based on the Data Analysis Process

14.2.1 From Conceptual Design to Modeling and Evaluation

The design of an application that requires adaptation is not a single step but an iterative process. Core of these process is the data analysis process generally considered to be highly iterative and consisting of different phases. At any stage of the process it may be necessary to go back to an earlier stage in order to guarantee appropriate modeling results. Figure 14.1 depicts a sketch of the most important ones included in this process.

The conceptual design of a smart adaptive system contains the analysis of

- the business needs and end-user user requirements,
- the business workflow (for development, application, maintenance),
- the transformation of the available data model to a data model necessary for analysis and modeling,

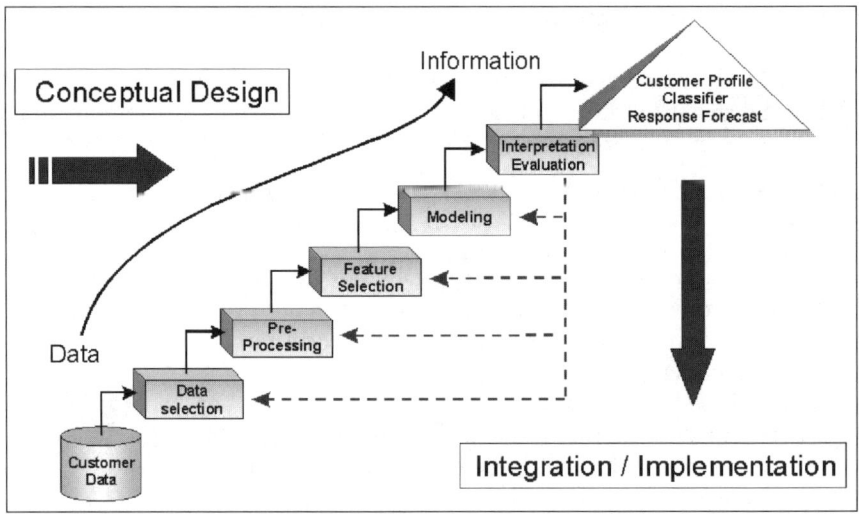

Fig. 14.1. The Data Analysis Process

- the available data sources and
- the existing software environment/technical infrastructure.

It actually starts with the acquisition and storage of data in a data base. Considering the vast amounts of data that can be acquired from business processes, the data to be analyzed has to be carefully selected.

In the pre-processing stage the data is filtered, missing values and outliers are treated, derived quantities are calculated from the basic input variables and all the variables are transformed and normalized.

This step is followed by a feature generation and selection procedure that is supposed to ensure that only the most relevant of the pre-processed variables are used for modeling. Typical feature generation examples are calculating the age (derived by date of birth) and aggregation and grouping (of similar products/customers). An appropriate feature selection is necessary because complete data sets without missing values are typically not always available, many features increase computing time for data pre-processing, modeling and recall and models containing many features are difficult to analyze and to interpret. Feature selection is influenced by the business content, confidential reasons and the modeling tasks to be carried out. The number of features has to be as large as necessary, as small as possible. An iterative modeling process comparing results of different feature sets can show the sensitivity if one feature is added or omitted.

In the modeling phase the above mentioned intelligent technologies are applied to the selected features in order to construct a model. This model has to be interpreted and its performance has to be evaluated before it is applied within a business automation environment. Typical modeling technologies are value distribution analysis, correlation analysis, application of expert knowledge or methods as shown in Fig. 14.2.

The intelligent technologies have in common that they imitate to some extend successful biological examples on life evolution, perception and reasoning. In practical these ideas are translated into algorithms and are implemented in computer systems like [4]. A detailed description of intelligent data analysis can be found in [1]. Depending on the business content, different data mining tasks (as shown in Fig. 14.3) have to be carried out. The available methods allow to chose between different technologies for the same task based on the user requirements.

14.2.2 Successful Modeling: Hints and Tips

Based on the practical experience of building different models in the finance applications some remarks on a proper design are given in the following. First, data is as well as important as the analysis part. Does the available input data cover the expected feature space the model should work properly? If you use data from external sources (geo-coded lifestyle and household information of customer addresses), had an evaluation shown the benefit to be expected?

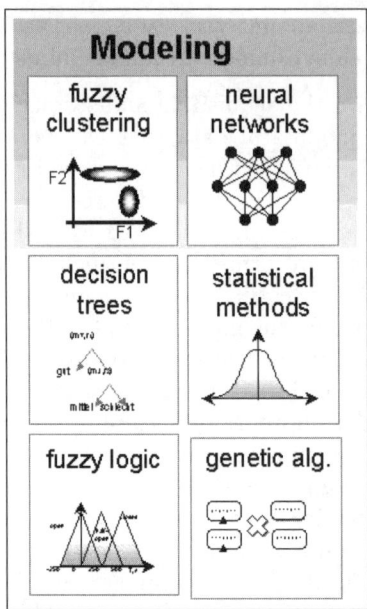

Fig. 14.2. Modeling Methods

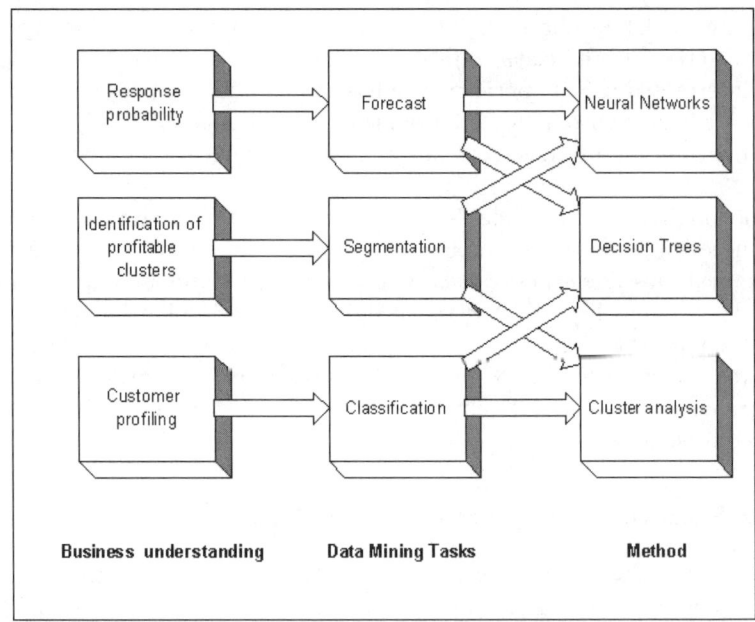

Fig. 14.3. Data Analysis Tasks and Methods

14 Intelligent Methods in Finance Applications   291

Will the used storage systems and interfaces (files, databases, ...) work under growing security and extensibility needs?

The selection of appropriate methods depends on the required technical performance (modeling results, computing time) and business needs, e.g. presentation of the results and interpretability of models. From the point of execution, most technologies can be compared and they perform well. But based on the available raw data, the model building process differs. For example, if no training data is available, a knowledge-based approach by a fuzzy rule base derived from an expert by questions is a possible choice. If training data is available, is there also test data available? That means, can we use a supervised methods for classification because we do know the classes to model, or do we have to find first the clusters by an unsupervised technology and label them by expert knowledge (Fig. 14.3)? In general, the rule-based methods like decision trees or fuzzy rule base models allow an easy understanding by using a model representation with readable rules and should be preferred if possible to present results to non-data analysis experts.

The following remarks and comments may help the inexperienced designer the choose the appropriate methods:

- Decision Tree: This method recursively partitions the data so as to derive a decision tree (a set of decision rules) for predicting outputs from inputs. The decision tree is constructed top-down, starting with a root node and incrementally adding variable-testing decision branches. The process continues until decisions can be added to the tip nodes correctly classifying all the given examples. Typically, decision trees have continuous or discrete inputs and discrete outputs. Training and test data has to be provided, and the model can be described in rules.
- Neural Network: The concept of artificial neural networks is to imitate the structure and workings of the human brain by means of mathematical models. Three basic qualities of the human brain form the foundations of most neural network models: knowledge is distributed over many neurons within the brain, neurons can communicate (locally) with one another and the brain is adaptable. In the case of supervised learning, in addition to the input patterns, the desired corresponding output patterns are also presented to the network in the training phase. The network calculates a current output from the input pattern, and this current output is compared with the desired output. An error signal is obtained from the difference between the generated and the required output. This signal is then employed to modify the weights in accordance with the current learning rule, as a result of which the error signal is reduced. The best-known and most commonly employed network model here is the multilayer perceptron with backpropagation learning rule. It has continuous or discrete inputs, continuous outputs. Training and test data has to be provided. In the case of unsupervised learning, the network is required to find classification criteria for the input patterns independently. The network attempts

to discover common features among the presented input patterns via a similarity comparison, and to adapt its weight structure accordingly. The neurons thus form independent pattern classes and become pattern detectors. This method is similar in effect to clustering algorithms or vector quantification methods. An example of this process is provided by Kohonen's self-organizing feature maps, which organize themselves with the aim of converting signal similarity into proximity between excited neurons. It has continuous or discrete inputs and discrete outputs. Training data has to be provided.

- Fuzzy Rule Base: In knowledge-based methods, fuzzy sets are employed primarily to carry out the formal, content-defined mapping of human knowledge. This makes it possible to process human empirical knowledge with electronic data-processing systems. This includes the following main functions
  – Knowledge representation: This is normally carried out by means of rules in the so-called knowledge base.
  – Contextual knowledge processing: This is normally carried out in an inference engine, which must be able to process linguistic knowledge in contextual form (and not in symbolic form).
  – Translation: On the input side, numerical information is translated, where possible, into linguistic information (fuzzification), and on the output side specific membership functions are translated either into numbers (defuzzification) or into linguistic expressions (linguistic approximation).

  In knowledge-based methods, a missing or inefficient algorithmic formulation is replaced by the use of human knowledge. It has continuous or discrete inputs and outputs.

- Cluster Analysis: Clustering procedures belong to the algorithmic methods of data analysis. Various different approaches have been proposed for the developing of classifiers by means of clustering, including iterative clustering, agglomerative hierarchical clustering and divisive hierarchical clustering. The first aim of clustering is to find structures contained within groups of data. These structures are usually classes to which objects from the data set are assigned. The result of the classification process is usually used as a classifier. Objects, which as of yet have no known class assignments, are assigned to classes using the classifier found through clustering. Classical clustering assigns each object to exactly one class, whereas in fuzzy clustering the objects are assigned different degrees of membership to the different classes. A fuzzy clustering model has continuous or discrete inputs and continuous outputs.

Sometimes it may be necessary to reduce a good performing solution to a linear model when a non-linear model is not accepted (e.g. for explanation reasons). Search especially for typical and non-typical patterns in the data

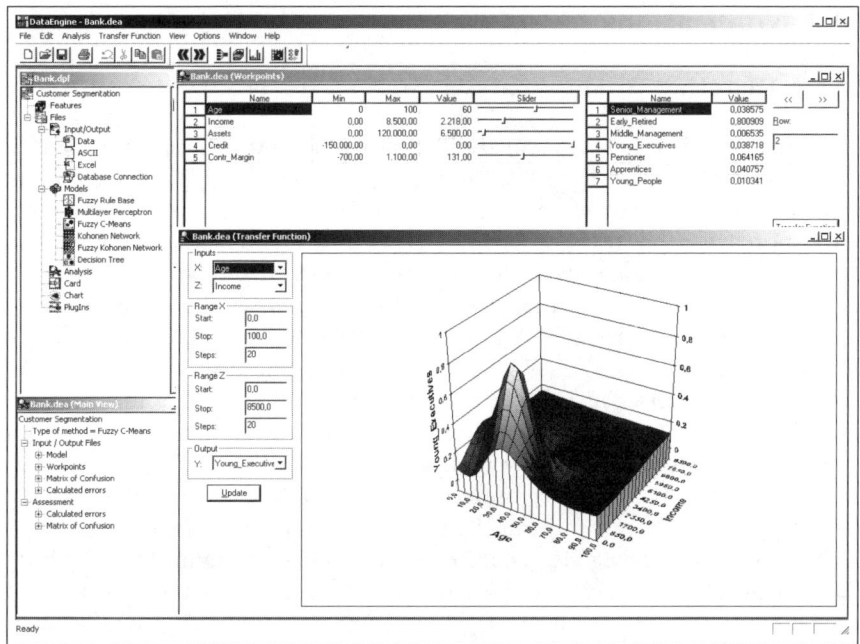

Fig. 14.4. Model Analysis

during the modeling phase (clusters and outliers). A log file helps to analyze consistency checks at any stage of the modeling process.

If you use new technologies, compare the results on known solutions with other (traditional) methods to make your customers trust in the technology. To find proper parameter settings for a special task, start from an existing solution with one dimensional parameter variation. Tools can support the model evaluation and analysis, see Fig. 14.4 and [4].

What is important? Success factors are a target-oriented conceptual design and detailed specifications, a real-world data base which allows to develop good-performing models using expert knowledge about the business process and modeling technologies as well as a teamwork of different computer scientists, data analysts and the business users.

After a successful modeling, the implementation and integration of the system has to consider security, safety and maintenance aspects.

In the next section some real world examples are presented.

14.3 Applications

14.3.1 Factoring Credit Analysis

Credit analysis is a field, in which different techniques to support automatic decisions were tested in the past. Conventional statistical approaches like discriminant analysis as well as neural networks have been examined so far. Such approaches, however, cannot model the way human experts derive their decisions about the solvency of the companies under investigation. Fuzzy techniques have advantages in areas where human knowledge has to be acquired and processed automatically. A fuzzy rule based decision support system for automated credit analysis has been integrated in a new software system for the factoring business at a large German business-to-business financial services company as decision support system for the credit managers ensuring that a flow of money follows every flow of merchandise. That means the financial services division takes over the entire handling of payments. Various service packages are put together here for special customer requirements.

The factoring package turns accounts receivable into immediate liquidity. That means claims on customers are transferred to a provider as factor, which immediately remits the equivalent. The buying and selling of accounts receivable or invoices is called factoring. The process is fairly common as a reason of cash flow problems for especially for young and fast growing businesses. The reason companies (factoring customers) sell their invoices to the factor is they do not want to wait (or can't afford to wait) for their customers to pay them in 30, 60, or 90 days. They need the money now. Sometimes they need the money to meet payroll expenses, rent, or other operating expenses. Sometimes they need the money to buy supplies needed to fill incoming orders, or they want advanced funds to take advantage of cash discounts offered by their suppliers. So regardless of when a customer pays, the supplier has immediate liquidity available. If required, security against bad debt losses is provided also.

The problem of sequential consideration of criteria can be handled by considering all relevant criteria simultaneously, as shown in Fig. 14.6 where the structure of the respective rulebase is sketched. For each branch in this tree, fuzzy rules have been formulated based on their terms. The entire system consists of rules which are processed by standard fuzzy inference procedures. Before the purchase price is paid to the factoring customer by the factor, the factor investigates the credit risk of the buyer (see Fig. 14.5). The buyer's credit application is checked via the fuzzy rule based system and the financial standing as output determines the given credit limit. For setting up the rule base, discussions with experts on credit analysis and comparisons with credit limit decisions done so far have been made to find out the best fitting parameter settings.

The automated on-line credit limit decision by the fuzzy rule base has been realized as a WindowsNT DLL in the framework of a large client server

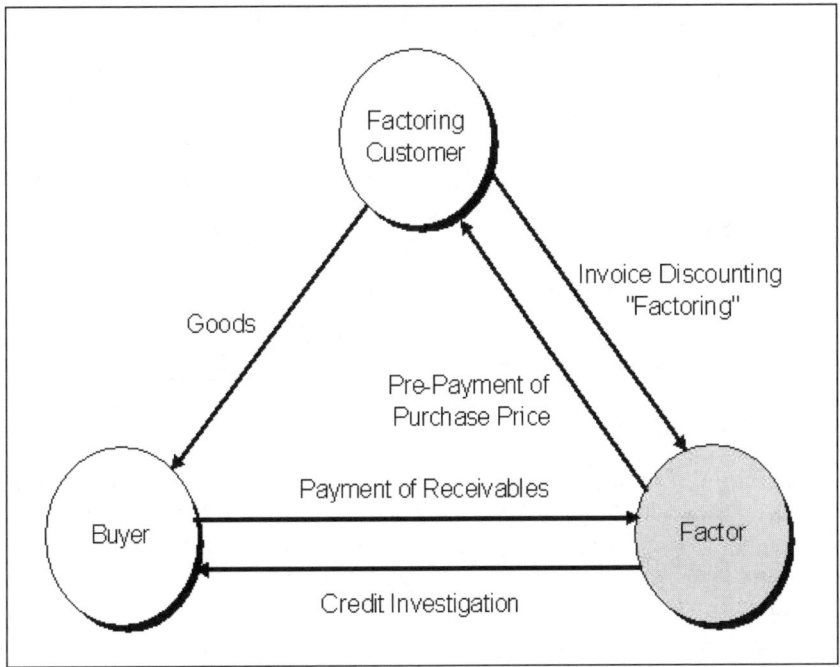

Fig. 14.5. The Credit Factoring Structure

application (including databases, workflow management, messaging for the credit managers, WWW access and so on).

Additionally the advanced user (credit manager) can analyze the financial standing decisions done by the fuzzy system using a graphical user interface working on a database. This enables the credit manager to see the reasons for a certain decision in detail (input values of a certain data set, rule base variables and terms, rules fired, financial standing) also in history. By modifying input parameters using the simulation part the what-if analysis shows the influence of current parameters to the output value. That helps for interpretation of the results and to check the rule base validity from time to time.

A lesson from industry learned during the project was to use a combined approach (data-based and knowledge-based) to find a proper parameter setting. The design of the system was mainly influenced by typical user requirements:

- 1:1 use of the existing knowledge of the credit professionals
- results easy to explain, especially by the system rejected risks
- extensibility for new criteria
- robustness against minor changes in the data range

Because of these requirements and the missing training data, a knowledge based model was preferred. A first model was build using a fuzzy rule base and

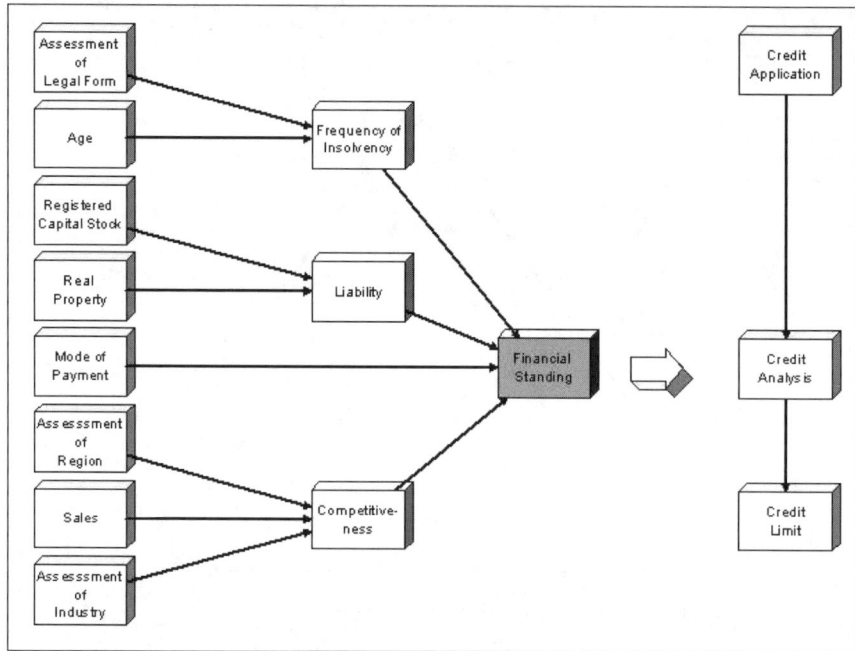

Fig. 14.6. Structure of Fuzzy Rulebase for Decision Support Systems

interviewing some credit managers to find out their decision procedures. The features and rules have been determined. The execution of the first rule set was not as expected, but after an analysis of the deviation on a small number of cases using a neural network, its connection weights helped to modify the rule base structure and parameters for a successful solution.

14.3.2 Customer Consultancy Assignment

The necessity for financial services providers is to comply with the private customers requirements for an individual support service and, at the same time, to make best use of the advisory resources; that means to put consultancy capacity where it is most cost effective. Usually, due to the present assignment of the clients, financial services providers are not operating at full capacity, and the available potentials are not fully exploited. In order to classify the customers into different customer advisory levels, a data based analysis of the clients by income, capital and securities is applied. Following the request of transparency and objectivity of the customer segmentation, an automated solution has been developed which applies intelligent technologies such as Fuzzy Set Theory for analyzing customer profiles and discovering new information from the data. The solution offers a flexible adaptation of all parameters and considers also the dynamic changes of a clients environment. Compared to

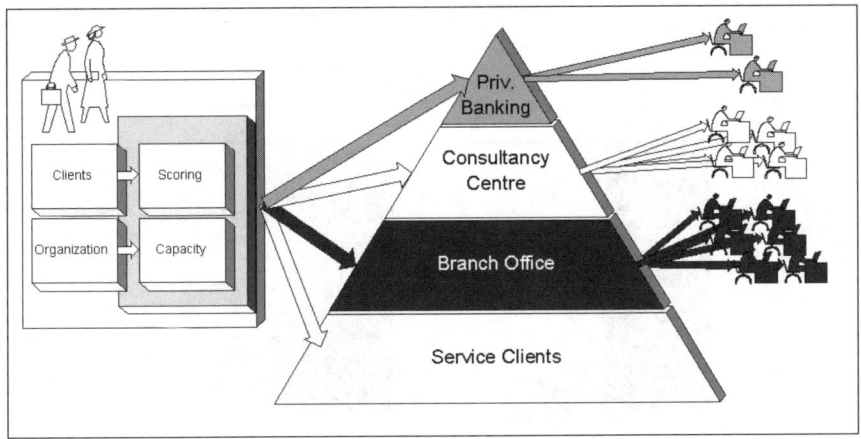

Fig. 14.7. Customer Advisor Management

the usual assignment of clients according to fixed categories of income and capital, the system provides a more flexible solution. The segmentation of the established clients identifies previously unknown and valuable information and helps to increase the efficiency of the customer service, e.g. by discovering clients, who are assigned marginally below the fixed limits of income and capital.

Only by applying a modified classification of customers, the number of attended clients can be increased by 15 percent. The capacity management of the customer advisory units (private banking, consultancy center, branch office,...) is managed by an overall evaluation of all features describing one client, and thus enables a customer segmentation coping with the available capacities. The customers are scored first, and assigned in a second step in descending order by their score to the different advisory units filling the capacities, see Fig. 14.7.

Following the request of transparency, objectivity and flexibility of a dynamic customer segmentation, an automated solution has been developed which applies intelligent technologies such as Fuzzy Set Theory [8] for the scoring of clients and their assignment to advisory units and consultants considering the available capacities and organization structure.

The solution offers different parameter settings and allows to monitor also the dynamic changes of a clients environment during his lifecycle. Compared to the usual assignment of clients according to fixed limits e.g. of income and assets, the mit.finanz.ks system provides a more flexible solution. It helps to increase the efficiency of the customer service e.g. by preferring clients, who have feature values marginally below the fixed limits of income and capital. In Fig. 14.8, client 1 would usually be selected in traditional way (monthly income more than 71.000 euro, capital 10.000 euro), whereas client 2 may offer more interesting asset volumes to work with (monthly income of 69.000

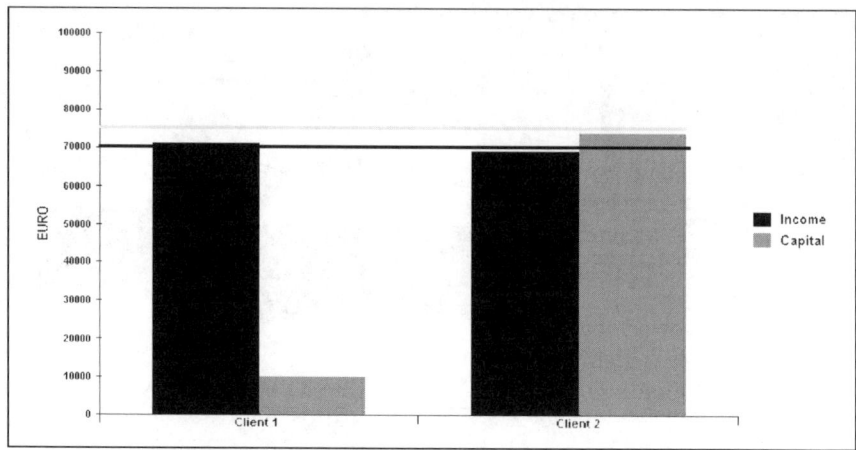

Fig. 14.8. Comparison of two customers

euro and capital of 74.000 euro). Typical common selections based on simple queries on the crisp limits (income more than 70.000 euro or capital more than 75.000 euro), whereas the designed scoring method combines both income and capital using a parameterized inference (adjustable between maximum and sum of the values for income and capital).

At the beginning, for each client different information is collected covering demographic, organizational and asset data. Master features like income, capital, securities, and others are derived from these. Single data sets belonging to members of the same family are associated to one client data set representing the family association. Depending on the structure of the family (number of adults, number of children) and the age, the different master features are scaled by parameterized functions. For example, an income of 3.000 EURO earned by a young executive can be treated in the same way as the income of 5.000 EURO of a middle-aged manager. Treating age and family status as input features for the model (and not as parameters) enables now to use one adaptive model for the scoring of different customer unions.

For each scaled master feature a partial score is calculated using a membership function by Fuzzy Set Theory. A client with a monthly income of 5.000 EURO will get a score of 0.7, a monthly income of 10.000 EURO will receive a score of 0.9, and a monthly income of 25.000 EURO ore more will be scored with a value of 1.0 (see example in Fig. 14.9). Each feature is related to a set of membership functions for different ages of customers. Based on the partial scores of each master feature, a final score for each client is calculated using a parameterized inference function. This score has a value between 0 an 100. The clients are assigned to the advisory levels and the related organization units depending on the score value and the available capacity for the consultancy service on that level; the system also offers the opportunity to use separate conditions to handle exceptions from standard assigning by score.

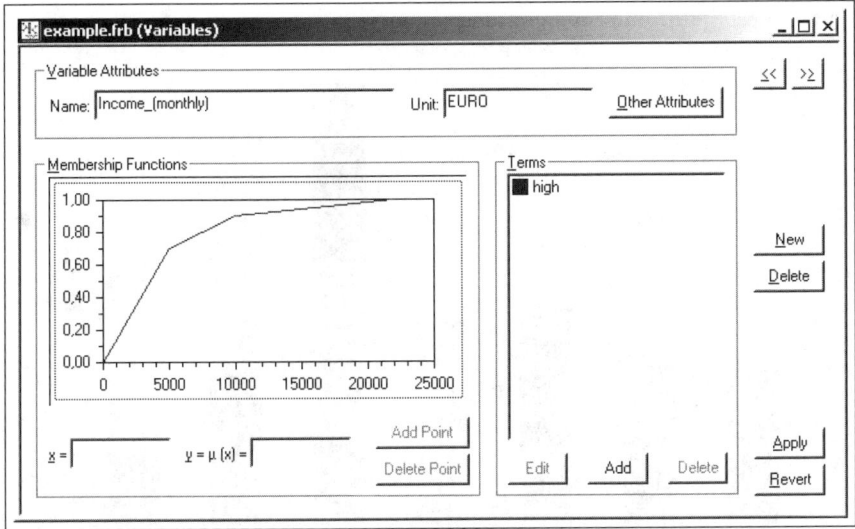

Fig. 14.9. Fuzzy Membershipfunction Income

The data analysis of the established clients identifies previously unknown and valuable information by applying cluster algorithms to each advisory level to offer a detailed look in the clients financial situation in a second step. Different input values (of income and capital) lead to the same final score (and the same advisory level). A detailed analysis of the customers score structure in the same advisory level is required to identify marketing activities and products for promotion.

Cluster analysis refers to a group of methods for recognition of structure in data sets. It can be applied if certain features of different objects (e.g. customers) can be observed, and if those objects are supposed to be divided according to their individual feature values. The entire set of objects will be clustered regarding all describing features in such a way that all objects belonging to one cluster are (possibly) similar. Vice versa, the objects belonging to different clusters should be different regarding their feature values. The clients have been clustered in different groups using the fuzzy c-means algorithm [2]. The first aim of clustering is to find structures contained within groups of data. These structures are usually classes to which objects from the data set are assigned. The result of the classification process is usually used as a classifier. Clients of a certain advisory level group are clustered using their partial score values. That allows to differ between different types of customers (though they have the same total score) and offer a more individual consultancy service. In the following figure, the result using the membership value as degree for the types e.g. of income; that means by having a monthly income of 10.000 EURO the degree for type of income is 0.9. From the profiles it can be derived that usually clients who have securities also have a respectively

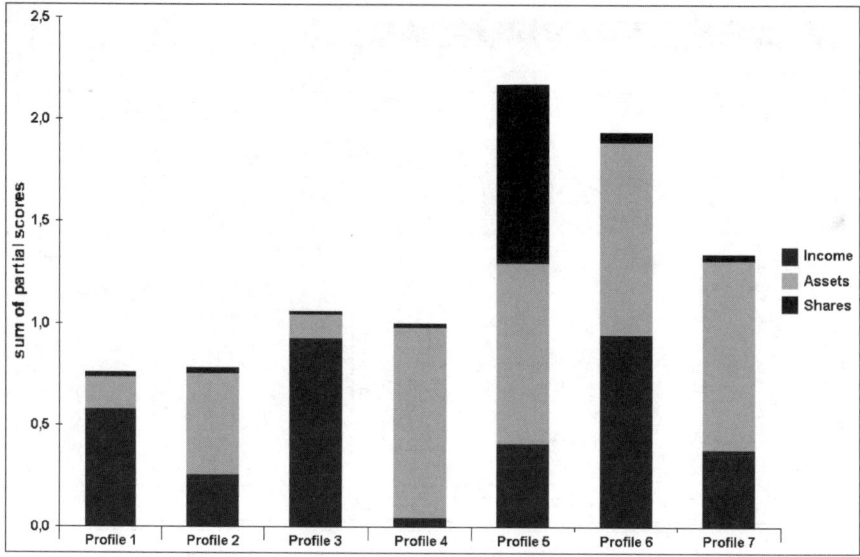

Fig. 14.10. Example Customer profiles

capital and income. Based on this, the estimated effort for the consultancy service is calculated and used during the assignment to optimize the capacity utilization.

The application is based on a distributed system architecture for offline use. It has two main components: a database with flexible interfaces to allow the data exchange with the financial services providers systems and the mit.finanz.ks software application for the data pre-processing, scoring, assignment and reporting. Two other components support the main system: for the data pre-processing a client association identification tool to generate additional association information and data administration support, and for the evaluation and use of the new assignment in the organization a web-based solution to support the consultants by executing the clients transition from one advisory level to another. The needs of financial services providers to comply with the private customers requirements for an individual support service and, at the same time, to make best use of the advisory resources in the framework of customer relationship management can be fulfilled with the mit.finanz.ks approach. The flexible adaptation of data selection, data pre-processing and data analysis followed by a reporting facility. The design of the interactive and user-friendly graphical interface enables an easy browser access for the end-users as confirmed by different financial services providers in Germany.

The main advantages to be noticed are:

- Efficient customer care capacities
- Increase of the number of attended clients and efficient use customer care capacities

- Flexible, transparent and objective allocation of customers to the different advisory levels
- Discovery of unknown potentials by identifying clients who have marginally values below the fixed limits of income and capital
- Determination of typical customer profiles for a more sophisticated consideration and further analysis of the customers
- Administration and comparison of different models for simulation of different parameter settings

The use of score values for a feature and the calculation of the final score orders the customer units. This order will be the kept when smaller changes of the data in the cause of time appear (parts of the assets values are influenced by a stock market index, income may be increase year by year) and this allows the adaptive customer segmentation.

14.3.3 Successful Applications: Hints and Tips

Generic lessons learned from these applications and the experience with the customers (system users) are discussed in the following. Is there a market place for smart adaptive systems? Yes! Looking at the systems and projects developed so far, the current situation for applications of smart adaptive systems in financial business is determined by headlines like "competition requires cost reductions" and "mass marketing is out of date, customers demand individual service". The marketing budgets are limited, but increasing the market share is only possible by detailed knowledge of market and customers. That is why solutions like

- customer segmentation
- customer scoring and identification of future buyers of a certain product
- support of sales representatives
- yield management
- response forecast for mail-order houses
- migration analysis/churn management
- product placement and basket analysis

are required. These solutions are a market place for system developers if they can point out the benefits of smart adaptive systems in those fields. The guiding principle of soft computing "exploit the tolerance for imprecision, uncertainty and partial truth to achieve tractability, robustness and low solution costs' can contribute because unlike the traditional, soft computing aims at an accommodation with the pervasive imprecision of the real world.

What was important? Teamwork right from the start and support of the whole process. A team of computer scientists, data analysts and business users has to work together, covering requirement analysis and definition of project goals, allocation of data, modeling, validation of the results, software integration, testing, training of staff and integration into daily workflow. Especially the integration of the business users should be as early as possible to avoid comments like "that is not what we expected", but technical details are not as important as access to first results and prototypes of graphical user interfaces.

What kind of problems have been identified? At the beginning, required data has not been delivered; the data quality in general caused some problems, e.g. fields like job information or date of birth are filled with default values. Determination of a customers income by analysis of transfer account data limited by confidential reasons is a great challenge. Take care on that because income is typically used in customer segmentation. During the modeling phase, lots models have been re-build and implemented because of former unknown or updated restrictions. The selection of rule-based methods was preferred to have a model description easy to understand, whereas the results have been analyzed by neural networks; use different technologies to verify and optimze your final model. Most of the problems occurred can been prevented with ongoing requirements verification and quality assurance practices.

14.4 Conclusions

The design of a successful smart adaptive system in finance applications is a complex iterative process. Having adaptation in the main focus here means that such systems should offer solutions which results are independent from (minor) changes of the existing environment covering

- data (providing, pre-processing, handling, storing),
- features and modeling techniques,
- interfaces/software design.

For a successful design of a smart adaptive system, you should answer yes to the following questions:

- Have you planned to set up a raw prototype for the whole system at an early stage covering all single steps of the whole process? Each step should be treated separately in detail when necessary as shown in the established rapid prototyping e.g. of automotive industry. Give an overview of the whole system as early as possible.
- Do the selected data sources will continue providing all required information in the future? If it is planned to build a model for periodic use, check if all the required input data will be available in the correct format, scale, period and data range.
- Do the interfaces offer a fast and flexible access and do they consider security aspects? Use encryption to make sure that information is hidden from anyone for whom it is not intended, even those who can see the encrypted data. If necessary, add an user administration to give access only for authorized people.

- Can you use model parameters as input features for a more general model to reduce manual adaptation efforts? Instead of building different models e.g. for car buying behavior in the four seasons, build one model and include the season as additional input parameter.
- Is the presentation of the results easy to understand also for non-experts? This is very important to show the benefit of a model and to convince customers to buy your system. Especially when risk evaluations or licensing procedures are carried out, a detailed explanation of the model for non-experts may be necessary.
- Does the evaluation phase in a real-world environment show the expected results and performance fulfilling the user requirements? You can deal with complex scientific modeling, colored graphics and nice user interfaces, but at any stage, focus on the problem to solve (management by objectives). Artificial approaches must learn from user experience in the system development process.

References

1. Berthold, M., Hand, D. J. (Eds.) (2000) Intelligent Data Analysis – An Introduction, 2nd rev. and ext. ed. 2003 XI, Springer Heidelberg
2. Bezdek, J. C. (1981) Pattern Recognition with Fuzzy Objective Function Algorithms. Plenum Press, New York, USA
3. EUNITE (2001) EUNITE – EUropean Network on Intelligent TEchnologies for Smart Adaptive Systems, funded by the Information Society Technologies Programme (IST) within the European Union's Fifth RTD Framework Programme, see http://www.eunite.org
4. MIT GmbH (2001), "Manual DataEngine 4.1", MIT GmbH, Aachen, Germany
5. Nelke M. (2002): Customer Relationship Management: A Combined Aproach by Customer Segmentation and Database Marketing. Hans-Juergen Zimmermann, Georgios Tselentis, Maarsten van Someren, Georgios Dounias (Eds.): Advances in Computational Intelligence and Learning: Methods and Applications. International Series in Intelligent Technologies Kluwer (2002): 281–290
6. Jostes M., Nelke M. (1999) Using Business Intelligence for Factoring Credit Analysis. In: Proceedings of ESIT 99 European Symposium on Intelligent Techniques, June 3–4 1999, Orthodox Academy of Crete, Greece
7. Nelke M., Tselentis G. (2002) Customer Identification and Advisory Management at Financial Services Providers. In: Proceedings of Eunite 2002 – European Symposium on Intelligent Technologies, Hybrid Systems and their implementation on Smart Adaptive Systems, 8.–21. September 2002 Albufeira, Portugal
8. Zimmermann H.-J., Zysno, P. (1983) Decisions and Evaluations by Hierarchical Aggregation of Information. Fuzzy Sets and Systems 10: 243–260
9. Zimmermann, H.-J. (1987) Fuzzy Sets, Decision Making, and Expert Systems. Kluwer Academic Publishers, Boston Dordrecht Lancaster 1987

15

Neuro-Fuzzy Systems for Explaining Data Sets

D.D. Nauck

Computational Intelligence Group
Intelligent Systems Lab
BT Exact
Adastral Park, Ipswich IP5 3RE
United Kingdom
detlef.nauck@bt.com

In this chapter we describe an application of a neuro-fuzzy approach to learning (fuzzy) rule-based explanations from travel data of a mobile workforce. Compared to rule bases for prediction we have different requirements when building explanatory rules. In order for explanations to be useful they must be simple and comprehensible. The accuracy of the rule base is obviously also important because incorrect explanations are useless. However, if there is a trade-off to be made between simplicity and accuracy we will favor simplicity as long as we can keep the error of the rule base within reasonable bounds.

The software platform for which we have implemented the explanation generation is called ITEMS – a system for the estimation, visualization and exploration of travel data of a mobile workforce. One key feature of ITEMS is the interactive exploration of travel data that is visualized on maps. Users cannot only see which journeys were late, on-time or early, but they can also request explanations why a journey was possibly late, for example. We have integrated a neuro-fuzzy learner based on the neuro-fuzzy approach NEFCLASS and a decision tree learner into ITEMS. Both learners generate explanatory rules for a selected data subset in real time and present them to the user. NEFCLASS creates fuzzy rules, and the decision tree builds crisp rules. The rules can help the user in understanding the data better and in spotting possible problems in workforce management and he can pick the type of rules he prefers.

In order to automatically pick an explanation, we show how the interpretability of a rule-base can be measured. This allows systems like ITEMS to employ several rule-learners in parallel and present the "better" explanation to the user both in terms of interpretability and accuracy.

15.1 Introduction

Our modern world is constantly growing in complexity. This complexity manifests itself, for example, in the form of information overload. Nowadays, we perceive a vast amount of information from channels like e-mail and other Internet-based services, hundreds of TV channels, telecommunication devices etc. Businesses have to deal with huge amounts of data they collect within their organization or from transactions with customers. It is paramount for businesses to put the collected data to good use [2]. Examples are dynamic workforce management based on internally collected data or customer relationship management based on transaction data gathered from Internet clickstreams or more mundane transactions like those at supermarket check-outs.

In order to transform data into valuable information an intelligent approach to data analysis is required. We do not simply require models for prediction which we would blindly follow and let them determine our business decisions. In order to properly facilitate our decision making processes we also require models for understanding, i.e. models that provide us with explanations. In this chapter we discuss how neuro-fuzzy systems can be used to generate explanations about data.

In data analysis applications an interpretable model of the data is especially important in areas

- where humans usually make decisions, but where machines can now support the decision making process or even take full responsibility for it,
- where prior knowledge is to be used in the data analysis process and the modification of this knowledge by a learning process must be checked,
- where solutions must be explained or justified to non-experts.

The application that we discuss in this chapter belongs to the last area. At BT Exact we have developed ITEMS, an intelligent data analysis system for estimating and visualizing the travel patterns of a mobile workforce [1, 5, 6]. The application allows users to track journey times and to generate explanations why a particular journey was possibly late, for example. We use the neuro-fuzzy approach NEFCLASS to generate such explanations.

In Sect. 15.2 we discuss some aspects of interpretable fuzzy systems and in Sect. 15.3 we consider the special circumstances of generating models for explanation instead of prediction. Section 15.4 explains how we use NEFCLASS to build explanatory rules and in Sect. 15.5 we present an example in the context of ITEMS before we conclude the chapter.

15.2 Interpretability in Fuzzy Systems

If we want to use a neuro-fuzzy method to generate explanatory rules for a given data set then we must consider the following issues.

- What does interpretability mean in the context of fuzzy sets?

- What kind of explanations does the user want and can understand?
- How can we build such rules quickly in interactive applications?
- How do we present the explanations?

In this chapter we only consider Mamdani-type fuzzy rules of the form
$$\text{if } x \text{ is } \mu \text{ and } \ldots \text{ and } x_n \text{ is } \mu_n \text{ then } \ldots,$$
i.e. fuzzy rules that use operators like "or" and "not" or linguistic hedges are not discussed here. The consequent of a fuzzy rule can be a linguistic term (regression) or a class label (classification). The following list describes a number of desirable features of an interpretable fuzzy rule base.

- Small number of fuzzy rules.
- Small number of variables used in each rule.
- Small number of fuzzy sets per variable (coarse granularity).
- Unambiguous representation of linguistic terms, i.e. only one fuzzy set for each linguistic term.
- No completely contradictory rules, i.e. there is no pair of fuzzy rules (A, C) and (A', C') with
$$A = A' \wedge C \neq C',$$
where A, A' are antecedents and C, C' are consequents. However, partial contradiction
$$(A \not\subseteq A') A \cap A' \neq \emptyset \wedge C \neq C'$$
is a feature of fuzzy rules and is therefore acceptable.
- No redundancy, i.e. there is no pair of rules (A, C) and (A', C') with
$$A \subseteq A' \wedge C = C'.$$
- Avoiding exceptions, i.e. there should not be a pair of rules (A, C) and (A', C') with
$$A \subseteq A' \wedge C \neq C'.$$

Although exceptions can be a way to reduce the number of rules they can make the rule base harder to read.
- Fuzzy partitions should be "meaningful", i.e. fuzzy sets should be normal and convex and must keep their relative positions during learning.

15.2.1 Hints and Tips

- There are two possible objectives when searching for an interpretable fuzzy rule base. We can look for global or local interpretability. Global interpretability means we can understand the (small) rule base completely, while local interpretability means we can at least understand a number of rules that apply at a time to a particular input situation. Depending on the application, local interpretability – i.e. understanding why we obtain a particular output from the rule base – may be sufficient.

- Interpretability is a subjective, vague and context-dependent notion. Not the designer, but only the user of a rule base is the judge in this matter.
- There is always a trade-off between interpretability and accuracy. You will typically not be able to achieve high degrees for both interpretability and accuracy.
- Interpretability in fuzzy systems is a controversial issue. For more information on this topic see, for example [3, 4, 8, 10, 13, 14, 15, 16, 17].

15.3 Explanations vs. Predictions

When neuro-fuzzy systems are applied in data analysis the usual target is to build a reasonably accurate predictor with a reasonably interpretable rule base. Obviously, there is a trade-off between accuracy and interpretability and depending on the application the solution will concentrate more on prediction than interpretability, or vice versa.

In order to obtain a model that can be used for prediction we have to avoid over-generalization. This means, we usually split up the available data in training and validation sets. In order to obtain a useful estimate of the error on unseen data this procedure is usually repeated several times (n-fold cross-validation).

When we build a model for explanation we have a different target. The model will not (and must not) be used for prediction. The idea of the model is to provide a summary of the currently available data in a meaningful way. In this sense an explanatory model is like a descriptive statistic of a sample, where we are only interested in describing the sample and not the underlying population.

That means to build an explanatory model we will use all available data and do not worry about over-generalization. Actually, we try to get the best possible fit to the data because that means we obtain the most accurate description of the data. We will also restrict the degrees of freedom for our model in order to obtain a small, simple and interpretable model. Obviously this leads to the same dilemma as for predictive models: the trade-off between accuracy and interpretability.

The demand for a predictive model usually does not impose strict time constraints on the model building process. The only real requirement is that the model is available before the first prediction is due. Explanations, however, are more likely to be demanded in a real time process. A typical scenario is that a user explores some data and demands an explanation for some findings. It is then not acceptable to wait a long time for a model. The model building process must be completed within seconds or minutes at most.

This time constraint forbids an extensive search for the smallest, most interpretable model with an acceptable accuracy. That means in a real world applications there is typically no time for building a model, checking its interpretability and then restart the model building with different parameters until we have reached the best result. However, it may be possible to do this in parallel, if the computing resources are available.

Therefore, we are interested in learning approaches that either try to build small models from the beginning or are capable of pruning a large model as part of the learning process.

Rule induction methods based on decision trees try to build small models by selecting the most promising variables first in the hope of not requiring all variables for the final model [7, 19]. Another option is to use a structure-oriented fuzzy rule learning method [16] as it is implemented in the neuro-fuzzy system NEFCLASS (see also Chap. 7). This approach first detects all rules that are induced by a data set, selects the best rules based on a rule performance measure, trains the fuzzy sets, and then prunes the rule base. For ITEMS we have implemented explanation facilities based on (crisp) decisions trees [20] and neuro-fuzzy rules. Users can select which version of rules they prefer.

If enough computing resources are available, it is possible to generate several different explanatory models in parallel. The system can then try to measure the interpretability based on approaches like the minimum description length principle [9] that can also be applied to neuro-fuzzy learning [8]. Another possibility to compare different rule based models is to consider the number of relevant parameters.

15.3.1 Measuring Interpretability

In the application that we discuss in this chapter we are interested in classification rules generated by different approaches. We want to enable the software to automatically select the more interpretable rule set and present it to the user first. Basically, we are looking for a measure that allows us to rank rule bases by interpretability [15] in the same way that results of an Internet search engine are ranked by their "usefulness".

We can measure the complexity comp of a classifier by

$$\text{comp} = m \left/ \sum_{i=1}^{r} n_i \right. ,$$

where m is the number of classes, r is the number of rules and n_i is the number of variables used in the ith rule. This measure is 1, if the classifier contains only one rule per class using one variable each and it approaches 0 the more rules and variables are used. We assume that at least one rule per class is required, i.e. systems with a default rule must be suitably recoded.

When we want to compare fuzzy rule bases we also want to measure the quality of the fuzzy partitions. Usually, a fuzzy partitioning is considered to be "good", if it provides complete coverage (i.e. membership degrees add up to 1 for each element of the domain) and has only a small number of fuzzy sets. If we assume that the domain X_i ($i \in \{1, \ldots, n\}$) of the ith variable is partitioned by p_i fuzzy sets $\mu_i^{(1)}, \ldots, \mu_i^{(p_i)}$ then we can measure the degree of coverage provided by the fuzzy partition over X_i by the coverage index

$$\mathrm{cov}_i = \frac{\int_{X_i} \hat{h}_i(x) dx}{N_i}, \quad \text{where} \tag{15.1}$$

$$\hat{h}_i(x) = \begin{cases} h_i(x) & if \ 0 \leq h(x) \leq 1 \\ \dfrac{p_i - h_i(x)}{p_i - 1} & \text{otherwise} \end{cases}, \quad \text{where}$$

$$h_i(x) = \sum_{k=1}^{p_i} \mu_i^{(k)}(x)$$

with $N_i = \int_{X_i} dx$ for continuous domains. For discrete finite domains we have $N_i = |X|$ and replace the integral in (15.1) by a sum.

We can see that $\mathrm{cov}_i \in [0,1]$, where $\mathrm{cov}_i = 0$ means that the variable is either not partitioned or that we have a crisp partition such that all $\mu^{(i)} = X_i$. Both extreme cases mean that the variable would be considered as "don't care" and would not appear in any rule. Complete coverage is indicated by $\mathrm{cov}_i = 1$. Note that a partition that covers only 10% of all values gets approximately the same score as a partition where 90% of the whole domain is completely covered by all sets. This feature may be disputable and has to be considered when applying the index.

In order to penalize partitions with a high granularity we can use the partition index

$$\mathrm{part}_i = \frac{1}{p_i - 1}$$

assuming that $p_i \geq 2$, because otherwise we would not consider that variable. We use $\overline{\mathrm{cov}}$ to denote the average normalized coverage and $\overline{\mathrm{part}}$ to denote the average normalized partition index for all variables actually used in the classifier.

Finally, we can use the interpretability index

$$I = \mathrm{comp} \cdot \overline{\mathrm{cov}} \cdot \overline{\mathrm{part}}$$

for measuring the interpretability of a fuzzy classifier. Obviously, we can apply I to any rule-based classifier that uses sets to partition variables, by simply representing crisp sets by corresponding fuzzy sets.

A classifier would only score $I = 1$ if it contains one rule per class, using only one variable per rule and providing complete coverage for each variable with two (fuzzy) sets and $m \leq 2n$ holds, where n is the number of variables. The value of I can give us an idea if we want to compare two rule-based classifiers for the same problem. It is less useful to compare the interpretability of classifiers for different domains.

15.3.2 Hints and Tips

- We should stress that for an explanatory model local interpretability can be more important than global interpretability. If, for example, an explanatory model is used for obtaining explanations only for individual patterns and not for obtaining an understanding of the data set as a whole, then the number of rules that fire for any given pattern are of interest and not the overall number of rules in the rule base. In this case the complexity index comp can be replaced by an average value based on the rules that fire for a selected pattern.
- It is important to note that the numbers produced by the interpretability index have little meaning by themselves and are only useful to rank rule bases that apply to the *same problem*. As we will see in Sect. 15.5 the numbers are usually small and do not provide an intuitive measure of interpretability.

15.4 Creating Explanatory Fuzzy Rule Bases

In this section we look at how a neuro-fuzzy system like NEFCLASS can be configured to produce interpretable rule-bases. The general ideas behind this procedure can probably be applied to other rule learners in a similar fashion.

The NEFCLASS structure and parameter learning algorithms are designed in a way that they can operate fully automatically if this is required (see also Chap. 7). Usually, if interpretable solutions are required from a neuro-fuzzy system, then user interaction is a desired feature, because only the user can decide what "interpretable" means in a certain application scenario. If, however, a fuzzy rule base shall be created within an application, then we cannot assume that the user is willing or capable of supervising a neuro-fuzzy learning process. Therefore it must be possible to create a rule base completely automatically while trying to obtain a high degree of interpretability. NEFCLASS uses the following strategies in order to achieve this goal.

- **Automatic best per class rule learning**: this feature creates a rule base that contains so many rules that all patterns of the training set are covered by at least one rule. In the first stage NEFCLASS creates a new rule each time it encounters a pattern that is not yet covered by a rule, i.e. if the pattern does not have a degree of membership of at least 0.5 with any rule antecedent currently stored in the rule base. During the second stage NEFCLASS selects for each class the best rules for the final rule base. It does that by selecting a rule for each class in a round-robin fashion until all patterns are covered by at least one rule. This algorithm also guarantees a similar number of rules for each class.
- **Automatic fuzzy set tuning**: the fuzzy set learning algorithm uses a heuristic to modify the membership function in order to reduce the sum of squared errors (SSE) and the number of misclassifications. A small overall

SSE usually indicates that the classifications are nearly crisp and a small number of misclassifications is an obvious target. In order to obtain meaningful fuzzy sets, the learning algorithm is constrained. For example, fuzzy sets must not change their relative position to each other and must always overlap.

- **Automatic exhaustive pruning**: NEFCLASS includes four pruning strategies that try to delete variables, rules, terms and fuzzy sets from the rule base. In order to obtain a rule base that is a small as possible all those four strategies are applied one after the other in an exhaustive way to all possible parameters. For the sake of speed the fuzzy sets are not retrained after each pruning step, but only once after pruning.

In order to start the learning process NEFCLASS requires initial fuzzy partitions for all variables. This part is not yet fully automated, because for numeric variables a number of fuzzy sets must be specified. For the discussed scenario this should be done by the application designer. We are currently working on automating this process as well. For symbolic variables, NEFCLASS can determine the initial fuzzy partitions automatically during rule learning [12].

When we execute NEFCLASS in an application environment with no user intervention, then we must try to balance the required computation time with the quality of the rule base. While rule learning and pruning are usually fast, fuzzy set learning can take a lot of time, because it requires many iterations through the training data set. Fuzzy set learning is influenced by some parameters (learning rate, batch/online learning, look ahead mode for trying to escape local minima) and by the already mentioned constraints for guaranteeing meaningful fuzzy sets.

For running fuzzy set learning automatically we can select a small learning rate (e.g. 0.1) together with batch learning to avoid oscillation and select a small look ahead value (10 epochs) which continues training beyond a local minimum in the hope to escape it again. The number of epochs are usually chosen in such a way that a minimum number of learning steps are computed (e.g. 100) and that learning is stopped after, for example, 30 seconds. This time has to be set according to the application scenario. If users of the application usually tend to inspecting other features of the requested data first before checking the fuzzy rules, learning may continue for longer. If the user mainly waits for the rules, learning may have to be shorter.

We also must balance the constraints we impose on the learning algorithm. We could enforce strict constraints like that the membership degrees for each element must add up to 1.0 [16, 18]. However, strict constraints tend to prevent the system from reaching an acceptable classification performance and usually require inspection of the learning outcome and repeated trials, for example, with different numbers of fuzzy sets. In an automated scenario this is not possible, therefore we use only the above-mentioned less strict constraints.

15.4.1 Hints and Tips

- Our objective was to embed a rule learner into a software platform such that it can perform fully automatically, is transparent to the user, and produces a rule base very quickly (in under a minute). If you have a similar objective, then look for approaches that have few parameters and simple learning algorithms. Decision tree learners are a good choice, for example.
- The data mining software package Weka [21] contains some rule learners that you can use in your own applications.
- Several NEFCLASS implementations are also available on the Internet [11].

15.5 Explaining Travel Patterns of a Mobile Workforce

In this section we will look at an application that requires the generation of explanations from data. After shortly introducing the application, we will give a short example and show how the interpretability of two generated rule base is measured.

15.5.1 ITEMS – The Application

Any organization with a large mobile workforce needs to ensure efficient utilization of its resources as they move between tasks distributed over a large geographical area. BT employs around 20000 engineers in the UK who provide services for business and residential customers such as network maintenance, line provision and fault repairs. In order to manage its resources efficiently and effectively, BT uses a sophisticated dynamic scheduling system to build proposed sequences of work for field engineers.

A typical schedule for a field engineer contains a sequence of time windows for travel and task. To generate accurate schedules the system must have accurate estimates for the time required to travel between tasks and estimates for task duration. At BT Exact – BT's research, technology and IT operations business – we have implemented a system that improves the accuracy of travel time estimates by 30% compared to the previous system. The system [1, 5, 6] contains an estimation module, a visual data mining module and an explanation facility.

Note that the travel time is calculated as the difference between the time a task is issued and the arrival time on site. Specifically, travel time includes the time required to leave the current site, walk to the car-park, start the car, drive to the destination site, park the car, and gaining access to the premises of the next customer. However, all these individual activities are not logged, only the start time of the journey and the start time of the actual work are available. It is obvious that we have to deal with huge differences between urban and rural areas, for example, just in finding a space to park.

The Intelligent Travel Estimation and Management System (ITEMS) is a web-enabled, Java-based software system that predicts, manages, visualizes and explains travel patterns of a mobile workforce. ITEMS is a tool for service industries like telecommunications, gas, water, electricity etc. that have to schedule jobs for large mobile workforces. Successful scheduling requires suitable estimates of inter-job times that are mainly determined by travel time. However, it is not sufficient to simply use routing software, because that cannot estimate the time that is required to find a parking space, to gain access to the site etc. It is also impossible for technicians to log detailed information about the routes they have taken, but they only log their actual travel (inter-job) time. Thus, it is not possible to compare actual travel data with recommendations from routing software.

Travel times that are estimated from historic travel data are more reliable, because they reflect the actual travel behavior of the workforce, even if only point-to-point information is available. However, point-to-point information is also the only kind of data a scheduler can suitably process, because it would be intractable for a scheduler to analyze different routes as well while computing a job schedule. Recorded inter-job times automatically reflect features of the areas between which the journey took place. Areas where it is difficult to find a parking space, for example, will automatically result in higher values for the estimated travel time. Because workforce travel displays distinct local patterns, only a fraction of all possible combinations of areas has to be considered by a model for travel estimation. If for some areas there is no travel data available, default estimates are used. For areas where travel occurs regularly, estimates will improve over time. Improved estimates for travel will improve job scheduling, which will finally result in a reduction of unnecessary travel and can thus create huge savings.

ITEMS has a learning component that constantly builds new models for travel time prediction. It compares the new model with the performance of the model currently used by the scheduler and recommends updating the scheduler if the new model performs significantly better. The learning component makes sure, that the scheduler will gradually adapt to changes in travel behavior. So if, for example, technicians are frequently late on specific journeys, the scheduler will obtain new travel time estimates for those journeys after a sufficient amount of data has been gathered. It can then compute better schedules and try to avoid those critical journeys. While updating the travel estimates may take time, managers can still quickly react to critical situations by making use of the visualization and explanation facility of ITEMS.

In addition to reliable estimates workforce managers also regularly need to analyze the travel behavior of their workforce in order to determine if improvements are required. ITEMS provides a color-coded geographical visualization of travel patterns. Managers can easily identify areas where travel is slow and can assess the performance of technicians on a weekly, daily and individual basis.

ITEMS contains an explanation facility based on decision trees and neuro-fuzzy systems that display rule-based information about individual journeys. The rules derived from travel data explain, for example, why a certain journey may have been late. The information of those rules can be used by managers to improve the overall system behavior. Let us, for example, assume that the automatically generated rules reveal that travel between two specific areas takes usually longer than predicted at a certain time of day. In this case, the scheduler can be advised and try avoid scheduling journeys between those two areas at that time of day.

The purpose of the explanatory rules is to provide resource managers with a tool to investigate workforce travel patterns. The rules provide a summary of the actual data and highlight influential variables. In order to be useful the rules must be simple and sparse. It must also be possible to create the rule base on the fly in a very short time. The user would select a specific set of journeys for which he requires explanations. The system must then create the rules completely automatically without user interaction.

For the explanation module we can use a decision tree learner or the NEFCLASS algorithms to generate rules. Figure 15.1 displays a screen shot of typical situation while using ITEMS. A user analyzes the travel pattern of some technician and has clicked an arrow in the displayed map. Additional windows display detailed information about the corresponding job. The top right window displays two fuzzy rules that match the travel information of the selected job. The user can see the degree of fulfillment of a rule and decide, if a particular rule is useful to explain the selected travel pattern, i.e. why the technician was late, early or on time.

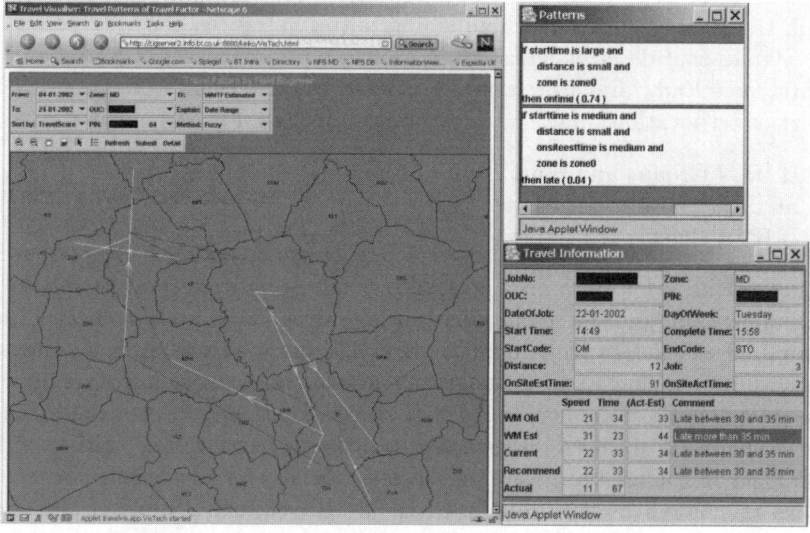

Fig. 15.1. Screen shot of ITEMS displaying two fuzzy rules

15.5.2 Example

In the following we present a short scenario about how ITEMS uses NEFCLASS and a decision tree learner to generate rules for explaining travel patterns. The data contains both numeric and symbolic data and we are using the algorithm described in [14]. For confidentiality reasons we can only reveal parts of the result.

As an example we take a closer look at one model for one organizational unit comprising 13 technicians, using three weeks of data. This is a typical set of data analyzed by a unit manager. The data contains 10 input variables, where five are symbolic. The data was classified into three classes: early, on time and late. After rule learning, fuzzy set tuning and pruning, NEFCLASS presented a rule base of seven rules using four variables (Technician, Time (hour), Start Location, Destination). The decision tree learner created a rule base with 66 rules with more than six variables on average.

The accuracy of the fuzzy rule base was 76.6% while the decision tree had and accuracy of 70%. On our interpretability index introduced in Sect. 15.3 the fuzzy rule base obtained a score of

$$I_{\text{fuzzy}} = 0.0066 \; (\text{comp} = 0.1875, \overline{\text{cov}} = 0.15, \overline{\text{part}} = 0.2375),$$

while the decision tree scored

$$I_{\text{dtree}} = 0.0025 \; (\text{comp} = 0.0076, \overline{\text{part}} = 0.33, \overline{\text{cov}} = 1).$$

Let us take a closer look at the more interpretable rule base. The class accuracies are early = 30%, late = 32%, on time = 97%. That means the rule base is better in predicting the majority class (on time) which is no surprise. Still, the rules are useful for explaining patterns. Let us concentrate on the fuzzy rules that describe late journeys for this particular set of jobs. Group_n, Start_n and End_n are fuzzy sets for the symbolic variables ID (technician), Start (start location) and End (destination), where n is an index.

- If ID is Group_1 and Start Hour is small and Start is Start_1 and End is End_1 then Late
- If ID is Group_4 and Start is Start_2 and End is End_3 then Late
- If ID is Group_5 and Start is Start_3 and End is End_4 then Late

Further analysis of the classification of individual jobs shows that 85% of the misclassified *late* patterns have almost equal membership with *late* and *on time*. In addition all *late* patterns have actually non-zero membership with the class *late*.

When the user clicks on the graphical representation of a journey in the graphical user interface, he will see all rules that fire for that particular record. If the user is particularly interested in late travel, at least one of the above presented rules will be displayed. Even if the pattern is misclassified, the rules can still provide a useful explanation.

Note, that the rules are not meant for prediction, but for explanation only. That is why we use all the data to generate the rules and do not use validation. The rules represent a rule-based summary of the data. For example, if there is a rule that correctly classifies a lot of late journeys, the manager can investigate, why this particular pattern is present in the data. Such a scenario can point to problems in the prediction module or to specific situations like on-going road works that are only indirectly reflected in the data. On the other hand, if there is a late journey that cannot be classified correctly, this means that it cannot be explained by the data and may be an outlier (exception) so that no action is required.

Most of the rules shown above contain only symbolic variables and therefore use fuzzy sets that are represented as a list of value/membership pairs. They are not as easily interpretable as, for example, the fuzzy set *small* for the variable Start Hour in the first rule. However, close inspection of the fuzzy sets reveals, that for example technicians who were frequently late all have high degrees of membership with the fuzzy sets $Group_1$, $Group_4$ and $Group_5$. This can help the manager in identifying, for example, individuals who may be require additional training or information about better routes, because they are, for example, new on the job.

15.5.3 Hints and Tips

- In our example we ranked rule bases only by interpretability. However, accuracy is certainly important as well. A way to compute a ranked list based on accuracy and interpretability is to compare any two rule bases R_i and R_j by computing a weighted sum of their relations of interpretability index I and accuracy A (using a weight $w \in [0, 1]$):

$$c = w \cdot \frac{I_i}{I_j} + (1 - w) \cdot \frac{A_i}{A_j} \ .$$

If $c > 1$ then R_i is preferred over R_j. Based on such pairwise comparisons a topological order of rule bases can be created.

15.6 Conclusions

Intelligent data analysis plays a crucial role in modern businesses [2]. They do not only require predictions based on data but also require a deep understanding of the data that is collected from internal or external sources. Rule based models that provide explanations can be a valuable tool in this area.

We feel that hybrid approaches like neuro-fuzzy systems are very well suited for this task, because they tend to provide more intuitive rules than crisp rule learners. For integrating any approach for explanation generation into applications it is also very important, that it works automatically and fast. For this reason and because of the scope of our application project we look only at NEFCLASS and decision trees.

While it is well known that decision trees are fast and run basically parameterless and automatically our project also showed that a neuro-fuzzy approach like NEFCLASS can compete in scenarios where explanatory rules are required.

The learning algorithms of NEFCLASS are capable of generating explanations about a data set selected by a user in a reasonably short time. In order to further automate the generation of fuzzy models by NEFCLASS we will study methods for automatically generating an appropriate number of fuzzy sets as they are described, for example, in [22].

References

1. Ben Azvine, Colin Ho, Stewart Kay, Detlef Nauck, and Martin Spott. Estimating travel times of field engineers. *BT Technology Journal*, 21(4): 33–38, October 2003.
2. Ben Azvine, Detlef Nauck, and Colin Ho. Intelligent business analytics – a tool to build decision-support sytems for eBusinesses. *BT Technology Journal*, 21(4): 65–71, October 2003.
3. Hugues Bersini, Gianluca Bontempi, and Mauro Birattari. Is readability compatible with accuracy? From neuro-fuzzy to lazy learning. In *Fuzzy-Neuro Systems '98 – Computational Intelligence. Proc. 5th Int. Workshop Fuzzy-Neuro-Systems '98 (FNS'98) in Munich, Germany*, volume 7 of *Proceedings in Artifical Intelligence*, pages 10–25, Sankt Augustin, 1998. infix.
4. J. Casillas, O. Cordon, F. Herrera, and L. Magdalena, editors. *Trade-off between Accuracy and Interpretability in Fuzzy Rule-Based Modelling*. Studies in Fuzziness and Soft Computing. Physica-Verlag, Heidelberg, 2002.
5. Colin Ho and Ben Azvine. Mining travel data with a visualiser. In *Proc. International Workshop on Visual Data Mining at the 2nd European Conference on Machine Learning (ECML'01) and 5th European Conference on Principles and Practice of Knowledge Discovery in Databases (PKDD'01)*, Freiburg, 2001.
6. Colin Ho, Ben Azvine, Detlef Nauck, and Martin Spott. An intelligent travel time estimation and management tool. In *Proc. 7th European Conference on Networks and Optical Communications (NOC 2002)*, pages 433–439, Darmstadt, 2002.
7. Cezary Z. Janikow. Fuzzy decision trees: Issues and methods. *IEEE Trans. Systems, Man & Cybernetics. Part B: Cybernetics*, 28(1): 1–14, 1998.
8. Aljoscha Klose, Andreas Nürnberger, and Detlef Nauck. Some approaches to improve the interpretability of neuro-fuzzy classifiers. In *Proc. Sixth European Congress on Intelligent Techniques and Soft Computing (EUFIT98)*, pages 629–633, Aachen, 1998.
9. I. Kononenko. On biases in estimating multi-valued attributes. In *Proc. 1st International Conference on Knowledge Discovery and Data Mining*, pages 1034–1040, Montreal, 1995.
10. Ludmilla I. Kuncheva. *Fuzzy Classifier Design*. Springer-Verlag, Heidelberg, 2000.

11. Detlef Nauck. NEFCLASS. Univ. of Magdeburg, WWW, 1998. http://fuzzy.cs.uni-magdeburg.de/nefclass.
12. Detlef Nauck. Using symbolic data in neuro-fuzzy classification. In *Proc. 18th International Conf. of the North American Fuzzy Information Processing Society (NAFIPS99)*, pages 536–540, New York, NY, 1999. IEEE.
13. Detlef Nauck. Adaptive rule weights in neuro-fuzzy systems. *Neural Computing & Applications*, 9(1): 60–70, 2000.
14. Detlef Nauck. Fuzzy data analysis with nefclass. In *Proc. Joint 9th IFSA World Congress and 20th NAFIPS International Conference*, pages 1413–1418, Piscataway, July 2001. IEEE.
15. Detlef Nauck. Measuring interpretability in rule-based classification systems. In *Proc. IEEE Int. Conf. on Fuzzy Systems 2003*, pages 196–201, St. Louis, MO, 2003. IEEE.
16. Detlef Nauck, Frank Klawonn, and Rudolf Kruse. *Foundations of Neuro-Fuzzy Systems*. Wiley, Chichester, 1997.
17. Detlef Nauck and Rudolf Kruse. How the learning of rule weights affects the interpretability of fuzzy systems. In *Proc. IEEE Int. Conf. on Fuzzy Systems 1998*, pages 1235–1240, Anchorage, May 1998.
18. Detlef Nauck and Rudolf Kruse. NEFCLASS-J – a Java-based soft computing tool. In Behnam Azvine, Nader Azarmi, and Detlef Nauck, editors, *Intelligent Systems and Soft Computing: Prospects, Tools and Applications*, number 1804 in Lecture Notes in Artificial Intelligence, pages 143–164. Springer-Verlag, Berlin, 2000.
19. J.R. Quinlan. Induction of decision trees. *Machine Learning*, 1: 81–106, 1986.
20. J.R. Quinlan. *C4.5: Programs for Machine Learning*. Morgan Kaufman, San Mateo, CA, 1993.
21. Ian H. Witten and Eibe Frank. *Data Mining: Practical Machine Learning Tools and Techniques with JAVA Implementations*. Morgan Kaufmann Publishers, San Francisco, CA, 2000. Software available at http://www.cs.waikato.ac.nz/~ml/.
22. Qingqing Zhou, Martin Purvis, and Nikola Kasabov. A membership function selection method for fuzzy neural networks. In *Proc. of Int. Conf. Neural Information Processing and Intelligent Systems ICONIP/ANZIIS/ANNES'97*, volume II, pages 785–788, Dunedin, 1997.

16

Fuzzy Linguistic Data Summaries as a Human Consistent, User Adaptable Solution to Data Mining

J. Kacprzyk[1,2] and S. Zadrożny[1]

[1] Systems Research Institute, Polish Academy of Sciences,
ul. Newelska 6, 01-447 Warsaw, Poland
[2] Warsaw School of Information Technology (WSISiZ)
ul. Newelska 6, 01-447 Warsaw, Poland
<kacprzyk,zadrozny>@ibspan.waw.pl

In this chapter fuzzy linguistic summaries of data (databases) in the sense of Yager (cf. Yager [1], Kacprzyk and Yager [3], and Kacprzyk, Yager and Zadrożny [4]) are presented as a flexible, user adaptable solution to data mining problem. The essence of this approach is that, for instance, if we have a (large) database on employees, then in case that we are interested in, say, age and qualifications, then the contents of the database in this respect may be summarized by, say, "most young employees are well qualified". We present the problem of deriving such linguistic summaries in the context of Zadeh's (cf. Zadeh and Kacprzyk [6]) computing with words and perceptions paradigm, and consider his recent idea of a protoform (cf. Zadeh [7]) that provides means to define and handle more general forms of summaries. We illustrate the approach on an a system developed for a small to medium computer retailer, and show how data from the Internet can qualitatively enhance the results obtained. We show that the approach presented may be viewed as an example of an inexpensive, human consistent, human friendly technology that is easily adaptable to changing interests and needs of users.

16.1 Introduction

In this chapter we address the problem that may be exemplified as follows. There is a small (or a small-to-medium, SME for short) company that – as all other companies and organizations – faces the problem of dealing with too large sets of data that are not comprehensible by the human user. They know that they need some data mining but they are fully aware of their limitations. Mainly, in comparison with larger and richer companies and organization, they need a simple and possibly inexpensive solution that is also as much human consistent as possible. They are aware that most of their employees are not qualified computer specialists, as they cannot afford to hire such people, and hence solutions adopted should be possibly human consistent and intuitive, basically as heavily as possible based upon the use of natural language. Such solutions have to offer at least a basic adaptability with respect to the interpretation of linguistic terms that are used to express data values and relations between data. Another dimension of the adaptability may be considered from the perspective of data sources taken into account. The primary data source for such data mining tasks is, of course, a database of the user. However, in order to discover some interesting phenomena in data it may be worthwhile to acquire some other data as well as no company operates in a vacuum, separated from the outside world. The Internet seems to be such a source of choice. Nowadays, it may be still difficult to get interesting, relevant data from the Internet without a careful planning and execution. However, as soon as promises of the Semantic Web become the reality, it should be fairly easy to arrange for automatic acquisition of data that is relevant for our problem but does not have to be identified in advance. In many cases such data may be easily integrated with our own data and provide the user with interesting results. For example, coupling the data on sales per day with weather information related to a given time period (that is not usually stored in sales databases) may show some dependencies important for running the business.

Generally, data summarization is still an unsolved problem in spite of vast research efforts. Very many techniques are available but they are not "intelligent enough", and not human consistent, partly due to the fact that the use of natural language is limited. This concerns, e.g., summarizing statistics, exemplified by the average, median, minimum, maximum, α-percentile, etc. which – in spite of recent efforts to soften them – are still far from being able to reflect a real human perception of their essence. In this chapter we discuss an approach to solve this problem. It is based on the concept of a *linguistic data (base) summary* and has been originally proposed by Yager [1, 2] and further developed by many authors (see, for instance, Kacprzyk and Yager [3], and Kacprzyk, Yager and Zadrożny [4]). The essence of such linguistic data summaries is that a set of data, say, concerning employees, with (numeric) data on their age, sex, salaries, seniority, etc., can be summarized linguistically with respect to a selected attribute or attributes, say age and salaries, by linguistically quantified propositions, say "almost all employees are well qualified", "most young employees are well paid", etc. Notice that such simple, extremely human consistent and intuitive statements do summarize in a concise yet very informative form what we may be interested in.

We present the essence of Yager's [1, 2] approach to such summaries, with its further extensions (cf. Kacprzyk and Yager [3], Kacprzyk, Yager and Zadrożny [4, 5]) from the perspective of Zadeh's computing with words and perception paradigm (cf. Zadeh and Kacprzyk [6]) that can provide a general theoretical framework which is implementable, as shown in works mentioned above. In particular, we indicate the use of Zadeh's concept of a protoform of a fuzzy linguistic summary (cf. Zadeh [7], Kacprzyk and Zadrożny [8]) that can provide a "portability" and "scalability" as meant above, and also some "adaptivity" to different situations and needs by providing universal means for representing quite general forms of summaries.

As an example we will show an implementation of the data summarization system proposed for the derivation of linguistic data summaries in a sales database of a computer retailer.

The basic philosophy of the approach and its algorithmic engine makes use of the computing with words and perception paradigm introduced by Zadeh in the mid-1990s, and best and most comprehensively presented in Zadeh and Kacprzyk's [6] books. It may be viewed as a new paradigm, or "technology" in the representation, processing and solving of various real life problems when a human being is a crucial element. Such problems are omnipresent. The basic idea and rationale of computing with words and perceptions is that since for a human being natural language is the only fully natural way of communication, then maybe it could be expedient to try to "directly" use (elements of) natural language in the formulation, processing and solution of problems considered to maintain a higher human consistence, hence a higher implementability. Notice that the philosophy and justification of the computing with words and perception paradigm are in line with the requirements and specifics of problems considered, and solution concepts adopted in this paper.

A prerequisite for computing with words is to have some way to formally represent elements of natural language used. Zadeh proposed to use here the PNL (precisiated natural language). Basically, in PNL, statements about values, relations, etc. between variables are represented by constraints. In the conventional case, a statement is, e.g., that the value of variable x belongs to a set X. In PNL, statements – generally written as "x is Z" – may be different, and correspond to numeric values, intervals, possibility distributions, verity distributions, probability distributions, usuality qualified statements, rough sets representations, fuzzy relations, etc. For our purposes, the usuality qualified representation will be of a special relevance. Basically, it says "x is usually Z" that is meant as "in most cases, x is Z". PNL may play various roles among which crucial are: the description of perceptions, the definition of sophisticated concepts, a language for perception based reasoning, etc.

Recently, Zadeh [7] introduced the concept of a protoform. For our purposes, one should notice that most perceptions are summaries. For instance, a perception like "most Swedes are tall" is some sort of a summary. It can be represented in Zadeh's notation as "most As are Bs". This can be employed for reasoning under various assumptions. For instance, if we know that "x is A", we can deduce that, e.g. "it is likely that x is B". We can also ask about an average height of a Swede, etc. One can go a step further, and define a protoform as an abstracted summary. In our case, this would be "QAs are Bs". Notice that we now have a more general, deinstantiated form of our point of departure "most Swedes are tall", and also of "most As are Bs". Needless to say that much of human reasoning is protoform based, and the availability of such a more general representation is vary valuable, and provides tools that can be used in many cases. From the point of view of the problem class considered in this chapter, the use of protoforms may be viewed to contribute to the portability, scalability and adaptivity in the sense mentioned above.

We discuss a number of approaches to mining of linguistic summaries. First, those based on Kacprzyk and Zadrożny's [9, 10] idea of an interactive approach to linguistic summaries in which the determination of a class of summaries of interest is done via Kacprzyk and Zadrożny's [11, 12] FQUERY for Access, a fuzzy querying add-on to Microsoft Access©. It is shown that by relating a range of types of linguistic summaries to fuzzy queries, with various known and sought elements, we can arrive at a hierarchy of protoforms of linguistic data summaries. Basically, there is a trade off between the specificity in respect to the summaries sought and the complexity of a corresponding mining process. In the simplest case, data mining boils down directly to a flexible querying process. In the opposite case, the concept of a linguistic association rule along with well known efficient mining algorithms may be employed. Also other approaches to linguistic summaries mining are briefly discussed in Sect. 16.3.

The line of reasoning adopted here should convince the reader that the use of a broadly perceived paradigm of computing with words and perceptions, equipped with a newly introduced concept of a protoform, may be a proper tool for dealing with situations when we have to develop and implement a system that should perform "intelligent" tasks, be human consistent and human friendly, and some other relevant requirements should also be fulfilled as, e.g., to be inexpensive, easy to calibrate, portable, scalable, being able to somehow adapt to changing conditions and requirements, etc.

16.2 Linguistic Data Summaries via Fuzzy Logic with Linguistic Quantifiers

The linguistic summary is meant as a natural language like sentence that subsumes the very essence (from a certain point of view) of a set of data. This set is assumed to be numeric and is usually large, not comprehensible in its original form by the human being. In Yager's approach (cf. Yager [1], Kacprzyk and Yager [3], and Kacprzyk, Yager and Zadrożny [4]) the following context for linguistic summaries mining is assumed:

- $Y = \{y_1, \ldots, y_n\}$ is a set of objects (records) in a database, e.g., the set of workers;
- $A = \{A_1, \ldots, A_m\}$ is a set of attributes characterizing objects from Y, e.g., salary, age, etc. in a database of workers, and $A_j(y_i)$ denotes a value of attribute A_j for object y_i.

A linguistic summary of data set D consists of:

- a summarizer S, i.e. an attribute together with a linguistic value (fuzzy predicate) defined on the domain of attribute A_j (e.g. "low salary" for attribute "salary");
- a quantity in agreement Q, i.e. a linguistic quantifier (e.g. most);
- truth (validity) T of the summary, i.e. a number from the interval $[0, 1]$ assessing the truth (validity) of the summary (e.g. 0.7); usually, only summaries with a high value of T are interesting;
- optionally, a qualifier R, i.e. another attribute together with a linguistic value (fuzzy predicate) defined on the domain of attribute A_k determining a (fuzzy subset) of Y (e.g. "young" for attribute "age").

Thus, the linguistic summary may be exemplified by

$$T(\text{most of employees earn low salary}) = 0.7 \qquad (16.1)$$

A richer form of the summary may include a qualifier as in, e.g.,

$$T(\text{most of young employees earn low salary}) = 0.7 \qquad (16.2)$$

Thus, basically, the core of a linguistic summary is a *linguistically quantified proposition* in the sense of Zadeh [13]. A linguistically quantified proposition, corresponding to (16.1) may be written as

$$Qy\text{'s are } S \qquad (16.3)$$

and the one corresponding to (16.2) may be written as

$$QRy\text{'s are } S \qquad (16.4)$$

Then, the component of a linguistic summary, T, i.e., its truth (validity), directly corresponds to the truth value of (16.3) or (16.4). This may be calculated by using either original Zadeh's calculus of linguistically quantified statements (cf. [13]), or other interpretations of linguistic quantifiers (cf. Liu and Kerre [14]), including Yager's OWA operators [15] and Dubois et al. OWmin operators [16]. The component of a linguistic summary that is a quantifier Q can also be interpreted from a more general perspective of the concept of a *generalized quantifier*, cf. Hájek and Holeňa [17] or Glöckner [18].

Using Zadeh's [13] fuzzy logic based calculus of linguistically quantified propositions, a (proportional, nondecreasing) linguistic quantifier Q is assumed to be a fuzzy set in the interval $[0, 1]$ as, e.g.

$$\mu_Q(x) = \begin{cases} 1 & \text{for } x \geq 0.8 \\ 2x - 0.6 & \text{for } 0.3 < x < 0.8 \\ 0 & \text{for } x \leq 0.3 \end{cases} \qquad (16.5)$$

Then, the truth values (from $[0, 1]$) of (16.3) and (16.4) are calculated, respectively, as

$$\text{truth}(Qy\text{'s are } S) = \mu_Q \left[\frac{1}{n} \sum_{i=1}^{n} \mu_S(y_i) \right] \quad (16.6)$$

$$\text{truth}(QRy\text{'s are } S) = \mu_Q \left[\frac{\sum_{i=1}^{n} (\mu_R(y_i) \wedge \mu_S(y_i))}{\sum_{i=1}^{n} \mu_R(y_i)} \right] \quad (16.7)$$

Both the fuzzy predicates S and R are assumed above to be of a rather simplified, atomic form referring to just one attribute. They can be extended to cover more sophisticated summaries involving some confluence of various attribute values as, e.g., "young and well paid". Clearly, when we try to linguistically summarize data, the most interesting are non-trivial, human-consistent summarizers (concepts) as, e.g.:

- productive workers,
- difficult orders, ... ,

and it may easily be noticed that their proper definition may require a very complicated combination of attributes as with, for instance: a hierarchy (not all attributes are of the same importance for a concept in question), the attribute values are ANDed and/or ORed, k out of n, most, ... of them should be accounted for, etc.

Recently, Zadeh [7] introduced the concept of a *protoform* that is highly relevant in this context. Basically, a protoform is defined as a more or less abstract prototype (template) of a linguistically quantified proposition. The most abstract protoforms correspond to (16.3) and (16.4), while (16.1) and (16.2) are examples of fully instantiated protoforms. Thus, evidently, protoforms form a hierarchy, where higher/lower levels correspond to more/less abstract protoforms. Going down this hierarchy one has to instantiate particular components of (16.3) and (16.4), i.e., quantifier Q and fuzzy predicates S and R. The instantiation of the former one consists in the selection of a quantifier. The instantiation of fuzzy predicates requires the choice of attributes together with linguistic values (atomic predicates) and a structure they form when combined using logical connectives. This leads to a theoretically infinite number of potential protoforms. However, for the purposes of mining of linguistic summaries, there are obviously some limits on a reasonable size of a set of summaries that should be taken into account. These results from a limited capability of the user in the interpretation of summaries as well as from the computational point of view.

The concept of a protoform may be taken as a guiding paradigm for the design of a user interface supporting the mining of linguistic summaries. It may be assumed that the user specifies a protoform of linguistic summaries sought. Basically, the more abstract protoform the less should be assumed about summaries sought, i.e., the wider range of summaries is expected by the user. There are two limit cases, where:

Table 16.1. Classification of protoforms/linguistic summaries

Type	Protoform	Given	Sought
0	QRy's are S	All	validity T
1	Qy's are S	S	Q
2	QRy's are S	S and R	Q
3	Qy's are S	Q and structure of S	linguistic values in S
4	QRy's are S	Q, R and structure of S	linguistic values in S
5	QRy's are S	Nothing	S, R and Q

- a totally abstract protoform is specified, i.e., (16.4)
- all elements of a protoform are specified on the lowest level of abstraction as specific linguistic terms.

In the first case the system has to construct all possible summaries (with all possible linguistic components and their combinations) for the context of a given database (table) and present to the user those verifying the validity to a degree higher than some threshold. In the second case, the whole summary is specified by the user and the system has only to verify its validity. Thus, the former case is usually more interesting from the point of view of the user but at the same time more complex from the computational point of view. There is a number of intermediate cases that may be more practical. In Table 16.1 basic types of protoforms/linguistic summaries are shown, corresponding to protoforms of a more and more abstract form.

Basically, each of fuzzy predicates S and R may be defined by listing its atomic fuzzy predicates (i.e., pairs of "attribute/linguistic value") and structure, i.e., how these atomic predicates are combined. In Table 16.1 S (or R) corresponds to the full description of both the atomic fuzzy predicates (referred to as linguistic values, for short) as well as the structure. For example:

$$Q \text{ young employees earn a high salary} \qquad (16.8)$$

is a protoform of Type 2, while:

$$\text{Most employees earn a "?" salary} \qquad (16.9)$$

is a protoform of Type 3.

In case of (16.8) the system has to select a linguistic quantifier (usually from a predefined dictionary) that when put in place of Q in (16.8) makes the resulting linguistically quantified proposition valid to the highest degree. In case of (16.9), the linguistic quantifier as well as the *structure* of summarizer S are given. The system has to choose a linguistic value to replace the question mark ("?") yielding a linguistically quantified proposition as valid as possible. Note that this may be interpreted as the search for a *typical* salary in the company.

Thus, the use of protoforms makes it possible to devise a uniform procedure to handle a wide class of linguistic data summaries so that the system can be easily adaptable to a variety of situations, users' interests and preferences, scales of the project, etc.

Usually, most interesting are linguistic summaries required by a summary of Type 5. They may be interpreted as fuzzy IF-THEN rules:

$$\text{IF } R(y) \text{ THEN } S(y) \tag{16.10}$$

that should be instantiated by a system yielding, e.g., a rule

$$\text{IF } y \text{ IS } young \text{ THEN } y \text{ EARNS } low\ salary \tag{16.11}$$

with a truth degree being a function of the two components of the summary that involve the truth (validity) T and the linguistic quantifier Q. In the literature (cf., e.g., Dubois and Prade [19]) there are considered many possible interpretations for fuzzy rules. Some of them were directly discussed in the context of linguistic summaries by some authors (cf. Sect. 16.3.3 in this chapter).

Some authors consider the concept of a *fuzzy functional dependency* as a suitable candidate for the linguistic summarization. The fuzzy functional dependencies are an extension of the classical crisp functional dependencies considered in the context of relational databases. The latter play a fundamental role in the theory of normalization. A functional dependency between two sets of attributes $\{A_i\}$ and $\{B_i\}$ holds when the values of attributes $\{A_i\}$ fully determine the values of attributes $\{B_i\}$. Thus, a functional dependency is a much stronger dependency between attributes than that expressed by (16.10). The classical crisp functional dependencies are useless for data summarization (at least in case of regular relational databases) as in a properly designed database they should not appear, except the trivial ones. On the other hand, fuzzy functional dependencies are of an approximate nature and as such may be identified in a database and serve as linguistic summaries. They may be referred to as extensional functional dependencies that may appear in a given instance of a database in contrast to intentionally interpreted crisp functional dependencies that are, by design, avoided in any instance of a database. A fuzzy functional dependency may be exemplified with

$$\text{AGE determines SALARY} \tag{16.12}$$

to be interpreted in such a way that "*usually* any two employees of a *similar* age have also *similar* salaries". Such a rule may be, as previously, associated with a certain linguistic quantifier (here: usually) and a truth qualification degree. Many authors discuss various definitions of fuzzy functional dependencies, cf., e.g., Bosc, Dubois and Prade [20].

16.3 Various Approaches to the Mining of Linguistic Summaries

The basic concept of a linguistic summary seems to be fairly simple. The main issue is how to generate summaries for a given database. The full search of the solution space is practically infeasible. In the literature a number of ways to solve this problem have been proposed. In what follows we briefly overview some of them.

The process of mining of linguistic summaries may be more or less automatic. At the one extreme, the system may be responsible for both the construction and verification of summaries (which corresponds to Type 5 protoforms/summaries given in Table 16.1). At the other extreme, the user proposes a summary and the system only verifies its validity (which corresponds to Type 0 protoforms/summaries in Table 16.1). The former approach seems to be more attractive and in the spirit of data mining meant as the discovery of interesting, unknown regularities in data. On the other hand, the latter approach, obviously secures a better interpretability of the results. Thus, we will discuss now the possibility to employ a flexible querying interface for the purposes of linguistic summarization of data, and indicate the implementability of a more automatic approach.

16.3.1 A Fuzzy Querying Add-on for Formulating Linguistic Summaries

Since we consider a problem that should be solved, and put to practice, we should find a proper way to implement the algorithmic base presented in the previous section. For this purpose we need first of all appropriate user interfaces since the tools involve many entities that should be elicited from the user, calibrated, illustratively displayed, etc.

In Kacprzyk and Zadrożny's [9, 10] approach, the interactivity, i.e. a user assistance, is in the definition of summarizers (indication of attributes and their combinations). This proceeds via a user interface of a fuzzy querying add-on. In Kacprzyk and Zadrożny [11, 12, 21], a conventional database management system is used and a fuzzy querying tool, FQUERY for Access, is developed to allow for queries with fuzzy (linguistic) elements. An important component of this tool is a *dictionary* of linguistic terms to be used in queries. They include fuzzy linguistic values and relations as well as fuzzy linguistic quantifiers. There is a set of built-in linguistic terms, but the user is free to add his or her own. Thus, such a dictionary evolves in a natural way over time as the user is interacting with the system. For example, an SQL query searching for *troublesome orders* may take the following WHERE clause (we make the syntax of a query to FQUERY for Access more self-descriptive in this example; examples of linguistic terms in italic):

WHERE *Most* of the conditions are met out of
  PRICE*ORDERED-AMOUNT IS *Low*
  DISCOUNT IS *High*
  ORDERED-AMOUNT IS *Much Greater Than* ON-STOCK

It is obvious that the condition of such a fuzzy query directly corresponds to summarizer S in a linguistic summary. Moreover, the elements of a dictionary are perfect building blocks of such a summary. Thus, the derivation of a linguistic summary of type (16.3) may proceed in an interactive (user-assisted) way as follows:

- the user formulates a set of linguistic summaries of interest (relevance) using the fuzzy querying add-on,
- the system retrieves records from the database and calculates the validity of each summary adopted, and
- a most appropriate linguistic summary is chosen.

Referring to Table 16.1, we can observe that Type 0 as well as Type 1 linguistic summaries may be easily produced by a simple extension of FQUERY for Access. Basically, the user has to construct a query, a candidate summary, and it is to be determined which fraction of rows matches that query (and which linguistic quantifier best denotes this fraction, in case of Type 1). Type 3 summaries require much more effort as their primary goal is to determine typical (exceptional) values of an attribute (combination of attributes). So, query/summarizer S consists of only one simple condition built of the attribute whose typical (exceptional) value is sought, the "=" relational operator, and a placeholder for the value sought. For example, using: $Q =$ "most" and $S =$ "age=?" we look for a typical value of "age". From the computational point of view Type 5 summaries represent the most general form considered: fuzzy rules describing dependencies between specific values of particular attributes.

The summaries of Type 1 and 3 have been implemented as an extension to Kacprzyk and Zadrożny's [22, 23, 24] FQUERY for Access.

16.3.2 Linguistic Summaries and Fuzzy Association Rules

The discovery of general rules as given by (16.10) (i.e. of Type 5) is essentially a difficult task. As mentioned earlier, some additional assumptions about the structure of particular fuzzy predicates and/or quantifier have usually to be done. One set of such assumptions leads to the idea of using *fuzzy association rules* as linguistic summaries.

Originally, the association rules were defined for binary valued attributes in the following form (cf. Agraval and Srikant [25]):

$$A_1 \wedge A_2 \wedge \ldots \wedge A_n \longrightarrow A_{n+1} \tag{16.13}$$

and note that much earlier origins of that concept are mentioned in the work by Hájek and Holeňa [17]).

Thus, such an association rule states that if in a database row all the attributes from the set $\{A_1, A_2, \ldots, A_n\}$ take on value 1, then also the attribute A_{n+1} is expected to take on value 1. The algorithms proposed in the literature for mining the association rules are based on the following concepts and definitions. A row in a database (table) is said to *support* a set of attributes $\{A_i\}_{i \in I}$ if all attributes from the set take on in this row value 1. The support of a rule (16.13) is the fraction of the number of rows supporting the set of attributes $\{A_i\}_{i \in \{1,\ldots,n+1\}}$ in a database (table). The *confidence* of a rule in a database (table) is the fraction of the number of rows supporting the set of attributes $\{A_i\}_{i \in \{1,\ldots,n+1\}}$ among all rows supporting the set of attributes $\{A_i\}_{i \in I}$. The well known algorithms (cf. Agrawal and Srikant [25] and Mannila et al. [26]) search for rules having values of the support measure above some minimal threshold and a high value of the confidence measure. Moreover, these algorithms may be easily adopted for the non-binary valued data and more sophisticated rules than one shown in (16.13).

In particular, *fuzzy association rules* may be considered:

$$A_1 \text{ IS } R_1 \wedge A_2 \text{ IS } R_2 \wedge \ldots \wedge A_n \text{ IS } R_n \longrightarrow A_{n+1} \text{ IS } S \qquad (16.14)$$

where R_i is a linguistic term defined in the domain of the attribute A_i, i.e. a qualifier fuzzy predicate in terms of linguistic summaries (cf. Sect. 16.2 of this chapter) and S is another linguistic term corresponding to the summarizer. The confidence of the rule may be interpreted in terms of linguistic quantifiers employed in the definition of a linguistic summary. Thus, a fuzzy association rule may be treated as a special case of a linguistic summary of type defined by (16.4). The structure of the fuzzy predicates R_i and S is to some extent fixed but due to that efficient algorithms for rule generation may be employed. These algorithms are easily adopted to fuzzy association rules. Usually, the first step is a preprocessing of original, crisp data. Values of all attributes considered are replaced with linguistic terms best matching them. Additionally, a degree of this matching may be optionally recorded and later taken into account. For example:

$$\text{AGE} = 45 \longrightarrow \text{AGE IS } medium \text{ (matching degree 0.8)} \qquad (16.15)$$

Then, each combination of attribute and linguistic term may be considered as a Boolean attribute and original algorithms, such as a priori [25], may be applied. They, basically, boil down to an efficient counting of support for all conjunctions of Boolean attributes, i.e., so-called itemsets (in fact, the essence of these algorithms is to count support for as small a subset of itemsets as possible). In case of fuzzy association rules attributes may be treated strictly as Boolean attributes – they may appear or not in particular tuples – or interpreted in terms of fuzzy logic as in linguistic summaries. In the latter case they appear in a tuple to a degree, as in (16.15) and the support counting should take that into account. Basically, a scalar cardinality may be employed (in the spirit of Zadeh's calculus of linguistically quantified propositions). Finally, each frequent itemset (i.e., with the support higher than a selected threshold) is split (in all possible ways) into two parts treated as a conjunction of atomic predicates and corresponding to the premise (predicate R in terms of linguistic summaries) and consequence (predicate S in terms of linguistic summaries) of the rule, respectively. Such a rule is accepted if its confidence is higher than the selected threshold. Note that such an algorithm trivially covers the linguistic summaries of type (16.3), too. For them the last step is not necessary and each whole frequent itemset may be treated as a linguistic summary of this type.

Fuzzy association rules were studied by many authors including Lee and Lee-Kwang [27] and Au and Chan [28]. Hu et al. [29] simplify the form of fuzzy association rules sought by assuming a single specific attribute (class) in the consequent. This leads to the mining of fuzzy classification rules. Bosc et al. [30] argue against the use of scalar cardinalities in fuzzy association rule mining. Instead, they suggest to employ fuzzy cardinalities and propose an approach for the calculation of rules' frequencies. This is not a trivial problem as it requires to divide the fuzzy cardinalities of two fuzzy sets. Kacprzyk, Yager and Zadrożny [4, 22, 23, 24, 31] advocated the use of fuzzy association rules for mining linguistic summaries in the framework of flexible querying interface. Chen et al. [32] investigated the issue of generalized fuzzy rules where a fuzzy taxonomy of linguistic terms is taken into account. Kacprzyk and Zadrożny [33] proposed to use more flexible aggregation operators instead of conjunction, but still in context of fuzzy association rules.

16.3.3 Other Approaches

George and Srikanth [34, 35] use a genetic algorithm to mine linguistic summaries. Basically, they consider the summarizer in the form of a conjunction of atomic fuzzy predicates and a void subpopulation. Then, they search for two linguistic summaries referred to as a *constraint descriptor* ("most specific generalization") and a *constituent descriptor* ("most general specification"), respectively. The former is defined as a compromise solution having both the maximum truth (validity) and number of covered attributes (these criteria are combined by some aggregation operator). The latter is a linguistic summary having the maximum validity and covering all attributes. As in virtually all other approaches, a dictionary of linguistic quantifiers and linguistic values over domains of all attributes is assumed. This is sometimes referred to as a *domain* or *background knowledge*. Kacprzyk and Strykowski [36, 37] have also implemented the mining of linguistic summaries using genetic algorithms. In their approach, the fitting function is a combination of a wide array of indices assessing a validity/interestingness of given summary. These indices include, e.g., a degree of imprecision (fuzziness), a degree of covering, a degree of appropriateness, a length of a summary, and yields an overall degree of validity (cf.

also Kacprzyk and Yager [3]). Some examples of this approach are presented and discussed in Sect. 16.4 of this chapter.

Rasmussen and Yager [38, 39] propose an extension, SummarySQL, to the SQL language, an industry standard for querying relational databases, making it possible to cover linguistic summaries. Actually, they do not address the problem of mining linguistic summaries but merely of verifying them. The user has to conceive a summary, express it using SummarySQL, and then has it evaluated. In [39] it is shown how SummarySQL may also be used to verify a kind of fuzzy gradual rules (cf. Dubois and Prade [40]) and fuzzy functional dependencies. Again, the authors focus on a smooth integration of a formalism for such rule expression with SQL rather than on the efficiency of a verification procedure.

Raschia and Mouaddib [41] consider the problem of mining hierarchies of summaries. Their understanding of summaries is slightly different than that given by (16.4). Namely, their summary is a conjunction of atomic fuzzy predicates (each referring to just one attribute). However, these predicates are not defined by just one linguistic value but possibly by fuzzy sets of linguistic values (i.e., fuzzy sets of higher levels are considered). It is assumed that both linguistic values as well as fuzzy sets of higher levels based on them form *background knowledge* provided by experts/users. The mining of summaries (in fact what is mined is a whole hierarchy of summaries) is based on a concept formation (conceptual clustering) process. The first step is, as usually, a translation of the original tuples from database into so-called *candidate tuples*. This step consists in replacing in the original tuples values of their attributes with linguistic values best matching them which are defined over respective domains. Then, candidate tuples obtained are aggregated to form final summaries of various levels of hierarchy. This aggregation leads to possibly more complex linguistic values (represented by fuzzy sets of a higher level). More precisely, it is assumed that one candidate tuple is processed at a time. It is inserted into appropriate summaries already present in the hierarchy. Each tuple is first added to a top (root), most abstract summary, covering the whole database (table). Then, the tuple is put into offspring summaries along the selected branch in the hierarchy. In fact, a range of operations is considered that may lead to a rearrangement of the hierarchy via the formation of new node-summaries as well as splitting the old ones. The concept of a linguistic quantifier does not directly appear in this approach. However, each summary is accompanied with an index corresponding to the number of original tuples covered by this summary.

16.4 Examples of Linguistic Summaries and Possible Extensions

Finally, to show the essence and virtues of the solution proposed we will briefly present an implementation of a system for deriving linguistic database

Table 16.2. The basic structure of the database

Attribute Name	Attribute Type	Description
Date	Date	Date of sale
Time	Time	Time of sale transaction
Name	Test	Name of the product
Amount (number)	Numeric	Number of products sold in the transaction
Price	Numeric	Unit price
Commission	Numeric	Commission (in %) on sale
Value	Numeric	Value = amount (number) × price, of the product
Discount	Numeric	Discount (in %) for transaction
Group	Test	Product group to which the product belongs
Transaction value	Numeric	Value of the whole transaction
Total sale to customer	Numeric	Total value of sales to the customer in fiscal year
Purchasing frequency	Numeric	Number of purchases by customer in fiscal year
Town	Test	Town where the customer lives or is based

summaries for a computer retailer. Basically, we will deal with its sales database, and will only show some examples of linguistic summaries for some interesting (for the user!) choices of relations between attributes.

The basic structure of the database is as shown in Table 16.2.

Linguistic summaries are generated using a genetic algorithm [36, 37]. We will now give a couple of examples of resulting summaries. First, suppose that we are interested in a relation between the commission and the type of goods sold. The best linguistic summaries obtained are as shown in Table 16.3.

Table 16.3. Linguistic summaries expressing relations between the group of products and commission

Summary
About 1/3 of sales of network elements is with a high commission
About 1/2 of sales of computers is with a medium commission
Much sales of accessories is with a high commission
Much sales of components is with a low commission
About 1/2 of sales of software is with a low commission
About 1/3 of sales of computers is with a low commission
A few sales of components is without commission
A few sales of computers is with a high commission
Very few sales of printers is with a high commission

16 Fuzzy Linguistic Data Summaries as a Human Consistent

Table 16.4. Linguistic summaries expressing relations between the groups of products and times of sale

Summary
About 1/3 of sales of computers is by the end of year
About 1/2 of sales in autumn is of accessories
About 1/3 of sales of network elements is in the beginning of year
Very few sales of network elements is by the end of year
Very few sales of software is in the beginning of year
About 1/2 of sales in the beginning of year is of accessories
About 1/3 of sales in the summer is of accessories
About 1/3 of sales of peripherals is in the spring period
About 1/3 of sales of software is by the end of year
About 1/3 of sales of network elements is in the spring period
About 1/3 of sales in the summer period is of components
Very few sales of network elements is in the autumn period
A few sales of software is in the summer period

As we can see, the results can be very helpful, for instance while negotiating commissions for various products sold.

Next, suppose that we are interested in relations between the groups of products and times of sale. The best results obtained are as in Table 16.4.

Notice that in this case the summaries are much less obvious than in the former case expressing relations between the group of product and commission. But, again, they provide very useful information.

Finally, let us show in Table 16.5 some of the obtained linguistic summaries expressing relations between the attributes: size of customer, regularity of customer (purchasing frequency), date of sale, time of sale, commission, group of product and day of sale.

Table 16.5. Linguistic summaries expressing relations between the attributes: size of customer, regularity of customer (purchasing frequency), date of sale, time of sale, commission, group of product and day of sale

Summary
Much sales on Saturday is about noon with a low commission
Much sales on Saturday is about noon for bigger customers
Much sales on Saturday is about noon
Much sales on Saturday is about noon for regular customers
A few sales for regular customers is with a low commission
A few sales for small customers is with a low commission
A few sales for one-time customers is with a low commission
Much sales for small customers is for nonregular customers

Notice that the linguistic summaries obtained do provide much of relevant and useful information, and can help the decision maker make decisions. It should be stressed that in the construction of the data mining paradigm presented we do not want to replace the decision maker but just to provide him or her with a help (support). This is clearly an example of the promising philosophy of decision support, i.e. to maintain user's autonomy and just to provide a support for decision making, and by no means to replace the user.

The system for deriving linguistic summaries developed and implemented for a computer retailer has been found useful by the user who has indicated its human friendliness, and ease of calibration and adaptation to new tasks (summaries involving new attributes of interest) and users (of a variable preparation, knowledge, flexibility, etc.). However, after some time of intensive use, the user has come to a conslusion (quite obvious!) that all summaries that could be derived by the system have been based on the own database of the company. Clearly, these data contain most relevant information on the functioning of the company. However, no company operates in a vacuum and some external data (e.g. on climate when the operation and/or results depend on climatic conditions, national and global economic indicators, etc.) can be of utmost importance and should be taken into account to derive more relevant summaries. Moreover, such external data do provide an easy and quick adaptation mechanism because they reflect what may be changing in the environment.

Following this rationale and philosophy, we have extended the class of linguistic summaries handled by the system to include those that take into account data easily (freely) available from Internet sources. These data are, on the one hand, most up to date so that their inclusion can be viewed as an obvious factor contributing to an efficient adaptation to most recent changes. A good example for the case cosnidered was the inclusion of data on wheather that has a cosniderable impact on the operation of the computer retailer. It is quite obvious that though such data are widely available because meteorological services are popular around the world, the Internet is the best source of such data. This is particularly true in the case of a small company that has limited funds for data, and also limited human resources to fetch such data. Data from the Internet may be therefore viewed as considerably contributing to an inexpensive technology that is so relevant for any small or medium company who has limited funds.

Using the data from meteorological commercial (inexpensive) and academic (free) services available through the Internet, we have been able to extend the system of linguistic database summarization described above.

For instance, if we are interested in relations between group of products, time of sale, temperature, precipitacion, and type of customers, the best linguistic summaries (of both our "internal" data from the sales database, and "external" meteorological data from an Internet service) are as shown in Table 16.6.

Table 16.6. Linguistic summaries expressing relations between the attributes: group of products, time of sale, temperature, precipitacion, and type of customers

Summary
Very few sales of software in hot days to individual customers
About 1/2 of sales of accessories in rainy days on weekends by the end of the year
About 1/3 of sales of computers in rainy days to individual customers

Notice that the use of external data gives a new quality to possible linguistic summaries. It can be viewed as providing a greater adaptivity to varying conditions because the use of free or inexpensive data sources from the Internet makes it possible to easily and quickly adapt the form and contents of summaries to varying needs and interests. And this all is practically at no additional price and effort.

16.5 Concluding Remarks

In this chapter we have presented an interactive, fuzzy logic based approach to the linguistic summarization of databases, and have advocated it as a means to obtain human consistent summaries of (large) sets of data. Such "raw" sets of data are incomprehensible by the human being, while they linguistic summaries are easily comprehensible. Our intention was to show it as an example of a simple inexpensive information technology that can be implementable even in small companies, and is easily adaptable to varying needs of the users, their preferences, profiles, and proficiency.

Moreover, through the use of Zadeh's computing with words and perceptions paradigm, and of protoforms we have attained the above characteristics to a higher extent and at a lower cost and effort.

References

1. R.R. Yager: A new approach to the summarization of data. Information Sciences, 28, pp. 69–86, 1982.
2. R.R. Yager R.R.: On linguistic summaries of data. In W. Frawley and G. Piatetsky-Shapiro (Eds.): Knowledge Discovery in Databases. AAAI/MIT Press, pp. 347–363, 1991.
3. J. Kacprzyk and R.R. Yager: Linguistic summaries of data using fuzzy logic. International Journal of General Systems, 30, 33–154, 2001.
4. J. Kacprzyk, R.R. Yager and S. Zadrożny. A fuzzy logic based approach to linguistic summaries of databases. International Journal of Applied Mathematics and Computer Science, 10, 813–834, 2000.
5. J. Kacprzyk, R.R. Yager and S. Zadrożny. Fuzzy linguistic summaries of databases for an efficient business data analysis and decision support. In W.

Abramowicz and J. Zurada (Eds.): Knowledge Discovery for Business Information Systems, pp. 129-152, Kluwer, Boston, 2001.
6. L.A. Zadeh and J. Kacprzyk (Eds.): Computing with Words in Information/Intelligent Systems. Part 1. Foundations. Part 2. Applications, Springer–Verlag, Heidelberg and New York, 1999.
7. L.A. Zadeh. A prototype-centered approach to adding deduction capabilities to search engines – the concept of a protoform. BISC Seminar, 2002, University of California, Berkeley, 2002.
8. J. Kacprzyk and S. Zadrożny. Protoforms of linguistic data summaries: towards more general natural-language-based data mining tools. In A. Abraham, J. Ruiz-del-Solar, M. Koeppen (Eds.): Soft Computing Systems, pp. 417–425, IOS Press, Amsterdam, 2002.
9. J. Kacprzyk and S. Zadrożny. Data Mining via Linguistic Summaries of Data: An Interactive Approach. In T. Yamakawa and G. Matsumoto (Eds.): Methodologies for the Conception, Design and Application of Soft Computing. Proc. of IIZUKA'98, pp. 668–671, Iizuka, Japan, 1998.
10. J. Kacprzyk and S. Zadrożny. Data mining via linguistic summaries of databases: an interactive approach. In L. Ding (Ed.): A New Paradigm of Knowledge Engineering by Soft Computing, pp. 325-345, World Scientific, Singapore, 2001.
11. J. Kacprzyk and S. Zadrożny. FQUERY for Access: fuzzy querying for a Windows-based DBMS. In P. Bosc and J. Kacprzyk (Eds.): Fuzziness in Database Management Systems, pp. 415-433, Springer-Verlag, Heidelberg, 1995.
12. J. Kacprzyk and S. Zadrożny. The paradigm of computing with words in intelligent database querying. In L.A. Zadeh and J. Kacprzyk (Eds.): Computing with Words in Information/Intelligent Systems. Part 2. Foundations, pp. 382–398, Springer–Verlag, Heidelberg and New York, 1999.
13. L.A. Zadeh. A computational approach to fuzzy quantifiers in natural languages. Computers and Mathematics with Applications. 9, 149–184, 1983.
14. Y. Liu and E.E. Kerre. An overview of fuzzy quantifiers. (I). Interpretations. Fuzzy Sets and Systems, 95, 1–21, 1998.
15. R.R. Yager and J. Kacprzyk (Eds.): The Ordered Weighted Averaging Operators: Theory and Applications. Kluwer, Boston, 1997.
16. D. Dubois, H. Fargier and H. Prade. Beyond min aggregation in multicriteria decision: (ordered) weighted min, discri-min,leximin. In R.R. Yager and J. Kacprzyk (Eds.): The Ordered Weighted Averaging Operators. Theory and Applications, pp. 181–192, Kluwer, Boston, 1997.
17. P. Hájek, M. Holeňa. Formal logics of discovery and hypothesis formation by machine. Theoretical Computer Science, 292, 345–357, 2003.
18. I. Glöckner. Fuzzy quantifiers, multiple variable binding, and branching quantification. In T.Bilgi c et al. IFSA 2003. LNAI 2715, pp. 135–142, Springer-Verlag, Berlin and Heidelberg, 2003.
19. D. Dubois and H. Prade. Fuzzy sets in approximate reasoning, Part 1: Inference with possibility distributions. Fuzzy Sets and Systems, 40, 143–202, 1991.
20. P. Bosc, D. Dubois and H. Prade. Fuzzy functional dependencies – an overview and a critical discussion. Proceedings of 3rd IEEE International Conference on Fuzzy Systems, pp. 325–330, Orlando, USA, 1994.
21. J. Kacprzyk and S. Zadrożny. Computing with words in intelligent database querying: standalone and Internet-based applications. Information Sciences, 134, 71–109, 2001.

22. J. Kacprzyk and S. Zadrożny. Computing with words: towards a new generation of linguistic querying and summarization of databases. In P. Sinčak and J. Vaščak (Eds.): Quo Vadis Computational Intelligence?, pp. 144–175, Springer-Verlag, Heidelberg and New York, 2000.
23. J. Kacprzyk and S. Zadrożny. On a fuzzy querying and data mining interface, Kybernetika, 36, 657–670, 2000.
24. J. Kacprzyk J. and S. Zadrożny. On combining intelligent querying and data mining using fuzzy logic concepts. In G. Bordogna and G. Pasi (Eds.): Recent Research Issues on the Management of Fuzziness in Databases, pp. 67–81, Springer–Verlag, Heidelberg and New York, 2000.
25. R. Agrawal and R. Srikant. Fast algorithms for mining association rules. Proceedings of the 20th International Conference on Very Large Databases, Santiago de Chile, 1994.
26. H. Mannila, H. Toivonen and A.I. Verkamo. Efficient algorithms for discovering association rules. In U.M. Fayyad and R. Uthurusamy (Eds.): Proceedings of the AAAI Workshop on Knowledge Discovery in Databases, pp. 181–192, Seattle, USA, 1994.
27. Lee J.-H. and H. Lee-Kwang. An extension of association rules using fuzzy sets. Proceedings of the Seventh IFSA World Congress, pp. 399–402, Prague, Czech Republic, 1997.
28. W.-H. Au and K.C.C. Chan. FARM: A data mining system for discovering fuzzy association rules. Proceedings of the 8th IEEE International Conference on Fuzzy Systems, pp. 1217–1222, Seoul, Korea, 1999.
29. Y.-Ch. Hu, R.-Sh. Chen and G.-H. Tzeng. Mining fuzzy association rules for classification problems. Computers and Industrial Engineering, 43, 735–750, 2002.
30. P. Bosc, D. Dubois, O. Pivert, H. Prade and M. de Calmes. Fuzzy summarization of data using fuzzy cardinalities. Proceedings of IPMU 2002, pp. 1553–1559, Annecy, France, 2002.
31. J. Kacprzyk and S. Zadrożny. On linguistic approaches in flexible querying and mining of association rules. In H.L. Larsen, J. Kacprzyk, S. Zadrożny, T. Andreasen and H. Christiansen (Eds.): Flexible Query Answering Systems. Recent Advances, pp. 475–484, Springer-Verlag, Heidelberg and New York, 2001.
32. G. Chen, Q. Wei and E. Kerre. Fuzzy data mining: discovery of fuzzy generalized association rules. In G. Bordogna and G. Pasi (Eds.): Recent Issues on Fuzzy Databases, pp. 45–66. Springer-Verlag, Heidelberg and New York, 2000.
33. J. Kacprzyk and S. Zadrożny. Linguistic summarization of data sets using association rules. Proceedings of The IEEE International Conference on Fuzzy Systems, pp. 702–707, St. Louis, USA, 2003.
34. R. George and Srikanth R. Data summarization using genetic algorithms and fuzzy logic. In F. Herrera and J.L. Verdegay (Eds.): Genetic Algorithms and Soft Computing, pp. 599–611, Springer-Verlag, Heidelberg, 1996.
35. R. George and R. Srikanth. A soft computing approach to intensional answering in databases. Information Sciences, 92, 313–328, 1996.
36. J. Kacprzyk and P. Strykowski. Linguistic data summaries for intelligent decision support. In R. Felix (Ed.): Fuzzy Decision Analysis and Recognition Technology for Management, Planning and Optimization – Proceedings of EFDAN'99, pp. 3–12, Dortmund, Germany, 1999.
37. J. Kacprzyk and P. Strykowski. Linguitic summaries of sales data at a computer retailer: a case study. Proceedings of IFSA'99, pp. 29–33, Taipei, Taiwan R.O.C, vol. 1, 1999.

38. D. Rasmussen and R.R. Yager. Fuzzy query language for hypothesis evaluation. In Andreasen T., H. Christiansen and H. L. Larsen (Eds.): Flexible Query Answering Systems, pp. 23–43, Kluwer, Boston, 1997.
39. D. Rasmussen and R.R. Yager. Finding fuzzy and gradual functional dependencies with SummarySQL. Fuzzy Sets and Systems, 106, 131–142, 1999.
40. D. Dubois and H. Prade. Gradual rules in approximate reasoning. Information Sciences, 61, 103–122, 1992.
41. G. Raschia and N. Mouaddib. SAINTETIQ: a fuzzy set-based approach to database summarization. Fuzzy Sets and Systems, 129, 137–162, 2002.

17

Adaptive Multimedia Retrieval: From Data to User Interaction

A. Nürnberger[1] and M. Detyniecki[2]

[1] University of Magdeburg, FIN IWS–IR Group, 39106 Magdeburg, Germany
nuernb@iws.cs.uni-magdeburg.de
[2] Laboratoire d'Informatique de Paris 6, CNRS 75015 Paris, France
Marcin.Detyniecki@lip6.fr

To improve today's multimedia retrieval tools and thus the overall satisfaction of a user, it is necessary to develop methods that are able to support the user in the retrieval process, e.g. by providing not only additional information about the search results as well as the data collection itself, but also by adapting the retrieval tool to the underlying data as well as to the user's needs and interests. In this chapter we give a brief overview of the state-of-the-art and current trends in research.

17.1 Introduction

During the last years several approaches have been developed that tackle specific problems of the multimedia retrieval process, e.g. feature extraction methods for multimedia data, problem specific similarity measures and interactive user interfaces. These methods enable the design of efficient retrieval tools if the user is able to provide an appropriate query. However, in most cases the user needs several steps in order to find the searched objects. The main reasons for this are on the one hand, the problem of users to specify their interests in the form of a well-defined query – which is partially caused by inappropriate user interfaces –, on the other hand, the problem of extracting relevant features from the multimedia objects. Furthermore, user specific interests and search context are usually neglected when objects are retrieved.

To improve today's retrieval tools and thus the overall satisfaction of a user, it is necessary to develop methods that are able to support the user in the search process, e.g. by providing additional information about the search results as well as the data collection itself and also by adapting the retrieval tool to the user's needs and interests.

In the following, we give an overview of methods that are used in retrieval systems for text and multimedia data. To give a guideline for the development of an integrated system, we describe methods that have been successfully used

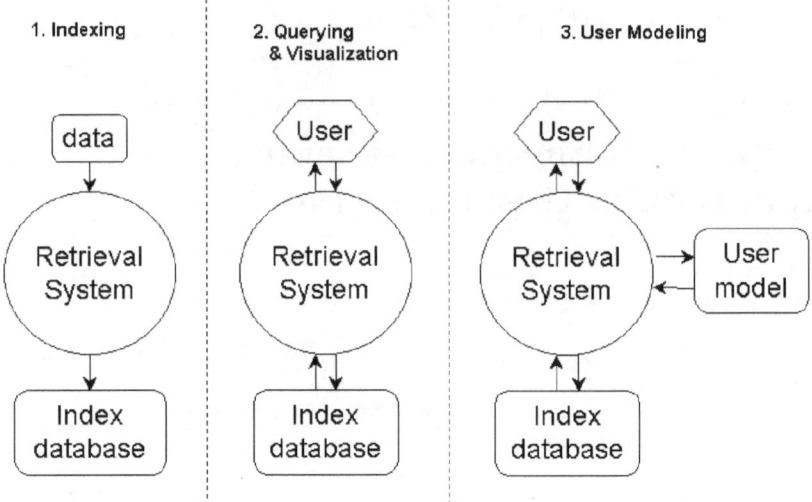

Fig. 17.1. Design Steps for Multimedia Retrieval Systems

by the authors to develop an adaptive retrieval system for multimedia data. Thus, we will all along this chapter follow the example of a system based on growing self organizing maps in order to illustrate the mentioned aspects. Furthermore, we give hints for improvement and point out related projects. The sections of this chapter, reflect the different steps necessary to develop a multimedia retrieval system, from initial feature extraction to interface design and user modelling.

The design of any retrieval system could and should be decomposed in three successive design steps (see Fig. 17.1): Preprocessing and indexing, querying and visualization, and finally interface design and user modelling.

In the first step features are extracted and indexes should be created. It is important to index, because on the one hand we work with large amounts of data and it is usually impossible to browse the whole data set in a reasonable time in order to locate what we are looking for. On the other hand, in case of non-textual-data, the indexing process aims to extract a semantic meaning of the visual or audio data available. This is particularly critical for the interaction with human users. The feature extraction is a crucial, but non-trivial process and therefore several research teams in the world focus on just this part. In Sect. 17.2 we present briefly the state of the art of automatic multimedia indexing tools and of optimized indexing techniques for text.

In the second step of the retrieval system design, we can start thinking on how to interact with the user. We distinguish three different ways of interaction: Querying, navigation through structures and visualization. In Sect. 17.3 we present the most simple and common way of interaction: Querying. This action can be decomposed in three successive steps. First the query formulation by the user, then how to find the information in the database and finally how to aggregate and rank the results. Each of these steps should be well though-out and designed. Another common way to help a user in the retrieval task is to structure the data set. Again, because of the volume of data, we are looking for automatic techniques. One family of algorithms known as clustering techniques allow to find groups of similar objects in data. In Sect. 17.4, after a short introduction and an overview of clustering techniques, we focus on one particular method: The growing self-organizing map. The third way of supporting a user in his task is visualization. Sophisticated visual interaction can be very useful, unfortunately it is rarely exploited. In Sect. 17.5, we first present briefly the state of the art of visualization techniques and then we focus on self-organizing maps, which are particularly suited for visualization.

The final step of retrieval system design is the integration and coordination of the different tools. This step is usually considered as interface design. We dedicate Sect. 17.6 to present some important aspects of it.

Once we have designed the retrieval system based on the above mentioned criteria, we can focus on personalization. The idea is to adapt the system to an individual user or a group of users. This adjustment to the user can be considered for every type of interaction. All personalization is based on a model of the user, which can be manually configured or learned by analyzing user behavior and user feedback. User feedback can be explicitly requested or – if the system is suitable designed – learned from the interaction between system and user. Finally, we discuss in Sect. 17.7 as exemplification, after a short state of the art of user modeling approaches, personalization methods for self-organizing maps.

17.2 Feature Extraction and Data Preprocessing

One of the major difficulties that multimedia retrieval systems face is that they have to deal with different forms of data (and data formats), as for instance text (e.g. text, html), hyperlinks (e.g. html, pdf), audio (e.g. waw, midi), images (e.g. jpeg, gif), video (e.g. mpeg, avi). Not only this polymorphism is a problem, but also the fact that the information is not directly available. In fact, it is not like a text file, where we can simply index by using the words appearing in the text. Usually, we have to index by annotating media data – manually or automatically –, in order to interact or work with them.

Formally, indexing is the process of attaching content-based labels to the media. For instance, existing literature on video indexing defines video indexing as the process of extracting the temporal location of a feature and its value from the video data. Currently indexing in the area of video is generally done manually. But since the indexing effort is directly proportional to the granularity of video access and since the number of videos available grows and new applications demand fine grained access to video, automation of the indexing process becomes more and more essential. Presently it is reasonable

1. Indexing

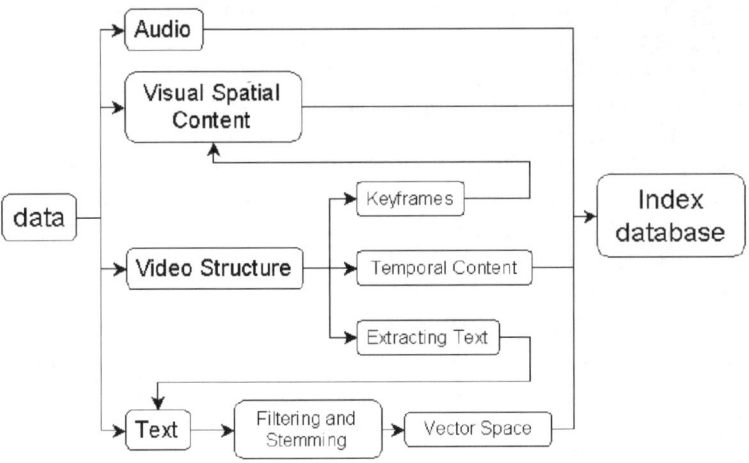

Fig. 17.2. The Process of Multimedia Indexing

to say that we can extract automatically the characteristics described in the following.

17.2.1 Visual Spatial Content

From an image, we can extract color, texture, sketch, shape, objects and their spatial relationships. The color feature is typically represented by image histograms. The texture describes the contrast, the uniformity, the coarseness, the roughness, the frequency, and the directionality. In order to obtain this features either statistical techniques are used (autocorrelation, co-ocurrence matrix) or spectral techniques as for instance detection of narrow peaks in the spectrum. The sketch gives an image containing only the object outlines and it is usually obtained by the combination of edge detection, thinning, shrinking and other transformations of this type. The shape describes global features as the circularity, eccentricity and major axis orientation, but also local ones such as for instance point of curvature, corner location, turning angles and algebraic moments. Note that the images can be already the result of an extraction, as for instance the key-frames, which are representative images from the visual stream (see video segmentation). Unfortunately, it is not yet possible to reliably extract a semantic content description of an image, even though several research groups are working on this problem.

17.2.2 Audio

From the audio stream, we can extract the following basic characteristics: The loudness as the strength of sound, determined by the amplitude of the sound wave. Using sound waves, loudness is dependant upon variations in pressure and the time rate at which sound energy flows through a defined area. The frequency translates what we perceive as pitch. The timbre is the characteristic distinguishing sounds from different sources. We can easily follow a particular timbre, without recognizing them.

More sophisticated characteristic can be obtained as for instance speaker tracking. To track a speaker means that we look for the person who speaks. Usually, at the beginning, a voice is learnt and then we recognize it during the whole audio stream. Most models for audio recognition use hidden Markov chains [1]. Another interesting feature is to recognize when noise, music, speech, or silence is predominant. Different approaches exist such as the use of expert systems [2], hidden Markov chains [3] or intelligent aggregation of specialized detectors.

17.2.3 Video Structure

We can simply extract all images and the sound track from a video or we can be more specific with respect to the video format and extract information regarding the structure by video segmentation and by extracting a representative image (called key-frame) from a segment. The purpose of video segmentation is to partition the video stream into basic units (shots) in order to facilitate indexing, browsing and to give some structure similar to paragraphs in a text document. Current techniques allow not only to find the shots, but also to describe the transition between them, as for instance fade in, fade out, dissolve or wipe.

Videos are stored either uncompressed or compressed, e.g. using MPEG or any other video compression method. The techniques for segmentation used in the uncompressed domain are based on pixel-wise or histogram comparison [4]. In the compressed domain [5] they are based either on coefficient manipulations as inner product or absolute difference or on the motion vectors. The key-frames are usually extracted to reduce the image processing to one image per shot. The idea is to find a representative frame (containing characteristic features) from the sequence of video frames. One simple method consists in extracting the first or the tenth frame [5]. More sophisticated methods look for local minima of motion or significant pauses [6]. Lately, expert systems having rules based on the camera motion have been proposed.

Video Visual Temporal Content

One important characteristic of the video is its temporal aspect. We can easily extract the following motions: The camera motion describes the real (translation and rotation) and factual movement of the camera (zoom in and out). It is usually obtained either by studying the optical flow by dividing the video in several regions [7] or by studying the motion vector [8]. The object motion describes the trajectory obtained by tracking one object on the screen [9, 10].

Extracting Text

Since we are able to use text very efficiently to retrieve objects from a collection (see the following Sect. 17.2.4), a lot of research is currently done in order to synchronize text with or to extract text from the video. One research direction is to synchronize the written script with the video [11]. Another one is to extract the text by doing a transcription of the audio channel [12]. This is usually done with complex speech recognition techniques on pre-processed data. Another interesting challenge is to extract the written information appearing on the screen [13, 14]. The idea is to locate the text on the screen and then to recognize it. These techniques are first attempts, but very often the synchronized text is available and can be exploited.

17.2.4 Text

For searching and navigation in large document collections it is necessary to pre-process the documents and store the information in a data structure, which is more appropriate for further processing than an unstructured text file. The currently predominant approaches are the vector space model [15], the probabilistic model [16], the logical model [17] and the Bayesian net model [18]. Despite of its simple data structure without using any explicit semantic information, the vector space model enables very efficient analysis of huge document collections and is therefore still used in most of the currently available document retrieval systems. The most popular retrieval methods that are based on this model are – besides the direct use of the vector description in an inverse index – Latent Semantic Indexing (LSI) [19], Random Projection [20] and Independent Component Analysis (ICA) [21]. The vector space description is also used for self-organizing maps. In the following we briefly describe the vector space model and appropriate document encoding techniques.

The Vector Space Model

The vector space model represents documents as vectors in t-dimensional space, i.e. each document i is described by a numerical feature vector $D_i = x_1, \ldots, x_t$. Thus, documents can be compared by use of simple vector operations and even queries can be performed by encoding the query terms similar to the documents in a query vector. The query vector can then be compared to each document and a result list can be obtained by ordering the documents according to the computed similarity [22]. The main task of the vector space representation of documents is to find an appropriate encoding of the feature vector.

Each element of the vector usually represents a word (or a group of words) of the document collection, i.e. the size of the vector is defined by the number of words (or groups of words) of the complete document collection. The simplest way of document encoding is to use binary term vectors, i.e. a vector element is set to one if the corresponding word is used in the document and to zero if the word is not. This encoding will result in a simple Boolean search if a query is encoded in a vector. Using Boolean encoding the importance of all terms for a specific query or comparison is considered as similar. To improve the performance usually term weighting schemes are used, where the weights reflect the importance of a word in a specific document of the considered collection. Large weights are assigned to terms that are used frequently in relevant documents but rarely in the whole document collection [23]. Thus a weight w_{ik} for a term k in document i is computed by term frequency tf_{ik} times inverse document frequency idf_k, which describes the term specificity within the document collection. In [22] a weighting scheme was proposed that has meanwhile proven its usability in practice. Besides term frequency and inverse document frequency – defined as $idf_k := log(N/n_k)$ –, a length normalization factor is used to ensure that all documents have equal chances of being retrieved independent of their lengths:

$$w_{ik} = \frac{tf_{ik} \log(N/n_k)}{\sqrt{\sum_{j=1}^{t} (tf_{ij})^2 (\log(N/n_j))^2}}, \qquad (17.1)$$

where N is the size of the document collection C and n_k the number of documents in C that contain term k.

Based on a weighting scheme a document i is defined by a vector of term weights $D_i = \{w_{i1}, \ldots, w_{ik}\}$ and the similarity S of two documents (or the similarity of a document and a query vector) can be computed based on the inner product of the vectors (by which – if we assume normalized vectors – the cosine between the two document vectors is computed), i.e.

$$S(D_i, D_j) = \sum_{k=1}^{m} w_{ik} \cdot w_{jk} . \qquad (17.2)$$

For a more detailed discussion of the vector space model and weighting schemes see, e.g. [15, 23, 24, 25].

Filtering and Stemming

To reduce the number of words and thus the dimensionality of the vector space description of the document collection, the size of the dictionary of words describing the documents can be reduced by filtering stop words and by stemming the words used. The idea of stop word filtering is to remove words that bear little or no content information, like articles, conjunctions, prepositions, etc. Furthermore, words that occur extremely often can be said to be of little information content to distinguish between documents, and also words that occur very seldom are likely to be of no particular statistical relevance and can be removed from the dictionary [26]. Stemming methods try to build the basic forms of words, i.e. strip the plural "s" from nouns, the "ing" from verbs, or other affixes. A stem is a natural group of words with equal (or very similar) meaning. After the stemming process, every word is represented by its stem in the vector space description. Thus a feature of a document vector D_i now describes a group of words. A well-known stemming algorithm has been originally proposed by Porter [27]. He defined a set of production rules to iteratively transform (English) words into their stems.

Automatic Indexing

To further decrease the number of words that should be used in the vector description also indexing or keyword selection algorithms can be used (see, e.g. [19, 28]). In this case, only the selected keywords are used to describe the documents. A simple but efficient method for keyword selection is to extract keywords based on their entropy. E.g. in the approach discussed in [29], for each word a in the vocabulary the entropy as defined by [30] was calculated:

$$W(a) = 1 + \frac{1}{\ln(m)} \sum_{i=1}^{m} p_i(a) \cdot \ln(p_i(a)) \text{ with } p_i(a) = \frac{n_i(a)}{\sum_{j=1}^{m} n_j(a)}, \quad (17.3)$$

where $n_i(a)$ is the frequency of word a in document i and m is the number of documents in the document collection. Here, the entropy gives a measure how well a word is suited to separate documents by keyword search. E.g. words that occur in many documents will have low entropy. The entropy can be seen as a measure of the importance of a word in the given domain context. As index words a number of words that have a high entropy relative to their overall frequency can be chosen, i.e. from words occurring equally often those with the higher entropy can be preferred. This procedure has empirically been found to yield a set of relevant words that are suited to serve as index terms [29].

17.2.5 Hints and Tips

- Feature extraction and data preprocessing are essential and fundamental steps in the design of a retrieval system. The feasibility of the following steps and the quality of the retrieval system directly depend on it.
- In general the difficulty and the time required to obtain good indexes is often underestimated. In particular, for real multimedia objects this task is even more complex, because we have not only very sophisticated indexing processes per media, but also we have to coordinate the cross-indexes. Therefore, very often multimedia retrieval systems are only focused on a particular kind of data, either on image or sound or video.
- Whenever text is available do not hesitate to use it. Query results using text are particularly good compared to other features.
- Even if it is possible to use not preprocessed text, the results are highly improved if preprocessing is done.

Given the current state of the art, reliable and efficient automatic indexing is only possible for the presented low-level characteristics. But is clear that any intelligent interaction with the multimedia data should be based on a higher level of description. For instance, in the case of a video we are more interested in finding a dialog, than in finding a series of alternated shots. In case of an image you can always query for a specific color and texture, but is this really what you want to do?

The current intelligent systems use high level indexing as for instance a predefined term index or even ontological categories. Unfortunately, the high level indexing techniques are based on manual annotation. So, these approaches can only be used for small quantities of new video and do not exploit intelligently the automatic extracted information.

In addition, the characteristics extracted by the automatic techniques are clearly not crisp attributes (color, texture) or they are defined with a degree of truth (camera motion: fade-in, zoom out) or imprecise (30% noise, 50% speech). Therefore, fuzzy techniques seem to be a promising approach to deal with this kind of data (see, for example, [31]).

17.3 Querying

In information retrieval it is usually distinguished between searching by example (or content based searching) and feature based querying. Content based searching means that the user has a sample and is looking for similar objects in a document collection. Content based searching is usually performed by preprocessing the provided sample object to extract the relevant features. The features are then used to perform a query. Therefore, content based and feature based searching differ only in the way the features for searching are defined: In feature based searching the user has to provide the features, in content based searching the features are automatically extracted from the sample given.

Searching based on given features is performed by retrieving the objects in the collection that have similar features. In case of Boolean queries the features of the objects have to match exactly the criteria defined in the query. Therefore, the result set cannot be sorted by similarity (since all objects match the criteria exactly), but by other criteria like date, size or simply by alphabetical order. If we have indexed a text document collection using the vector space model as described in the previous section, we can search for similar documents by encoding the query terms in form of a query vector similar to the documents. Then we can use the cosine measure (see 17.3) to compare the documents with the query vector and compute the similarity of each document to the query.

17.3.1 Ranking

Ranking defines the process of ordering a result set with respect to given criteria. Usually a similarity measure that allows to compute a numerical similarity value for a document to a given query is used. For text document collections ranking based on the vector space model and the $tf \times idf$ weighting scheme (17.1) discussed in the previous section has proven to provide good results. Here the query is considered as a document and the similarity is computed based on the scalar product (cosine; see 17.2) of the query and each document (see also [23]).

If fuzzy modifiers should be used to define vague queries, e.g. using quantifiers like *"most of (term1, term2,...)"* or *"term1 and at least two terms of (term2, term3,...)"*, ranking methods based on ordered weighted averaging operators (OWA) can be used [32]. If the document collection also provides information about cross-references or links between the documents, these information can be used in order to compute the "importance" of a document within a collection. A link is in this case considered as a vote for a document. A good example is the PageRank algorithm [33] used as part of the ranking procedure by the web search engine Google. It has proven to provide rankings that sorts heavily linked documents for given search criteria in high list ranks.

17.3.2 Hints and Tips

- Users are used to querying systems. Therefore any retrieval system should have a simple keyword query interface.
- Very often the raw results of the similarity (or distance) computation are directly used to rank from high to low. However, a short analysis of the ranking process can highly improve the quality of the system, especially if diverse features are combined.

17.4 Structuring Document Collections

A structure can significantly simplify the access to a document collection for a user. Well known access structures are library catalogues or book indexes. However, the problem of manual designed indexes is the time required to maintain them. Therefore, they are very often not up-to-date and thus not usable for recent publications or frequently changing information sources like the World Wide Web.

The existing methods to automatically structure document collections either try to find groups of similar documents (clustering methods) or assign keywords to documents based on a given keyword set (classification or categorization methods). Examples of clustering methods that are frequently used in information retrieval systems are k-means and agglomerative clustering [34, 35], self-organizing maps [36, 37] and fuzzy clustering methods [38].

To classify (or categorize) documents, we frequently find decision trees [34], probabilistic models [39] and neural networks [40, 41]. A recent softcomputing based approach to classify web documents was presented in [42]. For clustering and for classification non-hierarchical and hierarchical approaches exist, which enable the creation of a deep structure of the collection.

Hierarchical categorization methods can be used to organize documents in a hierarchical classification scheme or folder structure. To create a categorization model – if we neglect manual designed filters – a training set of already (e.g. manually) classified documents is required. If this information is not available we can use clustering methods to pre-structure the collection. Clustering methods group the documents only by considering their distribution in a document space (for example, a n-dimensional space if we use the vector space model for text documents). To each discovered group we can then assign a class label. However, this can be a tedious and time consuming task.

Meanwhile, methods are available that are able to use classified and unclassified documents to create a categorization scheme. The idea is to exploit the classification information of the pre-classified documents and the distribution of the document collections in document space. A recent approach – of these so-called semi-supervised clustering methods – for text categorization was presented in [43].

The approaches described above consider the documents itself as unstructured, i.e. they use only a feature vector created from each document. However, especially in web pages we could exploit structural information that marks headlines and paragraphs or even refers to other documents. The same holds for scientific contributions where usually references to different (additional or related) publications are given. These information can be provided to the user as additional information for navigation and search. An approach that exploits this idea was presented in [44].

The main problem of all these approaches is that they only consider information extracted from the underlying document collection and are not able to include user interests, e.g. based on keywords that define areas the user is interested in and that should be considered more important in the clustering process. We get back to this aspect in Sect. 17.7 when we discuss techniques for user modelling. In the following, we briefly describe self-organizing maps as an example for clustering methods.

17.4.1 An Example: Self-Organizing Maps

Self-organizing maps (SOMs) are trained in an unsupervised manner, i.e. no class information is provided, using a set of high-dimensional sample vectors. The network structure has two layers (see Fig. 17.3). The neurons in the input layer correspond to the input dimensions. The output layer (map) contains as many neurons as clusters needed. All neurons in the input layer are connected with all neurons in the output layer. The weights of the connection between

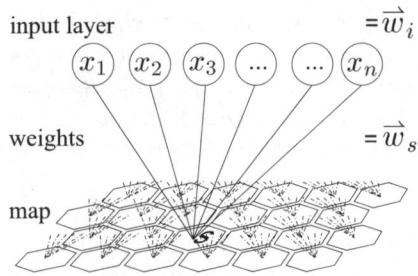

Fig. 17.3. Structure of a Self-Organizing Map Based on Hexagons

input and output layer of the neural network encode positions in the high-dimensional data space. Thus, every unit in the output layer represents a prototype.

Before the learning phase of the network, the two-dimensional structure of the output units is fixed and the weights are initialized randomly. During learning, the sample vectors are repeatedly propagated through the network. The weights of the most similar prototype w_s (*winner neuron*) are modified such that the prototype moves toward the input vector w_i. As similarity measure usually the Euclidean distance or scalar product is used. The weights w_s of the winner neuron are modified according to the following equation: $\forall i : w'_s = w_s + \delta \cdot (w_s - w_i)$, where δ is a learning rate.

To preserve the neighborhood relations, prototypes that are close to the winner neuron in the two-dimensional structure are also moved in the same direction. The strength of the modification decreases with the distance from the winner neuron. Therefore, the adaptation method is extended by a neighborhood function v:

$$\forall i : w_{s'} = w_s + v(c, i) \cdot \delta \cdot (w_s - w_i)$$

where d is a learning rate. By this learning procedure, the structure in the high-dimensional sample data is non-linearly projected to the lower-dimensional topology. Finally, arbitrary vectors (i.e. vectors from the sample set or prior "unknown" vectors) can be propagated through the trained network and are mapped to the output units. For further details on self-organizing maps see [45].

Unfortunately, the standard model of self-organizing maps requires a manual predefined map structure. Therefore, the size of the map is usually too small or too large to map the underlying data appropriately. If the size of the map was to small (the classification error for every pattern is usually very high and thus very dissimilar vectors are assigned to the same unit) or to large (similar vectors spread out on the map). Therefore, the complete learning process has to be repeated several times until an appropriate size is found. Growing self-organizing map approaches try to solve this problem by modifying the size (and structure) of the map by adding new units to the map,

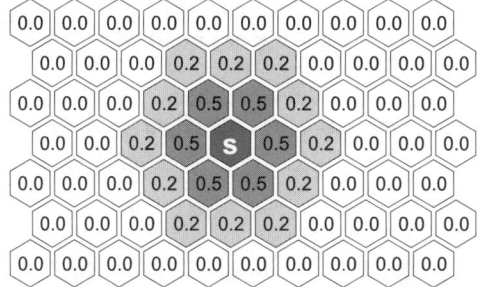

Fig. 17.4. Possible Neighborhood Function for a Self-Organizing Map

if the accumulated error on a map unit increases a specified threshold. Thus they adapt themselves to the structure of the underlying data collection. In the following we briefly describe the approach that we used in a prototypical retrieval system [46].

A Growing Self-Organizing Map Approach

The proposed method is mainly motivated by the growing self-organizing map models presented in [47, 48]. In contrast to these approaches we use hexagonal map structure and restrict the algorithm to add new units to the external units if the accumulated error of a unit exceeds a specified threshold value. The algorithm can be described as follows:

1. Predefine the initial grid size (usually 2 × 2 units)
2. Initialize the assigned vectors with randomly selected values. Reset error values e_i for every unit i.
3. Train the map using all inputs patterns for a fixed number of iterations. During training increase the error values of a winner unit s by the current error value for pattern i.
4. Identify the unit with the largest accumulated error.
5. If the error does not exceed a threshold value stop training.
6. Identify the external unit k with the largest accumulated error.
7. Add a new unit to the unit k. If more than one free link is available select the unit at the nearest position to the neighboring unit which is most dissimilar to k. Initialize the weights of the new unit with respect to the vectors of the neighboring units so that the new vector is smoothly integrated into the existing vectors (see Fig. 17.5).
8. Continue with step 3.
9. Continue training of the map for a fixed number of iterations. Reduce the learning rate during training.

This process creates an incremental growing map. Furthermore, it allows training the map incrementally by adding new data, since the training algorithm affects mainly the winning units to which new data vectors are assigned.

x_i, y_i: weight vectors
x_k: weight vector of unit with highest error
m: new unit
α, β: smoothness weights
Computation of new weight vector x_m for m:

$$x_m = [x_k + \alpha \cdot (x_k - y_k) + \sum_{\substack{i=0 \\ i \neq k}}^{n} (x_i + \beta \cdot (x_i - y_i))] \cdot \frac{1}{n+1}$$

Fig. 17.5. Insertion of a New Unit

If these units accumulate high errors, which means that the assigned patterns cannot be classified appropriately, this part of the map starts to grow. Even if the considered neuron is an inner neuron, than the additional data pushes the prior assigned patterns to outer areas to which new neurons had been created. This can be interpreted as an increase of the number of data items belonging to a specific area or cluster in data space, or if text documents are assigned to the map as an increased number of publications concerning a specific topic. Therefore also dynamic changes can be visualized by comparing maps, which were incrementally trained by, e.g. newly published documents [49].

17.4.2 Hints and Tips

- Structuring document collections can be very useful for a user. The question is which clustering algorithm to use. In order to choose an appropriate algorithm we have to consider the complexity of the algorithm. This depends on the amount of data that has to be processed and the time available for computation, which might be limited in interactive applications.
- Hierarchical methods are usually more appropriate to support a user in navigation, but its sometimes hard to use them in order to visualize structure or provide an overview of a set of documents.
- We have chosen self-organizing maps, because of their good visualization capabilities, good usability in interactive application and the simplicity of the algorithm (only a few parameters have to be defined). Nevertheless, the time required for the computation of a map usually prevents learning of SOMs online. Thus, a SOM has to be computed before it is used in an interactive application.

17.5 Visualization Techniques

Graphical visualization of information frequently provides more comprehensive and better and faster understandable information than it is possible by pure text based descriptions. Therefore, in information retrieval systems additional visual information can improve the usability to a great extend. Information that allow a visual representation comprises aspects of the document

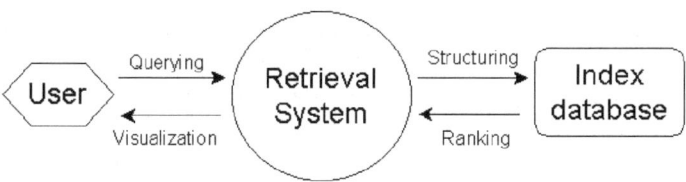

Fig. 17.6. Using Querying, Structuring and Visualization Techniques in a Multimedia Retrieval System

collection or result sets, keyword relations, ontologies or aspects of the search process itself, e.g. the search or navigation path in hyperlinked collections.

However, especially for text collections we have the problem of finding an appropriate visualization for abstract textual information. Furthermore, an *interactive* visual interface is usually desirable, e.g. to zoom in local areas or to select or mark parts for further processing. This results in great demands on the user interface and the hardware. In the following we give a brief overview of visualization methods that have been realized for information retrieval systems.

17.5.1 Visualizing Relations and Result Sets

Interesting approaches to visualize keyword-document relations are, e.g., the Cat-a-Cone model [50], which visualizes in a three dimensional representation hierarchies of categories that can be interactively used to refine a search. The InfoCrystal [51] visualizes a (weighted) boolean query and the belonging result set in a crystal structure. The lyberworld model [52] and the visualization components of the SENTINEL Model [53] are representing documents in an abstract keyword space. This idea was applied slightly modified to image databases in [54].

An approach to visualize the results of a set of queries was presented in [55]. Here, retrieved documents are arranged according to their similarity to a query on straight lines. These lines are arranged in a circle around a common center, i.e. every query is represented by a single line. If several documents are placed on the same (discrete) position, they are arranged in the same distance to the circle, but with a slight offset. Thus, clusters occur that represent the distribution of documents for the belonging query.

17.5.2 Visualizing Document Collections

For the visualization of document collections usually two-dimensional projections are used, i.e. the high dimensional document space is mapped on a two-dimensional surface. In order to depict individual documents or groups of documents usually text flags are used, which represent either a keyword or the document category. Colors are frequently used to visualize the density, e.g. the number of documents in this area, or the difference to neighboring documents, e.g. in order to emphasize borders between different categories. If three-dimensional projections are used, for example, the number of documents assigned to a specific area can be represented by the z-coordinate.

An Example: Visualization using Self-Organizing Maps

Visualization of document collections requires methods that are able to group documents based on their similarity and furthermore that visualize the similarity between discovered groups of documents. Clustering approaches that are frequently used to find groups of documents with similar content [35] – see also Sect. 17.4 – usually do not consider the neighborhood relations between the obtained cluster centers. Self-organizing maps, as discussed above, are an alternative approach which is frequently used in data analysis to cluster high dimensional data. They cluster high-dimensional data vectors according to a similarity measure. The resulting clusters are arranged in a low-dimensional topology that preserves the neighborhood relations of the corresponding high dimensional data vectors. Thus, not only objects that are assigned to one cluster are similar to each other, but also objects of nearby clusters are expected to be more similar than objects in more distant clusters. Usually, two-dimensional arrangements of squares or hexagons are used for the definition of the neighborhood relations. Although other topologies are possible for self-organizing maps, two-dimensional maps have the advantage of intuitive visualization and thus good exploration possibilities. In document retrieval, self-organizing maps can be used to arrange documents based on their similarity. This approach opens up several appealing navigation possibilities. Most important, the surrounding grid cells of documents known to be interesting can be scanned for further similar documents. Furthermore, the distribution of keyword search results can be visualized by coloring the grid cells of the map with respect to the number of hits. This allows a user to judge e.g. whether the search results are assigned to a small number of (neighboring) grid cells of the map, or whether the search hits are spread widely over the map and thus the search was – most likely – too unspecific.

A first application of self-organizing maps in information retrieval was presented in [56]. It provided a simple two-dimensional cluster representation (categorization) of a small document collection. A refined model, the WEBSOM approach, extended this idea to a web based interface applied to newsgroup data that provides simple zooming techniques and coloring methods [57, 58, 59]. Further extensions introduced hierarchies [60], supported the visualization of search results [37] and combined search, navigation and visualization techniques in an integrated tool [46]. A screenshot of the prototype discussed in [46] is depicted in Fig. 17.7.

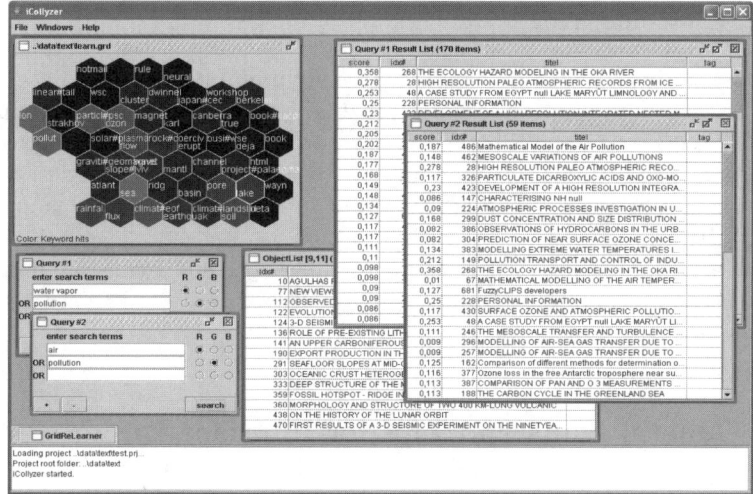

Fig. 17.7. A Prototypical Retrieval System Based on Self-Organizing Maps

Other Techniques

Besides methods based on self-organizing maps several other techniques have been successfully applied to visualize document collections. For example, the tool VxInsight [61] realizes a partially interactive mapping by an energy minimization approach similar to simulated annealing to construct a three dimensional landscape of the document collection. As input either a vector space description of the documents or a list of directional edges, e.g. defined based on citations of links, can be used. The tool SPIRE [62] applies a three step approach: It first clusters documents in document space, than projects the discovered cluster centers onto a two dimensional surface and finally maps the documents relative to the projected cluster centers. SPIRE offers a scatter plot like projection as well as a three dimensional visualization. The visualization tool SCI-Map [63] applies an iterative clustering approach to create a network using, e.g., references of scientific publications. The tools visualizes the structure by a map hierarchy with an increasing number of details.

One major problem of most existing visualization approaches is that they create their output only by use of data inherent information, i.e. the distribution of the documents in document space. User specific information can not be integrated in order to obtain, e.g., an improved separation of the documents with respect to user defined criteria like keywords or phrases. Furthermore, the possibilities for a user to interact with the system in order to navigate or search are usually very limited, e.g., to boolean keyword searches and simple result lists. Therefore, we describe in Sect. 17.7 fundamental methods to include user preferences and feedback in the visualization interface.

17.5.3 Hints and Tips

- Visualization Techniques can be extremely useful for the user.
- There is no visualization technique that is appropriate for all kinds of users. Some users are used to simple lists or trees and reject to use complex visual output. Others prefer complex interactive navigation capabilities. So far, no general design rule can be given, but to keep the output as simple and intuitive as possible.
- The coding of theoretical ideas for visualization into a running system is often complicated and time consuming. Therefore we recommend not to underestimate the workload of this step.

17.6 General Hints and Tips for Interface Design

In information retrieval it is usually distinguished between three types of input interfaces: Command language oriented (like SQL), form or menu based and (pseudo) natural language based. In most current information retrieval systems a combination of form and natural language based interface is used, which is usually extended by a set of operators to combine or refine queries (e.g. Boolean operators). This combination is accepted by users and has proven to be reasonably efficient [64, 65].

The search results are usually listed in a ranked list that is sorted according to the similarity to the query. Depending on the type of the query and the underlying data collection different ranking methods can be applied (see Sect. 17.3.1). Besides a document title and direct access to the document itself, usually some additional information about the documents content is automatically extracted and shown. For example, if the underlying document is a personal homepage it might be helpful to list besides the name of the person, the affiliation and phone number in order to avoid that the user needs to perform an additional interaction with the system to get the desired information. This is especially important, if the user can filter out documents based on the additional information without being required to scan the document and then going back to the result list, which is usually a time consuming task.

Furthermore, additional information on how a search can be refined can be given by providing additional keywords. This additional keywords can be derived based on thesauri or ontologies. Here research with respect to the semantic web might lead to further more refined techniques.

In order to improve the selection of the proposed keywords information about user interests can be used (for details see the following section). This can be done, for example, by selecting from a list of candidate keywords for refinement the words that have a high and words that have a very low importance in the user profile. The user profile information should also be used to change the order of the documents in the result list. Thus, the rank of a document should ideally be computed based on the degree of match with the query and information extracted from the user profile. Current web search engines usually sort hits only with respect to the given query and further information extracted from the web, e.g. based on a link analysis like PageRank (see [33]), a popularity weighting of the site or specific keywords. This results usually in an appropriate ordering of the result sets for popular topics and well linked sites. Unfortunately, documents dealing with special (sub) topics are ranked very low and a user has to provide very specific keywords in order to find information provided by these documents. Especially for this problem a user specific ranking could be very beneficial.

Another method to support a user in navigation is to structure result sets as well as the whole data collection using, e.g., clustering methods as discussed above. Thus, groups of documents with similar topics can be obtained (see, for example, [37]). If result sets are grouped according to specific categories, the user gets a better overview of the result set and can refine his search more easily based on the proposed classes.

17.7 User Modelling

The main objective of user modelling in the area of information retrieval is to extract and store information about a user in order to improve the retrieval performance. A model of the user usually consist of at least a list of keywords to which relevance degrees are assigned. More complex models distinguish between different search contexts, e.g. private and business, or store additionally relations between keywords in order to model a more expressive search context. A simple profile of user interests can be obtained, e.g., by extracting keywords from the queries performed or documents read by the user in the past.

User models are currently extensively used in the area of intelligent agents. For example, agents that adapt the content of a website by changing links or contents based on the classification of a user in a specific user group like "fashion-conscious" or "pragmatic". The belonging agents that analyze the user behavior based on the web pages the user visits, are called profiling agents. By analyzing the content of the pages and the sequence in which the user visited the pages (click stream analysis), it is very often possible to identify the interests of a user. A good overview of methods in this area is given, for example, at http://www.dfki.de/AgentSurvey/ and in [66].

17.7.1 User Feedback

User feedback is any response of a user to an output of the system. In information retrieval we are especially interested in feedback that can be used to learn something about the user (user profiling as described above) or feedback that can directly be used to improve the retrieval performance. For example, we can provide a user the possibility to mark relevant and not relevant objects

3. User Modeling

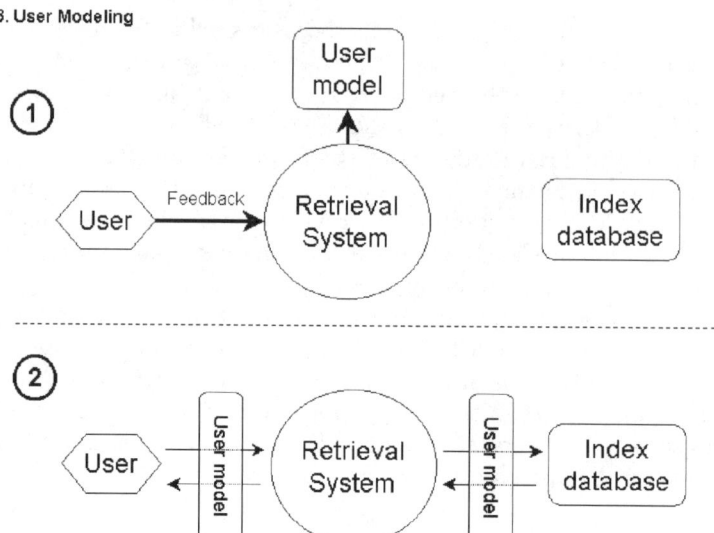

Fig. 17.8. User Profiling and Application of User Models in Retrieval Systems

of a result set. This is called *relevance feedback*. Based on this information a modified query can be derived automatically that makes use of information extracted from the tagged objects. This can be done by increasing the importance of keywords that are used in documents marked as relevant and decreasing the importance of keywords used in documents that are marked as not relevant. Thus, the result set can be iteratively refined. For details see Rochio [67] for the vector space model or [68] for the probalistic model. Almost all current approaches for relevance feedback are based on the ideas presented in these papers.

A model that utilizes user feedback to obtain a user profile was presented in [69]. This approach is also able to distinguish between long term interests of a user and ad hoc queries. Thus it avoids the undesired modification of queries that can not appropriately be described by the information stored in the user profile. The model was specifically designed for a Newsreader (NewsDude) that filters relevant information from different news sources. A Bayesian network is used in order to compute the user specific relevance of documents. The approach discussed in [70] follows a similar motivation in order to derive interest categories for users by an incremental clustering approach.

The tool WebWatcher [71] was designed to support a user by highlighting potentially relevant links on a web site. This approach uses a reinforcement learning approach to learn a profile based on the link usage of a group of users.

More general ideas concerning user and context modelling can be found in [72, 66]. As a more detailed example for the use of user feedback we discuss in the following an approach that has been integrated in a clustering and visualization tool for multimedia object collections.

17.7.2 Incorporating Weighting and User Feedback: An Exemplification Using Self-Organizing Maps

Self-organizing maps perform clustering of the objects in an unsupervised manner. Therefore it depends on the choice of describing feature vectors and the definition of similarity between them whether the resulting groups meet the users' expectations. This is especially true for multimedia data, where a general definition of significant features is rather hard and strongly depends on the application. However, the user has often a good intuition of a "correct" clustering. Therefore it seems to be very important to incorporate user feedback information to optimize the classification behavior by gathering information about the desired similarity measure and hereby modifying, e.g., the importance of features using weights. To allow the user to give feedback information, a user can be allowed to drag one or several objects from one node on the map to another that in his opinion is more appropriate for the considered objects. Furthermore, the user can mark objects that should remain at a specific node, thus preventing the algorithm from moving them together with the moved object after re-computing the groups on the map. In the following, three user-feedback models are described that have been designed to solve specific clustering problems. All approaches implement some kind of semi-supervised learning, as they use classification information about some of the objects as given by the user. The first two approaches modify the underlying similarity measure by increasing or decreasing the importance of individual features. The idea of the third approach is to fine-tune or re-compute parts of the self-organizing map to guide the learning process in how the high-dimensional feature space should be folded (i.e. non-linearly projected) into the two dimensions of the map. The methods are described in more detail in the following (see also [73, 74, 75]).

Learning a Global Feature Weighting

For the implementation of a global feature weighting scheme, we replaced the Euclidean similarity function used for the computation of the winner nodes by a weighted similarity measure. Therefore, the distance of a given feature vector to the feature vectors of the prototypes is computed by

$$e_s = \left(\sum_i w^i \cdot (x_s^i - y_k^i)^2 \right)^{\frac{1}{2}} \tag{17.4}$$

where w is a weight vector, y_k the feature vector of an document k and x_s the prototypical feature vector assigned to a node s.

We update the global-weight vector w based on the differences of the feature vectors of the moved document and the vectors of the origin node and of the target node. The goal is to increase the weights of similar features between the document and target node and to decrease the weights of similar features between the document and its current node. And symmetrically decrease and increase the weights of dissimilar features.

Let y_i be the feature vector of an document i, s be the source and t the target node, x_s and x_t be the corresponding prototypes, then w is computed as described in the following. First we compute an error vector e for each object based on the distance to the prototypes

$$e_{ji}^k = |d_{ji}^k| \, \forall k \,, \quad \text{where} \quad d_{ji} = \frac{y_i - x_j}{\|y_i - x_j\|} \,. \tag{17.5}$$

If we want to ensure that an object is moved from the source node to the target node using feature weights, we have to assign higher weights to features that are more similar to the target than to the source node. Thus for each object we compute the difference of the distance vectors

$$f_i = e_{si} - e_{ti} \,. \tag{17.6}$$

The global weight vector is finally computed iteratively. For the initial weight vector we choose $w^{(0)} = w_1$, where w_1 is a vector where all elements are equal one. Then we compute a new global weight vector $w^{(t+1)}$ by doing a by element multiplication:

$$w^{k(t+1)} = w^{k(t)} \cdot w_i \,, \forall k \text{ with } w_i = (w_1 + \eta \cdot f_i) \,, \tag{17.7}$$

where η is a learning rate. The global weight is modified until – if possible – all moved objects are finally mapped to the target node. A pseudocode description of this approach is given in Fig. 17.1.

Obviously, this weighting approach also affects the assignments of all other documents. The idea is to interactively find a feature weighting scheme that improves the overall classification performance of the map. Without a feature weighting approach the map considers all features equally important.

The global weights can also be used to identify features that the user considers important or less important. If, for example, text documents are used where the features represents terms, then we might get some information about the keywords that the user seems to consider important for the classification of the documents.

Learning a Local Weighting Scheme

The global weighting scheme emphasizes on general characteristics, which support a good overall grouping of the data collection. Unfortunately, this may lead to large groups of cells with quite similar documents. In this case some features – which are of less importance on a global scope – might be useful for distinguishing between local characteristics. Thus, modifying locally the weights assigned to these features might improve the assignment of the documents to more specific *local* classes.

Table 17.1. Pseudocode Description of the Computation of a Global Weight

> Compute the weight vectors w_i;
> If the global weight vector w is undefined
>   create a vector and initialize all elements to one;
> $cnt = 0$;
> Repeat until all documents are moved or $cnt > max$
> $cnt + +$;
>   For all documents i to be moved do
>   Compute the winning node n for i;
>   if $N \neq t_i$ (target for i) then
> $w^k := w^k \cdot w_i^k, \forall k$;
>   normalize w;
>   end if;
>   end for;
> end repeat;

The proposed learning method used is quite similar to the method described above. However, instead of modifying the global weight w, we modify local weights assigned to the source and the target nodes (noted here w_s and w_t). As before, we first compute an error vector e for each document based on the distance to the prototypes, as defined in (17.2). Then we set all elements of the weight vectors w_s and w_t to one and compute local document weights w_{si} and w_{ti} by adding (subtracting) the error terms from the neutral weighting scheme w_1. Then we compute the local weights iteratively similar to the global weighting approach:

$$w_s^{k(t+1)} = w_s^{k(t)} \cdot w_{si} \, \forall k , \quad \text{with} \quad w_{si} = w_1 + \eta \cdot e_{si} \quad (17.8)$$

and

$$w_t^{k(t+1)} = w_t^{k(t)} \cdot w_{ti} \, \forall k , \quad \text{with} \quad w_{ti} = w_1 - \eta \cdot e_{ti} , \quad (17.9)$$

where η is a learning rate. The weights assigned to the target and source node are finally normalized such that the sum over all elements equals the number of features in the vector, i.e.

$$\sum_k w_s^k = \sum_k w_t^k = \sum_k 1 . \quad (17.10)$$

In this way the weights assigned to features that achieved a higher (lower) error are decreased (increased) for the target node and vice versa for the source node.

A Generalized Learning Model

With the local approach we just modified weighting vectors of the source and target nodes. However, as adjacent map nodes should ideally contain

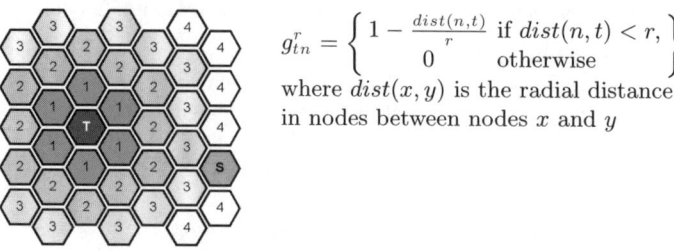

$$g_{tn}^r = \begin{Bmatrix} 1 - \frac{dist(n,t)}{r} & \text{if } dist(n,t) < r, \\ 0 & \text{otherwise} \end{Bmatrix},$$

where $dist(x,y)$ is the radial distance in nodes between nodes x and y

Fig. 17.9. Neighborhood Function Centered on Target Node (decreasing to zero for r)

similar documents, one could demand that the weights should not change abruptly between nodes. Thus, it is a natural extension of this approach to modify the weight vectors of the neighboring map units accordingly with a similar mechanism as in the learning of the map. Depending on the radius r of the neighborhood function, the result would lie between the local approach (r = 0) and the global approach (r = ∞). In the following, we present such an extension.

As for the local approach we have a weighting vector per node. Then – as before – we start by computing an error vector e for each object based on the distance to the prototypes, as defined in (17.2). Based on the error vectors e weight vectors of each node n are computed iteratively. For the initial weight vector $w_n^{k(0)}$ we choose vectors where all elements are equal to one. We then compute a new local weight vector for each node by an elementwise multiplication:

$$w_n^{k(t+1)} = w_n^{k(t)} \cdot w_{ni} \, , \, \forall k \, , \quad \text{with} \quad w_{ni} = w_1 + \eta \cdot (g_{sn}^r \cdot e_{si} - g_{tn}^r \cdot e_{ti}) \quad (17.11)$$

where η is a learning rate and where g_{sn}^r and g_{tn}^r are weighting values calculated using a neighborhood function.

Because two similar prototypes can be projected in distant cells of the map, the neighborhood function should be based on the actual topology of the map. Therefore a linear decreasing function for g_{sn}^r, which equals one for the source node and equals zero at the hull defined by the radius r can be used. The same holds for the target node and r (see also Fig. 17.2). Notice that more refined functions can be used as for instance Gaussian-like functions.

As above, all weights vectors are modified until – if possible – all moved objects are finally mapped to the target node.

This weighting approach affects the assignments of documents of neighboring cells. The influence of the modification is controlled by the neighborhood function. The idea is that a local modification has a more global repercussion on the map. In this way we can interactively find a feature weighting scheme that improves the classification performance of the map.

17 Adaptive Multimedia Retrieval 365

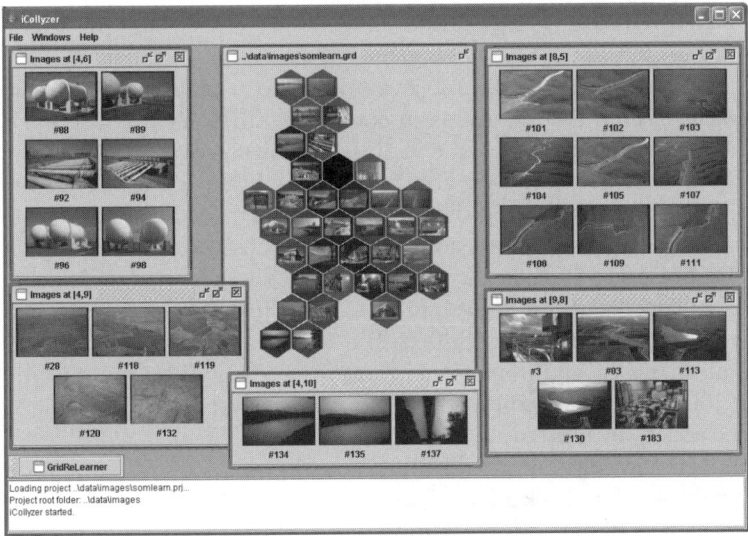

Fig. 17.10. A Prototypical Image Retrieval System: Overview of the Image Collection by a Map and Selected Clusters of Images

We note here that the general approach has by construction the local and global approach as limiting models. In fact, depending on the radius r of the neighborhood function, we obtain the local approach for $r \to 0$ and the global approach for $r \to \infty$.

The methods described above have been integrated into a prototype for image retrieval shown in Fig. 17.10. Further details can be found in [74].

17.7.3 Hints and Tips

- The user modelling step should be integrated at the very end of the design process. All other functionalities of the system should be tested and evaluated before.
- A user model should contain at least a list of features ot keywords describing the user interests.
- The user should have access to his profile in order to increase his confidence in the system.
- Particular care should be given on how to obtain user feedback. Should the user be requested directly or should we "spy" on him? One hint to answer this question is to take into account that most users do not like to be disturbed too much by questions. This particular aspect has been the reason of several failures of besides carefully designed systems.

17.8 Concluding Remarks

The creation of a multimedia retrieval system is a difficult process that requires considerable efforts. As we pointed out all along this chapter, these efforts are often underestimated and this is particularly true for some crucial steps. It also important to note that a well thought out design of the retrieval system, for instance following the methodology proposed here, is the key of a successful system. We recommend in particular considering all the possible interactions presented in this chapter. Because of these hidden difficulties, very often a multimedia retrieval system focuses on just one media like image or audio, eventually combined with text. But it is clear that in the future the use of several media types in one single retrieval system will show its synergies, and the joined use of several media types is anyway required for the design of improved video retrieval systems.

Although the construction of a multimedia retrieval system is a difficult task, it represents a fascinating challenge. The results are always gratifying, because the new designed tools help to search through data that is richer and its meaning subtler than that of pure text.

References

1. Bonastre, J.F., Delacourt, P., Fredouille, C., Merlin, T., Wellekens, C.J.: A speaker tracking system based on speaker turn detection for nist evaluation. In: Proc. of ICASSP 2000, Istanbul (2000)
2. Santo, M.D., Percannella, G., Sansone, C., Vento, M.: Classifying audio streams of movies by a multi-expert system. In: Proc. of Int. Conf. on Image Analysis and Processing (ICIAP01), Palermo, Italy (2001)
3. Montacié, C., Caraty, M.J.: A silence/noise/music/speech splitting algorithm. In: Proc. of ICSLP, Sydney, Australia (1998)
4. Idris, F., Panchanathan, S.: Review of image and video indexing techniques. Journal of Visual Communication and Image Representation **8** (1997) 146–166
5. Zhang, H.J., Low, C.Y., Smoliar, S.W., Wu, J.H.: Video parsing, retrieval and browsing: an integrated and content-based solution. In: Proc. of ACM Multimedia 95 – electronic proc., San Franscisco, CA (1995)
6. Yu, H.H., Wolf, W.: A hierarchical multiresolution video shot transition detection scheme. Computer Vision and Image Understanding **75** (1999) 196–213
7. Sudhir, G., Lee, J.C.M.: Video annotation by motion interpretation using optical flow streams. Journal of Visual Communication and Image Representation **7** (1996) 354–368
8. Pilu, M.: On using raw mepg motion vectors to determine global camera motion. Technical report, Digital Media Dept. of HP Lab., Bristol (1997)
9. Lee, S.Y., Kao, H.M.: Video indexing – an approach based on moving object and track. SPIE **1908** (1993) 81–92
10. Sahouria, E.: Video indexing based on object motion. Master's thesis, UC Berkeley, CA (1997)

11. Ronfard, R., Thuong, T.T.: A framework for aligning and indexing movies with their script. In: Proc. of IEEE International Conference on Multimedia & Expo (ICME 2003), IEEE (2003)
12. Potamianos, G., Neti, C., Luettin, J., Matthews, I.: Audio-visual automatic speech recognition: An overview. In Bailly, G., Vatikiotis-Bateson, E., Perrier, P., eds.: Issues in Visual and Audio-Visual Speech Processing. MIT Press (2004)
13. Jain, A.K., Yu, B.: Automatic text location in images and video frames. Pattern Recognition **31** (1998) 2055–2076
14. Wu, V., Manmatha, R., Riseman, E.M.: Textfinder: An automatic system to detect and recognize text in images. IEEE Transactions on Pattern Analysis and Machine Intelligence **21** (1999) 1224–1229
15. Salton, G., Wong, A., Yang, C.S.: A vector space model for automatic indexing. Communications of the ACM **18** (1975) 613–620 (see also TR74-218, Cornell University, NY, USA)
16. Robertson, S.E.: The probability ranking principle. Journal of Documentation **33** (1977) 294–304
17. van Rijsbergen, C.J.: A non-classical logic for information retrieval. The Computer Journal **29** (1986) 481–485
18. Turtle, H., Croft, W.B.: Inference networks for document retrieval. In: Proc. of the 13th Int. Conf. on Research and Development in Information Retrieval, New York, ACM (1990) 1–24
19. Deerwester, S., Dumais, S.T., Furnas, G.W., Landauer, T.K.: Indexing by latent semantic analysis. Journal of the American Society for Information Sciences **41** (1990) 391–407
20. Kaski, S.: Dimensionality reduction by random mapping: Fast similarity computation for clustering. In: Proc. of the International Joint Conference on Artificial Neural Networks (IJCNN'98). Volume 1., IEEE (1998) 413–418
21. Isbell, C.L., Viola, P.: Restructuring sparse high dimensional data for effective retrieval. In: Proc. of the Conference on Neural Information Processing (NIPS'98). (1998) 480–486
22. Salton, G., Allan, J., Buckley, C.: Automatic structuring and retrieval of large text files. Communications of the ACM **37** (1994) 97–108
23. Salton, G., Buckley, C.: Term weighting approaches in automatic text retrieval. Information Processing & Management **24** (1988) 513–523
24. Baeza-Yates, R., Ribeiro-Neto, B.: Modern Information Retrieval. Addison Wesley Longman (1999)
25. Greiff, W.R.: A theory of term weighting based on exploratory data analysis. In: 21st Annual International ACM SIGIR Conference on Research and Development in Information Retrieval, New York, NY, ACM (1998)
26. Frakes, W.B., Baeza-Yates, R.: Information Retrieval: Data Structures & Algorithms. Prentice Hall, New Jersey (1992)
27. Porter, M.: An algorithm for suffix stripping. Program (1980) 130–137
28. Witten, I.H., Moffat, A., Bell, T.C.: Managing Gigabytes: Compressing and Indexing Documents and Images. Morgan Kaufmann Publishers, San Francisco (1999)
29. Klose, A., Nürnberger, A., Kruse, R., Hartmann, G.K., Richards, M.: Interactive text retrieval based on document similarities. Physics and Chemistry of the Earth, Part A: Solid Earth and Geodesy **25** (2000) 649–654

30. Lochbaum, K.E., Streeter, L.A.: Combining and comparing the effectiveness of latent semantic indexing and the ordinary vector space model for information retrieval. Information Processing and Management **25** (1989) 665–676
31. Detyniecki, M.: Browsing a video with simple constrained queries over fuzzy annotations. In: Flexible Query Answering Systems FQAS'2000, Warsaw (2000) 282–287
32. Yager, R.R.: A hierarchical document retrieval language. Information Retrieval **3** (2000) 357–377
33. Brin, S., Page, L.: The anatomy of a large-scale hypertextual web search engine. In: Proc. of the 7th International World Wide Web Conference, Brisbane, Australia (1998) 107–117
34. Manning, C.D., Schütze, H.: Foundations of Statistical Natural Language Processing. MIT Press, Cambridge, MA (2001)
35. Steinbach, M., Karypis, G., Kumara, V.: A comparison of document clustering techniques. In: KDD Workshop on Text Mining. (2000) (see also TR 00-034, University of Minnesota, MN)
36. Nürnberger, A.: Clustering of document collections using a growing self-organizing map. In: Proc. of BISC International Workshop on Fuzzy Logic and the Internet (FLINT 2001), Berkeley, USA, ERL, College of Engineering, University of California (2001) 136–141
37. Roussinov, D.G., Chen, H.: Information navigation on the web by clustering and summarizing query results. Information Processing & Management **37** (2001) 789–816
38. Mendes, M.E., Sacks, L.: Dynamic knowledge representation for e-learning applications. In: Proc. of BISC International Workshop on Fuzzy Logic and the Internet (FLINT 2001), Berkeley, USA, ERL, College of Engineering, University of California (2001) 176–181
39. Weigend, A.S., Wiener, E.D., Pedersen, J.O.: Exploiting hierarchy in text categorization. Information Retrieval **1** (1999) 193–216
40. Ruiz, M.E., Srinivasan, P.: Hierarchical text categorization using neural networks. Information Retrieval **5** (2002) 87–118
41. Wermter, S.: Neural network agents for learning semantic text classification. Information Retrieval **3** (2000) 87–103
42. Teuteberg, F.: Agentenbasierte informationserschließung im www unter einsatz von künstlichen neuronalen netzen und fuzzy-logik. Künstliche Intelligenz **03/02** (2002) 69–70
43. Benkhalifa, M., Mouradi, A., Bouyakhf, H.: Integrating external knowledge to supplement training data in semi-supervised learning for text categorization. Information Retrieval **4** (2001) 91–113
44. Vegas, J., de la Fuente, P., Crestani, F.: A graphical user interface for structured document retrieval. In Crestani, F., Girolami, M., van Rijsbergen, C.J., eds.: Advances in Information Retrieval, Proc. of 24th BCS-IRSG European Colloqium on IR Research, Berlin, Springer (2002) 268–283
45. Kohonen, T.: Self-Organization and Associative Memory. Springer-Verlag, Berlin (1984)
46. Nürnberger, A.: Interactive text retrieval supported by growing self-organizing maps. In Ojala, T., ed.: Proc. of the International Workshop on Information Retrieval (IR 2001), Oulu, Finland, Infotech (2001) 61–70
47. Fritzke, B.: Growing cell structures – a self-organizing network for unsupervised and supervised learning. Neural Networks **7** (1994) 1441–1460

48. Alahakoon, D., Halgamuge, S.K., Srinivasan, B.: Dynamic self-organizing maps with controlled growth for knowledge discovery. IEEE Transactions on Neural Networks **11** (2000) 601–614
49. Nürnberger, A., Detyniecki, M.: Visualizing changes in data collections using growing self-organizing maps. In: Proc. of International Joint Conference on Neural Networks (IJCNN 2002), Piscataway, IEEE (2002) 1912–1917
50. Hearst, M.A., Karadi, C.: Cat-a-cone: An interactive interface for specifying searches and viewing retrieval results using a large category hierarchie. In: Proc. of the 20th Annual International ACM SIGIR Conference, ACM (1997) 246–255
51. Spoerri, A.: InfoCrystal: A Visual Tool for Information Retrieval. PhD thesis, Massachusetts Institute of Technology, Cambridge, MA (1995)
52. Hemmje, M., Kunkel, C., Willett, A.: Lyberworld – a visualization user interface supporting fulltext retrieval. In: Proc. of ACM SIGIR 94, ACM (1994) 254–259
53. Fox, K.L., Frieder, O., Knepper, M.M., Snowberg, E.J.: Sentinel: A multiple engine information retrieval and visualization system. Journal of the American Society of Information Science **50** (1999) 616–625
54. Pu, P., Pecenovic, Z.: Dynamic overview technique for image retrieval. In: Proc. of Data Visualization 2000, Wien, Springer (2000) 43–52
55. Havre, S., Hetzler, E., Perrine, K., Jurrus, E., Miller, N.: Interactive visualization of multiple query result. In: Proc. of IEEE Symposium on Information Visualization 2001, IEEE (2001) 105–112
56. Lin, X., Marchionini, G., Soergel, D.: A selforganizing semantic map for information retrieval. In: Proc. of the 14th International ACM/SIGIR Conference on Research and Development in Information Retrieval, New York, ACM Press (1991) 262–269
57. Honkela, T., Kaski, S., Lagus, K., Kohonen, T.: Newsgroup exploration with the websom method and browsing interface. Technical report, Helsinki University of Technology, Neural Networks Research Center, Espoo, Finland (1996)
58. Honkela, T.: Self-Organizing Maps in Natural Language Processing. PhD thesis, Helsinki University of Technology, Neural Networks Research Center, Espoo, Finland (1997)
59. Kohonen, T., Kaski, S., Lagus, K., Salojärvi, J., Honkela, J., Paattero, V., Saarela, A.: Self organization of a massive document collection. IEEE Transactions on Neural Networks **11** (2000) 574–585
60. Merkl, D.: Text classification with self-organizing maps: Some lessons learned. Neurocomputing **21** (1998) 61–77
61. Boyack, K.W., Wylie, B.N., Davidson, G.S.: Domain visualization using vxinsight for science and technology management. Journal of the American Society for Information Science and Technologie **53** (2002) 764–774
62. Wise, J.A., Thomas, J.J., Pennock, K., Lantrip, D., Pottier, M., Schur, A., Crow, V.: Visualizing the non-visual: Spatial analysis and interaction with information from text documents. In: Proc. of IEEE Symposium on Information Visualization '95, IEEE Computer Society Press (1995) 51–58
63. Small, H.: Visualizing science by citation mapping. Journal of the American Society for Information Science **50** (1999) 799–813
64. Nielsen, J.: Usability Engineering. Morgan Kaufmann Publishers (1994)
65. Shneiderman, B., Byrd, D., Croft, W.B.: Sorting out searching: A user-interface framework for text searches. Communications of the ACM **41** (1998) 95–98

66. Klusch, M., ed.: Intelligent Information Agents. Springer Verlag, Berlin (1999)
67. Rochio, J.J.: Relevance feedback in information retrieval. In Salton, G., ed.: The SMART Retrieval System. Prentice Hall, Englewood Cliffs, NJ (1971) 313–323
68. Robertson, S.E., Jones, K.S.: Relevance weighting of search terms. Journal of the American Society for Information Science **27** (1976) 129–146
69. Wong, S.K.M., Butz, C.J.: A bayesian approach to user profiling in information. Technology Letters **4** (2000) 50–56
70. Somlo, G.S., Howe, A.E.: Incremental clustering for profile maintenance in information gathering web agents. In: Proc. of the 5th International Conference on Autonomous Agents (AGENTS '01), ACM Press (2001) 262–269
71. Joachims, T., Freitag, D., Mitchell, T.M.: Webwatcher: A tour guide for the world wide web. In: Proc. of the International Joint Conferences on Artificial Intelligence (IJCAI 97), San Francisco, USA, Morgan Kaufmann Publishers (1997) 770–777
72. Jameson, A.: Modeling both the context and the user. Personal and Ubiquitous Computing **5** (2001) 29–33
73. Nürnberger, A., Klose, A., Kruse, R.: Self-organising maps for interactive search in document databases. In Szczepaniak, P.S., Segovia, J., Kacprzyk, J., Zadeh, L.A., eds.: Intelligent Exploration of the Web. Physica-Verlag, Heidelberg (2002) 119–135
74. Nürnberger, A., Klose, A.: Improving clustering and visualization of multimedia data using interactive user feedback. In: Proc. of the 9th International Conference on Information Processing and Management of Uncertainty in Knowledge-Based Systems (IPMU 2002). (2002) 993–999
75. Nürnberger, A., Detyniecki, M.: User adaptive methods for interactive analysis of document databases. In: Proc. of the European Symposium on Intelligent Technologies (EUNITE 2002), Aachen, Verlag Mainz (2002)

Printing: Krips bv, Meppel
Binding: Stürtz, Würzburg